Total Construction
Project Management

Other McGraw-Hill Titles of Interest

Total Construction Project Management

George J. Ritz

McGraw-Hill, Inc.

New York   San Francisco   Washington, D.C.   Auckland   Bogotá
Caracas   Lisbon   London   Madrid   Mexico City   Milan
Montreal   New Delhi   San Juan   Singapore
Sydney   Tokyo   Toronto

Library of Congress Cataloging-in-Publication Data

Ritz, George J.
    Total construction project management / George J. Ritz.
        p.   cm.
    Includes bibliographical references.
    ISBN 0-07-052986-8 (alk. paper)
    1. Construction industry—Management.   I. Title.
TH438.R54   1994
624'.068—dc20   93-27856
                                                         CIP

1 2 3 4 5 6 7 8 9 0   DOC/DOC   9 9 8 7 6 5 4 3

ISBN 0-07-052986-8

*The sponsoring editor for this book was Larry Hager, the editing
supervisor was Kimberly A. Goff, and the production supervisor was
Suzanne W. Babeuf. This book was set in Century Schoolbook. It
was composed by McGraw-Hill's Professional Book Group
composition unit.*

Printed and bound by R. R. Donnelley & Sons Company.

This book is printed on acid-free paper.

This book is dedicated to all those laborers, crafts-people, foremen, engineers, architects, builders, owners, contractors, managers, academicians, and administrators who have, over the past millennia, brought the construction industry to its present state of sophistication. Fortunately for future generations, they have left plenty of room for further growth.

Contents

Appendixes

Preface

This book presents my personal thoughts and opinions on the management aspects of total construction management garnered over 45 years of executing a broad range of capital projects around the world. It stresses the *practical* side of construction management normally encountered in completing a construction contract.

The goal of the book is to focus on the *mental side* of construction management rather than the technology side of the business. In my review of the literature on construction management, I have found few, if any, books devoted to the everyday management of construction projects. With the possible exception of the computer, construction management is relatively independent of technology. Actually, developing construction management and technology skills are two independent but parallel paths to total construction management.

The material is generic enough for the reader to adapt the practices to any of the diverse facets of the construction environment. The management methods espoused in this book may call for learning new skills, techniques, styles, and changing your way of thinking acquired from past practices. They are designed to give a construction manager the essential tools to survive and prosper in a highly competitive environment. Remember, no manager, athletic coach, or businessperson ever succeeded without drilling *the fundamentals* into his or her team.

I discovered in writing this book that it is virtually impossible to cover a subject as broad as construction management in one moderately-sized volume. Many of the subjects discussed here need a volume of their own to cover the subject in detail. My only recourse was to address the subject from the construction manager's viewpoint, and refer readers to other books for more detail on the subject. Those readers who have not yet started a management library are strongly advised to start building one, beginning with some of the titles referenced at the end of each chapter in this book.

As the term *construction manager* is used here, it is intended to mean the person in responsible charge of the construction work at the job site. There are a number of names for that person, such as site superintendent, project manager, field manager, or construction engineer, so don't let your title confuse you as to what your responsibilities really are. Anyone in responsible charge of the field activities must practice *total construction project management* to meet their project and personal goals.

I hope the information passed along to you in this book will help you to enjoy managing in the construction industry as much as I have. Remember, none of us will ever know all there is to know about construction management, so keep an open mind. Good luck on your projects!

George J. Ritz

1

The Construction Management Environment

If one is going to practice management within an industry, it is a good idea to define the arena in which the management techniques will be applied. We really need to know just what business we are in to evaluate our present goals, find out where we have been, and where we hope to go from here.

The theme of this book is the practice of management across all facets of construction project execution. Therefore, construction technology will be introduced only as it bears on total construction project management. It is a given that you must master the basic technology applicable to your specialized field of construction before you can effectively take on the total management of a construction project.

The "Total" part of this book's title means that I am addressing the application of construction management practices in an integrated systems context. One must apply all the techniques presented here over the life of the construction project to be successful. The construction project manager's genuine commitment to Total Construction Project Management must be absolute. Superficial utilization will not meet the goal of effective construction project management.

Some History of Construction Management

Construction has had a long history—closely paralleling the development of human civilization. We can start with prehistoric do-it-your-selfers piling up some rocks in the Rift Valley to mark a good hunting ground. Or with the first people to build a fireplace, piling some stones around the fire to improve its overall performance and enhance their cooking. The lever and a strong back were the construction technology of that era.

Technology moved rather slowly in those early years, but later it produced some impressive projects, such as the rudimentary but precise stone observatory at Stonehenge in England. How the Druid construction people erected those huge stone structures without heavy equipment is most impressive. We must also consider the management abilities of the people who conceived and supervised the construction. I would love to have sat in on the constructibility analysis meetings on that project!

The temple and tomb builders of the Mediterranean Civilization left many high-profile construction projects of their era. Construction technology had moved along with the development of the wedge, lever, sledges, rollers, and the inclined plane to make these monuments possible. At this point the wheel, the block and tackle, or the derrick had not yet been invented.

Even with the new technology, those projects were still highly labor-intensive. I remember seeing a stack of 6-in bell and spigot pipe on a tour of the ancient city of Ephesis. On closer examination, I noted that each pipe had been chiseled out of a solid piece of stone! The quality of the workmanship was superb, and all the pipe appeared to be made to precise tolerances—certainly as good as equivalent present-day products. One can readily imagine the problems of producing pipe under those conditions, rather than phoning your order to the local builder supply yard.

Much of the heavy construction work in those days was done with slave labor. I think that we can safely assume that the construction managers had minimal labor problems. Also, they probably used more stick than carrot in their approach to labor relations.

The Egyptian work, however, was done with free labor. They may have been among the first to practice personnel management on their construction projects. Considering the scope and quality of their projects, and the amount of manpower required over normal 20-year project schedules, keeping the work force motivated was a monumental construction management accomplishment in itself.

The next quantum leaps in construction technology occurred in the 1800s, as a result of the industrial revolution. Construction people quickly adapted the tools and machinery from the workshops and mines to construction use. Cranes, derricks, hoists, and shovels were converted to steam power. Earth-moving equipment converted from horse and mule power to steam. Less labor-intensive operations made field productivity soar. Construction managers had to adapt their thinking to get the most from the new technology. The new technology also made new forms of facilities possible. We could now build larger power dams, skyscrapers, transportation systems, bridges, and the like, to serve the burgeoning economy.

The availability of electricity, the internal combustion engine, and electrical motors at the turn of the century replaced steam to make construction tools and equipment even more mobile and efficient. Construction technology continued to race ahead along a broad front in the twentieth century as well. We learned to pump concrete, use slip forms, improve welding, and apply the steady stream of new products and materials made available by modern research and industry.

Even management made some strides during the twentieth century. We adapted Gantt bar-charting techniques for scheduling construction projects. We even adapted some budgeting and cost-controlling procedures from business and industry to improve our project financial performance. Finally, in the early sixties, the computer burst on the construction scene. We suddenly had a new tool with which to crunch all of those payroll, budget, and scheduling numbers in record time.

The computer as a construction management tool arrived just as construction projects were getting much larger and more complex after World War II. The increased size and complexity of the construction projects caused scheduling and cost performance to suffer. Field productivity improvement started to wane and even decrease in the face of the new technology advances.[1]

Suddenly our quiet little construction field offices were inundated with data in the form of interminable computer printouts from the home office. In the early computer age the flow of data was largely ignored by the field people, because they were not trained to use it. This was a slight oversight by the early computer gurus. And that "slight oversight" almost caused the application of computers in construction management to fail.

It also took some time and training to make the new computer technology effective in the field. First the commercial software simplified the payroll and accounting end of the field operations. Later, scheduling and cost-control applications were developed and proven out in the field.

Where do construction managers come from?

Until the arrival of the computer, the development of field construction managers was pretty much left to its own devices. Those foremen who had some business smarts and a dash of charisma rose through the craft ranks to become field superintendents and construction managers. Many were long on construction technology and know-how, but short on management skills. The advent of more sophisticated construction technology, along with larger and more complex projects, brought with it a need for construction managers with more technical and business skills.

That in turn meant that present-day construction managers needed to start with a four-year bachelor's degree in a field related to construction just to master the basic technology. They would then start in lower-management, on-the-job training assignments to build their construction know-how skills in their specialized field.

However, that was still not the total solution to improving construction project performance, because the management training received along with the technical degree was so limited. The priority given to the technical training left little or no room for business and management courses. Would-be construction managers still need courses and training to learn and sharpen the necessary management skills to become effective on the management side of construction project execution. The number of universities offering construction management training has been growing in recent years, but more are needed to raise the level of training.

I think this brief review of history tells us a few things about the construction industry:

- Construction has a long tradition of creating structures and facilities promoting the development of humankind.

- Construction has provided a world civilization with a huge infrastructure, ranging from basic shelter to facilities in outer space.

- Construction productivity and efficiency have improved greatly over the centuries.

However, the construction industry cannot afford to rest on its past laurels. The U.S. construction industry has been experiencing slackening growth in overall productivity and efficiency in the face of improving productivity in other businesses. Considering the size and dollar volume of the construction industry, we can't let that condition persist in a developing economy. Construction must join the other industries in improving productivity and performance to survive in the world economy. Improving construction management practices appears to offer the surest route to meeting the improvement goals.

Construction Industry Cost Effectiveness

The Business Roundtable (BRT), an association of the chief executives of 200 major corporations in America, was founded 20 years ago in 1972. As part of its work, BRT has published a series of investigative reports focusing on some of the problem areas faced by the construction industry. Those reports are listed in Appendix A.

At that time the construction industry was the largest business segment in the U.S. economy, representing about 10 percent of the gross national product (GNP). However, construction's percentage of GNP has since eroded by about half, to about 5 percent of GNP.

The BRT reports indicated a continuing decline in construction industry cost effectiveness, so it formed a special committee called the Construction Industry Cost Effectiveness (CICE) Task Force. The CICE task force includes several hundred experts from all branches of the construction industry including owners, contractors, labor unions, trade associations, academia, government, and the like.

After five years of studying the problems underlying the decline in construction industry performance, CICE issued in 1983 a series of reports[1] that tried to come to grips with the problems. The resulting series of reports is available free from the Business Roundtable, and is an excellent source of construction information useful to a construction manager. The pamphlets are available on a no-cost basis, so they should be in every CM's reference library.

One of the main areas that CICE arrived at as a possible area for improving construction industry performance is construction management. Because the reports are so construction-management-intensive, I will refer to them often as examples of problem areas. It is also only fair to say that not all construction management practitioners feel that the CICE reports are an entirely fair representation of the construction industry as it exists today. In my opinion there is enough evidence in the reports to indicate a need for improvement, even if some people think that the facts may be somewhat overstated.

The most productive and immediate means of correcting the adverse trends lies in improving both the construction management and the construction technology areas. The industry urgently needs to move forward in both areas. The purpose of this book is to offer ways to improve our construction management practices in order to immediately improve overall construction performance. Including the technology area in this book would greatly exceed the planned page allotment, so it has been left to other existing and future volumes to suggest improvements in that area.

How large is the construction industry?

The construction industry has long been a major component of the economies of the world's industrialized nations. Statistical analysis of the industry, even in the U.S., is very difficult. The U.S. Department of Labor keeps statistics, as well as private groups such as *Engineering News Record* (ENR) and various contractor associations. The ENR-generated figures are generally given more credence in the industry. The actual numbers are not that critical as long as one uses them simply to evaluate industry trends. Here, we are only trying to evaluate the relative size of the construction industry and its annual performance trends.

In the United States, for example, ENR estimated nonresidential construction turnover at $125 to $150 billion, even in the midst of the prolonged 1990–1993 recession. That represents approximately 5 percent of U.S. gross national product (GNP). In former years the percentage of GNP ran as high as 10 percent. That gives the reader some idea of how closely construction contract value tracks the nation's economic cycle.

Regardless of the doubtful accuracy of the industry's actual reported numbers, the trends indicated by the numbers do give an indication of overall performance direction. The trends of total construction volume usually trail the U.S. general economic cycle by about 6 to 12 months. This was true of the prolonged recession of the early 1980s. Unfortunately, construction productivity figures have eroded over the past several decades and have not yet reversed the trend.[1] Even though contractors try to reduce costs during recessions, field productivity seems to respond better to an active market.

Although the estimated total volume of construction is an impressive number, one has to consider the structure of the industry when evaluating the use of management practices. The makeup of the construction industry is extremely fragmented.[1] At the top are about 25 major contractors, each with a turnover of $1 billion or more per year. Those giants are followed by about 750,000 smaller contractors represented by 70 national contractor associations.

Adding such other ingredients as a large customer base, labor organizations, suppliers, subcontractors, architects, engineers, and government to the mixture further complicates the business structure. The fragmentation on the labor side is even more impressive considering the number of craft unions and the open shop options available in today's construction labor environment.

To top off the stew, many of the organizations cited are in a constant state of confrontation with each other, further aggravating the atmosphere of managing a construction project. Industry leaders have little hope that this condition will improve in the foreseeable future, given the strong links to its past history and the independent thinking of the participants.

Even in its fragmented and depressed economic condition in the early nineties, construction still qualifies as a major industry. Robert G. Zilly gives an excellent review of the current status of the construction industry at the beginning of the 1990s.[2] Certainly it is too large to ignore the benefits of effective management practices at the cutting edge of the industry's efforts in the field. Construction project managers do indeed face a serious challenge in practicing their profession. That's the bad news! The good news is that they have overcome many greater challenges in the past centuries, so there is always hope for improvement in the future. However, modern construction managers can no longer ignore the use of improved management practices if they hope to succeed.

Defining Construction Project Management

It is impossible to define a complex operation such as construction management in a simple one-sentence definition. We will have to dissect the term and define its many facets.

What is a construction project?

The term *construction project* means different things to different people. It can mean building a house, a high-rise building, a dam, an industrial plant, an airport, or even remodeling or upgrading a facility. To summarize the various major categories of construction projects I have developed Table 1.1, delineating the diverse parts of the industry. The chart groups construction projects into the various specialty areas based on the markets served by the industry. My intent is to make the list representative rather than exhaustive. You should be able to fit your type of construction projects into a suitable class.

The Types of Construction Activities list, under each category, gives an idea of the complexity of each type of project. The construction technology tends to become less complex as the categories move from left to right. Donald S. Barrie gives a good description of the major categories of construction in his book *Directions in Managing Construction*.[6]

Although the types of projects shown in Table 1.1 differ, they do have at least four traits in common:

TABLE 1.1   A Matrix of Construction Project Characteristics

Process-type Projects			
Liquid/gas processing plants (1)	Liquid/solid processing plants (2)	Solids processing plants (3)	Power plants (4)
Refineries & petro-chemical plants Organic & inorganic chemicals Monomers Basic chemicals LNG & industrial gases Nat. gas cleanup Etc.	Pulp and paper Mineral & ore dressing plants Polymers & plastics Pigments & paints Synthetic fibers Food, beverage & pharmaceuticals Specialty chemicals Soaps & detergents Films & adhesives Etc.	Cement, clays, & rock Mining & smelting Iron and steel Nonferrous metals Fertilizers & agricult. chemicals Glass & ceramics Rubber and polymer extrusion Activated carbon & carbon black Etc.	Fossil fuel & hydroelectric plants Power transmission lines & substations Cogeneration plants Coal gasification & hot-gas clean-up Flue gas scrubbers Solid waste burning Nuclear power* Etc.
% of TIC** 12%	10%	9%	7–8%
Types of construction activities			
Complex process piping systs., with much process equip. using exotic alloys Major computerized process contr. systs. Extensive power & control wiring systems Major utility equip. & distribution systems Piling & heavy equip. foundations Extens. light & heavy structural steel Thermal insulation Extens. safety systs. Major pollution & waste contr. systs. Minor bldgs. & enclosed spaces—min. architect. treatment Minor HVAC & plmbg.	Average process piping systems with mixture proc. & mech. equip.—some in exotic alloys Major computerized proc. contr. systs. Extensive power & control wiring systs. Major utility equip. & distribution systems Piling & heavy equip. foundations Heavy & light structural steel Thermal insulation Exten. safety systs. Major pollution & waste control systs. Major industrial & support buildings Sanitary construction	Minor process piping systems with heavy mech. equip. Some exotic alloys & finishes Moderate computer proc. control systs. Extensive wiring & elect. systems Major utility equip. & distribution systems Piling & heavy equip. foundations Heavy & light structural steel Thermal insulation Exten. safety systs. Major pollution & waste-control systs. Major industrial & support buildings Sanitary construction	Minimal proc. piping systems, but large hi-temp. steam & gas piping systs. Heavy boiler & turbogenerator equip. Some alloys & special finishes Major boiler control & minor proc. contr. Major wiring & elect. substation work Major utility systs. Piling & heavy equip. foundations Heavy structural steel & stacks Thermal insulation Major industrial & support buildings Some architectural finished areas
Applicable codes: OSHA, MSHA, ASME, ANSI, EPA, ASTM, and local building, plumbing, and electrical codes.			
Key craft labor used			
Major: Pipefitter-welders, ironworkers, riggers, boilermakers, millwrights, equipment operators, electricians, instrument technicians, and concrete workers. Minor: Insulators, painters, specialty brick masons, carpenters, laborers, and some architectural trades			

*Nuclear jobs are exponentially more complex due to NRC regulations.

Nonprocess-type projects				
Manufacturing plants (5)	Civil works projects (6)	Support facility projects (7)	Commercial and A&E (8)	Miscellaneous projects (9)
Automotive & heavy equip. assy. plants Light manufacturing Electronic component mfg. & assembly plants Aerospace manufacturing plants Etc.	Dams & irrigation Highways & bridges Public transport systems Water & sewerage treatment systems Port & marine projects Airports Public works infrastructure projects Oil, gas, & water transmission lines Etc.	Laboratories Test facilities Aerospace test facilities R&D facilities Pilot plants Etc.	Office buildings High-rise buildings Shopping malls Health care facilities Institutions, schools, banks & prisons Multiple-family housing units Multiple-unit housing schemes Military facilities Etc.	Plant turn-arounds Revamp projects Restoration projects Single-family housing units Etc.
7%	6%	6%	4–6%	Variable
Types of construction activities				
Site devel. per size & local conditions Major indust. bldgs. with architect. treatmt. Major HVAC, plmbg., utility & some minor process piping systems Light to medium assembly line & matl. handling systems Clean room assembly & laboratory areas Computer rooms Fire protection systs. Moderate electrical distrib. systems Heavy pollution control & waste treatmt.	Heavy earth-moving, blasting, & site preparation Large underground piping systems Major forming, rebar instl., & conc. pours Heavy concrete & asphalt paving Some industrial & passenger bldgs. Fuel, water, & waste treatment, & mech. equip. installation Railroad track installation Signalling & communications systs. Use of pre- & post-stressed concrete Sheet piling, coffer dams, and marine construction	Mixture of industrial & institutional bldgs. Some specialized equip. installation Moderate utility systems and minor process systems Heavy HVAC with fume-control systs. Some special finishes on bldgs. and equip. Moderate to heavy elect. wiring systems Average site dev. & concrete work per site conditions Heavy plumbing & waste treatment Interior & exterior architectural trades	Heavy architectural treatments & high-rise building construction technology Major HVAC & plumbing systems Major-use masonry & precast wall units Major use of window-wall installations Use of high productivity techniques for duplicate floors High-rise structural steel del. & erect. Installation of people-movers Heavy foundns. for high-rise bldgs. Smart bldg. control systems	Moderate to heavy demolition work Safety requirements for working in an operating facility Tight schedules Specialty craft requirements for restorations Other input is highly variable depending on type of project

Applicable codes: OSHA, DOT, ASME, ANSI, EPA, ASTM, and local building, plumbing, and electrical codes.

Key craft labor used
Major: Ironworkers, equipment operators, riggers, electricians, plumbers, cement workers, laborers, sheetmetal workers, architectural trades, brick masons, welders, and carpenters Minor: Pipefitters, instrument techs, boilermakers, and insulators.

**Design cost as a percentage of total installed cost (TIC).

1. Each project is unique and not repetitious.

2. A project works against schedules and budgets to produce a specific result.

3. The construction team cuts across many organizational and functional lines that involve virtually every department in the company.

4. Projects come in various shapes, sizes, and complexities.

Looking at the matrix of construction projects depicted in Table 1.1, one can readily see that each type of project is unique and not repetitious. Rarely do we find that even two single-family homes are built exactly the same. The individuality of the owners assures that all construction projects will have some degree of uniqueness about them.

Figure 1.1 is a visual presentation of the organizational lines that construction managers may cross on any given day. Because not all the contacts have the same goals as the construction team, the CM must handle each contact with the utmost care—as we will later learn.

What are the project variables?

Some unique project variables such as size, complexity, and life cycle also occur in the construction project environment. Usually, large projects become complex just by reason of problems of size and scope of work. You may want to refer to Table 1.2 to get a feel for the effects of project size. We will discuss Table 1.2 in more detail later.

Small projects also can become complex by reason of new technologies, remote location, tight schedules, or other unusual factors in the project. Small projects are often more difficult to execute than large ones, so don't take them lightly. One particular type of complex small project is the plant maintenance turnaround project done on an extremely tight, hourly-based schedule.

All projects go through a typical life-cycle curve, as shown in Fig. 1.2. The project starts with a *conceptual* phase, then passes through a *definition* phase into an *execution* phase. Finally the work tapers off into a *turnover* phase, and the owner accepts the project. The project execution phase accounts for most of the project resources, so it becomes the focal point of the life-cycle curve.

In most cases, a single project manager does not handle a project completely through the full life cycle. Most owners have specialists who develop the conceptual phase and other specialists who handle the execution phase. The conceptual phase has a heavy accent on research and development (R&D), market analysis, licensing, finance,

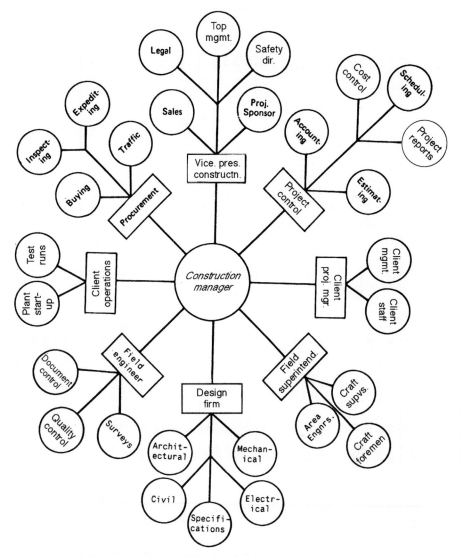

Figure 1.1 Construction manager coordination interfaces.

and economics. The early phase requires a set of skills different from those needed in the project execution phase.

The life-cycle curve gains most of its vertical growth during the detailed design and procurement phases, and peaks during the construction phase. The major commitment of financial and human resources occurs during that part of the project. That is also the time when the project master plan must be in place, because pres-

TABLE 1.2 A Matrix of Project Size Characteristics

Size TIC, $ in thousands	No. of buildings or units	Utility systems	Tagged equipment items	Purchase orders	Subcontractors	A&E hours inc. field followup	Field labor hours	Control systems Cost control	Control systems Schedule control	Construction management team
Small 1000 to 5000	1 or 2	Tie into existing systems	10 to 50	10 to 50	8 to 12	2500 to 12,500	15,000 to 75,000	Manual or PC PC with Spreadsheet	Man. barchart or PC with commercial SW	SP/FT with CM/PT to CM/SP/FT and PT to FT CM
Medium 6000 to 25,000	3 or 4	Tie-ins or some new systems	60 to 200	60 to 150	15 to 20	15,000 to 62,500	90,000 to 400,000	PC with commercial SW	PC with commercial SW	SP/FT with CM/PT to CM/FT with SP, AEs, and CP full time
Large 26,000 to 70,000	5 or 6	Moderate to major new systems	225 to 350	160 to 200	20 to 30	65,000 to 175,000	422,000 to 1,000,000	PC or MC with commercial or proprietary SW	PC or MC with commercial software	CM, SP, AEs and CP full time
Super 80,000 to 200,000	8 to 15	Major new systems and offsites	375 to 600	200 to 500	30 to 35	80,000 to 500,000	1,000,000 to 3,250,000	MC with commercial or proprietary software	PCs for SCs MC or MF for GC with Primavera or equal SW	CM/FT with SP, AEs, and CP full time
Nega 201,000 to 999,000 and over	15 to 25	Major new systems and infrastructure	600 to 1000	500 to 900	15 major and 30 to 175 minor	500,000 to 2,500,000	3,250,000 to 16,500,000	Several MCs networked to GC's MF with compatible SW	Several MCs networked to GC's MF with compatable SW	PD managing multiple CMs and SCs with full field organizations

PD = project director
CM = const. manager
SP = const. supt.
SC = subcontractor
GC = general contractor

AE = area engineer
CP = control people
FT = full-time
PT = part-time

PC = personal computer
MC = mini computer
MF = mainframe computer
SW = software
TIC = total installed cost

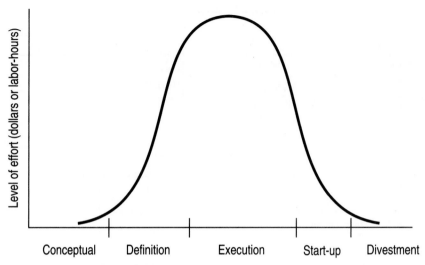

Figure 1.2 Project life-cycle curve.

sure on budgets and schedules increases during that phase, especially during construction.

We have defined a project; now let's define a manager. The dictionary tells us that a manager is "a person charged with the direction of an institution, business or the like." Synonyms listed are "administrator," "executive," "supervisor," and "boss." At one time or another the construction manager will perform all of those functions in executing the project.

Going one step further, we can define the construction management system. It is a centralized system of planning, organizing, and controlling the field work to meet the goals of schedule, cost, and quality.

Figure 1.3 graphically illustrates the definition by showing the key functions of the construction management system. The construction team uses the inputs of people, money, plans, specifications, and materials and organizes them to deliver the desired facility. Figure 1.3 also shows the feedback and control loop that ensures that the construction project goals are met.

Project Goals

The primary goal of the construction team is to *finish projects as specified, on schedule and within budget*. I have dubbed this message

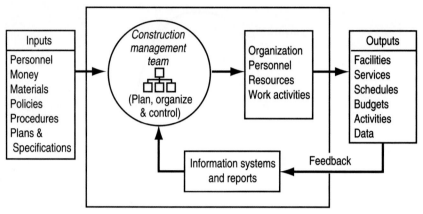

Figure 1.3   The construction management process.

"The Construction Manager's Creed." It is so important a motto that I recommend you enlarge it and hang it on your wall. It will serve as a reminder of your goals when the water in the swamp is rising and the alligators are snapping at you from every side!

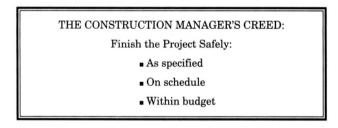

The whole system of construction management exists to ensure that we meet those goals. Obviously, the construction management system has not met its goals every time. Otherwise we wouldn't continue to hear about projects that failed because they were poorly managed, late, over budget, or abandoned.[3]

I designed this book to present the fundamental methods that construction management practitioners must know and use for effective total construction project management. Fortunately there is no particular mystique involved, just good common sense. When construction projects fail, it is not from the lack of a complex management system. More likely, it happened because one of the basic principles that we will be discussing in this book was overlooked.

Goal-oriented project groups

Let us look for a moment at the key goal-oriented groups involved in the execution of any capital construction project, namely:

Group	Description
Owner client	The one who commissions and owns the completed facility.
Project team	The project manager(s) and key staff members.
A/E firm	The entity responsible for design of the facility.
Construction team	The entity responsible for delivering the finished product.

The underlying thought for all capital project and construction management people must be that they are creating facilities for a client who seeks to earn a profit or to provide services at a reasonable cost. The first case applies to the business sector, the second to government and service agencies. Even the single-family home builder is providing living services to the home buyer/owner.

In addition to the functionality of the facilities, we must consider the aesthetics, safety, and the public welfare. The final tradeoff decisions on this group of project requirements often involve the construction manager.

Owner/client goals

The owners create the need for the facilities and raise the necessary financial resources for their creation, so they certainly rate top billing in the goals department. Actually, the owners' needs are quite simple:

Need	Reason
1. The best facility for the money	To maximize profits or services at a reasonable cost
2. On-time completion	To meet production or service schedules and financial goals
3. Completion within budget	To meet financial plans for the facility and return capital
4. A good project safety record	To meet the owner's safety standards

In their search for tools to maximize profits or services at a reasonable cost, owners want reliability, efficiency, safety, and good on-stream time. We hope they do not want goldplating and overdesign, which add unnecessary costs. All these conditions should add up to an edge over the competition or successful acceptance of the public services.

On-time completion allows owners to meet production quotas and schedules, while avoiding high cost of added interest and start-up costs. On-time completion usually results in the added bonus of a smooth start-up and rapid acceptance of the facility.

Finishing within budget avoids nasty surprises, which can upset the owner's financial plan. Project overruns lead to slower payout, and negate the chance for early return on the owner's investment. On very large projects, these financial jolts are often felt right down to the owner's bottom line. For example, SOHIO's major participation in the Trans-Alaskan Pipeline in the 1970s adversely affected the firm's general financial performance for several years.

The owner sets the minimum safety standards and responsibilities in the contract. Good construction safety performance leads to overall lower cost for the owner. We will discuss this factor in more detail in Chapter 10.

The project team's goals

The project management team is responsible to the owner for the direction and coordination of all facets of the project. The team is often made up of representatives from the owner, the design firm, and the constructor depending on the contractual arrangements. Their main responsibilities include overall project schedule, budget, quality, and performance to contract. Because this group is a composite of interests, their goals are the total goals covered in this section.

Architect/engineer (A/E) goals

The A/E group is the entity responsible for delivering design documents for the facility. It can be an outside contracting group such as an A/E firm or the owner's captive central engineering department. In the latter case, the ultimate client is the corporate division or service group owning the facility. It is desirable to have at least a quasi-contractual understanding between the internal groups to ensure that the operating division's project goals are met.

The design team's goals are to:

Need	Reason
1. Make a profit on each project.	This applies only to A/E entities. CEGs must minimize cost only.
2. Finish on time.	To satisfy the owner/client and meet contractual requirements.
3. Design within budget.	Ensure goal 1 and satisfy owner.
4. Furnish quality per contract.	Ensure goal 1, satisfy owner; meet contractual requirements.
5. Get repeat business.	Maintain company reputation and reduce selling expense.

The construction team's goals

The construction team's assignment is to deliver the finished facility ready for acceptance by the owner. This may be as a separate construction contractor or as a member of a design-build team.

Need	Reason
1. Make a profit on each project.	This applies to constructors or design-build teams.
2. Finish on time.	To satisfy the owner/client and meet contractual requirements.
3. Build within budget.	Ensure goal 1 and satisfy owner.
4. Furnish quality per contract.	Ensure goal 1, satisfy owner; meet contractual requirements.
5. Finish the job safely.	Meet company's and owner's safety goals.
6. Get repeat business.	Maintain company reputation and reduce selling expense.

Contracting firms must strive to make a profit on every project performed or they will not be able to stay in business. Because of the fixed duration of the project life cycle, the contracting entity has only one chance to make its profit. Owners may suffer some financial setback on a project with unmet goals, but they can often make up the losses later than originally anticipated. That one-time profit requirement places additional pressure on construction managers to meet their company and personal project goals.

The owner/client may not want to see the contracting entity lose money, but the owner/client does not have the contractor's profit goal strongly in mind. That is the only major point of difference among the goal-oriented groups involved in executing most capital projects. As we will discuss in Chapter 12 in relation to human factors, the contractor's profit motive represents a major stumbling block to maintaining good client-contractor relations.

Finishing projects on time and within budget has several attractions for the contracting entity. This is especially true when the contract involves an incentive clause or a fixed-price bid. In that case, any budget overruns are net losses to the contractor. The contractor's general management considers such losses a catastrophic event! On the other hand, finishing early and under budget increases profits and/or reduces costs, which are happy events for everyone in the contractor's organization.

Finishing on time and under budget are important considerations when the owner is evaluating the performance of the contractor and deciding whether to provide repeat business. The possibility that it may withhold repeat business is the best leverage a client has to ensure good contractor performance. Owners should use that leverage to the fullest extent when a contractor's performance falls below par.

Another key goal of both contractor and owner, which must be met for a successful project, is to finish the project with a good safety record. A good safety performance has a strong bearing on being able to accomplish all the other goals listed.

Building the facility to the specifications is an important goal, because it affects the overall cost of the facility and thus the budget. It also affects the quality of the facility and its ability to carry out its mission. To keep that goal in perspective, it is critical to develop a thorough and sound project definition before starting any work. Many project failures result directly from inadequate project definition before the project was begun.[3]

The construction team's personal goals

The last goal-oriented group to consider is the construction team. Here we are talking about people much like ourselves, so the matter becomes even more personal. The project teams goals are to:

Need	Reason
1. Earn a financial reward.	This can be in the form of a bonus or promtion.
2. Identify with a successful project.	Build a reputation and gain personal satisfaction.
3. Finish on time and within budget.	Attain goals 1 and 2, and satisfy client and company managements.
4. Maintain one's reputation in the business.	Personal satisfaction and professionalism.

It should come as no surprise that personal satisfaction and financial reward top the list of our human needs. Behavioral scientists have been telling us that for years.[5] With the life cycles of most projects running from six months to four years, our personal satisfaction comes relatively quickly compared with those working in a manufacturing environment. Since the results of our labors are usually visible and permanent, we remember them whenever we see or hear about them. People who do not take pride or personal satisfaction in their work will never become successful construction managers.

Completing a difficult project on schedule and under budget gives one additional satisfaction, that of having met a challenge. Having both client and management satisfied at the end of a project is indeed much more satisfying than the reverse would be.

Financial rewards can come in several forms: a bonus, a promotion, or both. Many contracting firms put their project teams on an incentive program, in which the rewards are tied to successful performance. Commercial and industrial clients also favor incentives, because these ensure, at minor additional cost, that their project goals will be met.

It is extremely important for a CM to build and maintain a reputation in the field. The number of capital project construction management practitioners is relatively small, particularly within each specialized area of practice. News of an unsuccessful project travels through the industry quite rapidly, and can detract from one's professional reputation. It takes several successful project performances to overcome the adverse press from one bad project.

Looking at all the goals for the contributing groups, we see that they are almost identical. The only goal not common to all the key players is that of the contracting entity making a profit on the project. It is difficult to understand how that single difference in goals can cause so many client-contractor relations problems. We will explore these further in later chapters.

Assuming that the project has gone through sound conceptual and design phases, it is now ready for construction. The CM and his or her team now hold the keys to delivering a successful facility and satisfying all the goal-oriented groups. All parties will be able to meet their project goals if they use the total construction project management principles presented in this book.

Basic Construction Project Management Philosophy

My basic construction project management philosophy is simply stated in three words: *plan, organize, and control.* I call that my Golden Rule of Construction Management. It should hang right alongside the Construction Manager's Creed. Practicing the Golden Rule will deliver the goals in the creed!

Many of you may recognize my Golden Rule as a distillation of a general management philosophy of plan, organize, execute, coordinate, and control. To ensure its easy application by busy CMs, I have boiled it down to the three essentials.

Total construction project management requires that you practice the Golden Rule on your projects in general, as well as for each project activity. By that I mean that you must plan, organize, and control every activity on the project. A good example of applying the rule to a piece of the project is your approach to project meetings. Plan your meeting (or your part in it); organize the logistics for the meeting; and control the meeting if you are its leader. I will cover handling meetings in more detail in a later chapter on project communications.

As a corollary to the Golden Rule I want to add the KISS principle, to further stress the need for a simple, uncluttered approach to construction project management. KISS is the acronym for the phrase "Keep It Simple, Stupid," which turned up in a joke some years ago. Some people are a little stuffy about using the word *stupid* in connection with managers. However, I believe that any unintended derogation is well worth the benefits gained by remembering to keep it simple! At the risk of spoiling your office decor, I recommend hanging a KISS sign. It's a valuable reminder for a serious construction management practitioner.

My aim in introducing simplicity is to point out that overdetailed plans, budgets, schedules, and control systems do not automatically equate to more successful projects. In fact the reverse is true: Overkill systems have hurt more small projects than underpowered ones have hurt large projects. There will be more on this subject as we progress through the book.

What Are We Going to Plan, Organize, and Control?

The key construction activities that we need to plan, organize, and control lie at the very heart of the total construction management system. They also establish the format of the project execution portion of this book. They are the basic building blocks of the Total Construction Project Management approach.

Project planning activities

The planning activities for a typical capital project are:

- Construction execution plan
- Time plan—field schedules
- Money plans—construction budget and cashflow
- Resources plan—people, materials, systems, and money

The *construction execution plan* is the master plan for executing the field work, from bidding to completion and turnover to the owner. The master plan must operate within the restraints of overall project financing, strategic schedule dates, allocation of project resources, and contracting procedures.

The *time plan* is the itemized working plan for the project execution, which results in the detailed construction schedule. The construction team makes a work breakdown structure by breaking the project scope into major work activities. They then assign completion dates for each operation.

The *money plans* consist of a detailed project budget based on a sound construction cost estimate. The *cashflow plan* results from the budget and the schedule, and forecasts how the budgeted funds will be spent.

The last major planning effort is forecasting the human, material, and systems resources required to execute a construction project according to the master plan and the schedule.

Organizational activities

The activities in the organizational activities area cover deployment of the human resources and systems required to meet the master plan and the project schedule. In this area managers must

- Prepare organization charts and personnel loading curves
- Write key position descriptions

- Issue site operating procedures
- Mobilize and motivate the field staff
- Arrange site facilities and systems
- Issue and start control procedures

Plan the field organization based on the scope of work and the personnel plan. Issue detailed construction procedures promptly to instruct and guide the new organization. Mobilize the staff and suitable facilities to kick off the project on schedule. Start the cost controls immediately; you're already spending money!

Control activities

Controlling the construction plans and activities extends over the life of the project. Control is the most vital of the three total construction project management steps. Major areas of control are:

- Quality—field engineering, materials, and construction
- Time—measured by the construction project schedule
- Money—measured by construction budget and cashflow plan
- Measuring physical progress and productivity
- Project reporting

The controlling function must monitor the quality of all phases of the work, to meet the universal goal of building the project as specified. Monitor time by checking physical progress (not labor-hours expended) against the schedule. Monitor costs through a cost-control system based on the project budget. The project-reporting system regularly informs key project players as to the status of project activities and the results of the project-control systems in detail.

The lists just given constitute the skeleton of an effective method for the successful execution of capital projects. In later chapters we will fill in the flesh, sinew, and blood, to give you a complete body of knowledge for total construction project management!

Construction Manager's Job Description

A discussion of the project environment would not be complete without addressing the duties of the construction manager. I have included a comprehensive construction manager's job description in Appendix B—a hybrid of several versions that I have gathered over the years.

Writing an industrywide job description suitable for the total CM population would result in a glut of models to cover all cases. Therefore I have selected a model for a contractor's environment, because much of the CM population works in that area. Readers working in noncontractor environments should substitute "contractor" for "client" in the model given.

Section 1.0, Concept, sets the stage for granting the CM's charter and delegates responsibility to complete the project safely, as specified, on time, and within budget. This is also where top management makes its commitment to the construction management system. If management's charter statement is weak at the outset, little can be done later to strengthen it.

Section 2.0, Foreword, suggests some good construction management techniques for successful construction project execution: goal setting; effective delegation; planning, organizing, and controlling the work; and leadership. These suggestions are vital to successful construction management.

Sections 3.0, Duties and Responsibilities, gets down to defining the on-the-job operating techniques required to execute a project. It gives a step-by-step approach to the planning, organizing, and controlling of a capital construction project. Section 3.0 serves as a good checklist to follow as you proceed through the project life cycle, even if you don't use it as a job description.

Section 4.0, Authority, expands on the CM's authority in relation to other company activities. Strength in the authority area is important to the creation of the CM's power base in the company hierarchy. Weakness here can seriously hobble the construction manager in staffing projects and marshaling the firm's construction resources to execute the projects.

Section 5.0, Working Relationships, gives a few suggestions about the human relations involved in using the power granted under Section 4.0 when operating across departmental lines. You can expand on this list to include other interfaces in your particular environment.

Section 6.0, Leadership Qualities, lists some of the leadership qualities that are most useful in motivating the construction team and others involved in the project. I will expand on these valuable qualities in later chapters.

The job description applies within the context of a contractor's environment, which usually grants a stronger construction management charter than that of a typical owner. This is primarily due to the competitive nature of the contracting business in which the profit motive discussed earlier is of the essence.

If you are working in an owner's organization, you may have difficulty getting a strong version of your job description approved. In any event it's worth a try, since it is often doubtful whether your management will volunteer a strong construction management charter. The main difference between the owner and contractor CM's job descriptions is found in section 3.0, Duties and Responsibilities. The owner CM has the same duties and responsibilities, but delegates their execution to the contractor's CM whenever a contractor is engaged to execute the work. The owner CM's responsibility is to see that the contractor is performing according to contract and plan.

If you do use the job description in Appendix B as a model for your own position, you can adapt the parts that apply to your work. Also add any new material describing your particular situation. It makes good sense to write your charter as broadly as possible and let your management cut it back. You might as well see how far management is willing to go. But remember this: If you are given total construction project responsibility to meet the project goals, you must have the authority necessary to do the job!

I hope that you noticed the application of my Golden Rule of Construction Management in the layout of Section 3.0. We divided the duties and responsibilities into subsections for the planning, organizing, and controlling functions that are so essential to effective construction project management.

The job description is quite general, as it must be without a specific project assignment. Therefore you should prepare a written list of project-specific goals to measure your performance and that of your staff. You should in turn incorporate those specific project goals into your staff's job descriptions. This will allow you to take a Management by Objectives (MBO) approach to the execution of the project. We will go into more detail on MBO systems in later chapters.

Project Size

The one project variable that needs more detailed discussion is project size. Most people seem to have more difficulty with size than with any other aspect of construction management discussions.

The matrix of the project characteristics chart shown in Table 1.2 highlights the differences in various size projects. The size divisions in the left-hand column represent my opinion as to project size categories. Opinions will vary somewhat among the different types of capital projects listed in Table 1.1. The common denominators for judging project size are scope of services, numbers of drawings, major equipment items, bulk material quantities, and the home office and field labor-hours. It behooves CMs to give these size factors some careful consideration when reviewing their project assignments for the first time. Making a quick review of the project's major size indicators will give you a better idea of how best to approach your new assignment.

Since a high percentage of capital projects fall into the small-project group, some readers may feel overwhelmed when looking at the numbers for the larger projects. However, larger projects do come along occasionally, so you should know how to deal with them. I have found that doing a single large project is much easier than running several small ones simultaneously.

The important thing to remember when reading the chart in Table 1.2 is that the same principles of total construction project management covered in this book apply to projects of any size. Only the systems, tools, and procedures vary to suit the size. Smaller projects require that you keep the "KISS" principle uppermost in your mind. Use simple, uncomplicated methods that don't get in the way of the work. Larger projects demand more complicated methods, which in turn require use of automated management-by-exception techniques to control the larger number of work activities.

Notice in the right-hand column of Table 1.2 that staffing and systems become more sophisticated as the projects increase in size. As projects become larger, they are broken down into smaller, more manageable units (small projects). An area engineer or manager is responsible for each unit and reports to the overall CM. On large projects, we may also sublet complete units to subcontractors who specialize in executing that type of work. Making them responsible for those pieces of the project reduces your input on the managerial side of the project through effective delegation.

The most important differences in handling small and large projects are the systems and tools used for project control. We can still effectively control small projects (under $10,000,000) by using manual systems to control schedule and cost. Now we have personal computers (PCs) and commercial software to assist us with these chores even on small projects. Notice that I say *commercially available* software. Creating custom software for small projects becomes too involved, expensive, and time-consuming, so be wary of it. Commercially available spreadsheets and database programs can do the job very well.

As projects become larger, computers become more productive as a means of processing the more complex budgets, material lists, schedules, cost reports, etc. Remember, computers are tools capable of sorting and crunching large volumes of data, but they do not add any *intelligence* to your input data. We will discuss computer applications in later chapters.

Project size ratios

There are several good rule-of-thumb ratios between labor-hours and total project cost that are useful for rough cost estimates early in the project. The simplest one is the ratio between home office labor-hours and field labor-hours. On complex projects, field labor-hours are from five to seven times the design hours. Thus, each design labor-hour generates five to seven construction labor-hours. Those projects on the right half of the matrix shown in Table 1.1 are likely to have a lower ratio, so you will have to establish the actual ratio numbers for your own construction specialty area.

In Table 1.1, we see ratios of home office costs to total installed cost (TIC) ranging from 6 to 12 percent, depending on the type of facility. By using the applicable percentage and an average home office labor-hour cost, one can readily get from the home office hours to the total project cost. The same applies for the field labor-hours.

Today, average overall home office hourly rates run from $35 to $60 per hour inclusive of overhead and fees. For example, by using a medium-size project with 150,000 home office hours and a $40.00 per average hourly rate, we can roughly estimate the total project cost as follows:

$$150,000 \text{ h} \times \$40/\text{h} = \$6,000,000 \text{ (design cost) } \$6,000,000 \text{ / } 0.10$$
$$(\% \text{ TIC}) = \$60,000,000 \text{ approx. total project cost}$$

Using the above numbers, we can approximate the field labor-hours from the design hours, assuming a 6 to 1 ratio of field to design labor-hours as follows:

$$150,000 \text{ design hour} \times 6 = 900,000 \text{ field labor hour}$$

If we multiply the field labor-hours by an average field labor-hour rate of say $30 per hour, we get a field labor cost of about $27 million.

One should use the above method of factoring costs only in the early phases of a project, before more accurate and reliable estimating data has become available. Owners use these cost and labor-hour relationships in checking contractors' proposals. A&E firms and contractors also use them in the early stages of proposal preparation. You should develop a similar set of numbers for your projects from previous project cost reports by averaging them to get your own rule-of-thumb ratios.

Handling megaprojects

The largest projects on the list are the megaprojects, which are the World Series and Super Bowls of construction project management. Some of us secretly dream of participating in such a major undertaking.

Let's look at one such project that is now in the early stages: the new Department of Energy's Superconducting Supercollider (SSC) project being built in a small community just south of Dallas, Texas. The project will further extend America's frontiers in that most recent high-tech area, superconductivity. The SSC is about a 10-times scale-up of the already impressive Enrico Fermi linear accelerator facility, located outside of Chicago.

The 1993 budget forecast for this project is now at $8.25 billion, after starting at $4.4 billion, which is about par for the course on most government megaprojects. The new research center will include about 53 miles of accelerator tunnel, an operations center, research buildings, cryogenic cooling units, and support facilities. As was just noted, the facility is to be located around a small town south of Dallas, Texas, and will thus call for major additions to the local infrastructure. The $8.25 billion budget figure does not include any off-project infrastructure cost.

Assume you are the owner's construction project manager on this project. How would you go about getting the construction of such a mind-boggling operation organized? Remember, your mission is to bring this venture to a successful conclusion as specified, safely, on time, and within budget!

As construction manager for the owner, you would be serving on a project team headed by a project director who has the final say on the contracting plan for executing the project. The project director will be depending on you for the construction recommendations.

After the initial shock of your assignment has passed, I expect you to remember to divide the monster construction project into manageable packages. That will allow you to bring the world's best available resources to bear on the project. Because of the specialized nature of the process and equipment involved in the supercollider, this is not your everyday type of facility. There is probably not a firm in the world experienced in all the diverse technical details of constructing this facility.

The project environment also gives us some restraints on the formulation of a construction plan. This is a government-owned project, so federal laws and regulations will require public bidding procedures. Design-build contracts will not be considered, except for the more sophisticated scientific areas of the project. Other factors to consider are the small business and minority firm set-asides that will affect the construction contracting plan.

What are some of the natural dividing lines we could use to subdivide this megaproject into workable units? Some major areas that quickly come to mind are:

1. Site development—site planning, site preparation, drainage, roads, utilities, security, and so on

2. The accelerator tunnel

3. The accelerating equipment and associated electrical work

4. The control building and operations center

5. The research center and laboratories

6. The support facilities—engineering building, shops, power plant, food services, and so on

7. Supporting infrastructure, offsite facilities, and so on

With our limited view of the project, these are seven major identifiable areas available to start a project breakdown. Let's also assume that a preliminary design study has been completed to set the overall parameters and budget for the project. We are also early in the project life cycle and are just starting to commit major funding.

In reviewing the list of major areas, we find they readily fall into several areas of design and construction specialization. Tunneling, for example, is a highly specialized construction skill handled only by a few firms with that expertise. It is a large dollar-value piece of the work that we can delegate to a qualified tunneling specialist. They will require our close schedule coordination and quality-control supervision to ensure good performance. Contract administration will also be a major concern for that amount of work.

The accelerator equipment design and build is highly specialized, and limited to the capability of only a few firms in the world. The last report I read said that the magnets are to be produced by a firm located in Russia. How's that for a major switch in the world political climate! With an owner's staff of specialists overseeing the work, we can effectively delegate to the Russian supplier a design, build, and installation supervision subcontract. With the selected supplier being located in Russia, we will have to be especially careful about standards, quality, delivery, and the political situation in the country.

Site development is a specialty area where we could use an experienced consultant to help in the planning and construction aspects of this task. Site development is basically civil-oriented work, so we will need a qualified civil-engineering-oriented construction management contractor to supervise it. In turn, the site development manager can subdivide this major piece of the construction budget into work packages suitable for smaller local contractors.

Other areas of the project involving buildings can be broken down into related bid packages and let to qualified construction firms. In that way we get the benefit of firms who specialize in the various types of buildings involved. Most of these building projects will be large enough to warrant construction managers to supervise and coordinate their integration into the construction of the overall project.

We have farmed out most of the construction work to a group of highly qualified specialty contractors. All that remains is the overall coordination of the owner's construction responsibilities. For that you will need a staff of specialist engineers, planners and schedulers, cost engineers, contract administrators, accountants, legal advisers, procurement people, inspectors, and a clerical staff to handle the paperwork. The construction manager will bring all the exceptional items that cannot be handled at that level to the attention of the project director.

Aside from the sheer numbers involved, handling a job in this fashion is not much different from the work of a construction manager on a large construction project. The only other difference is that the stakes are much higher and the risk of failure much greater. A steely set of nerves and a cool head are essential for survival in the megaproject arena. We will get into that part of the business in Chapter 12, Human Factors in Construction Management.

This chapter has been a discussion of some major points that I have found to be important in practicing total construction management. There are many more things you should know about the subject. Therefore I recommend that you expand your viewpoint through an outside reading program. Select applicable material from the references listed at the end of each chapter, and from other bibliographies available in the literature.

I think that you will also find it particularly valuable to expand your interest into general management books such as those by Peter Drucker and other management gurus. They will give you an insight into how construction and project management fits into the global management scheme. Construction managers usually do a lot of traveling or living away from home. The time you otherwise waste in airplanes and airports can be put to good use by reading some of these books. Remember, the most likely upward path from construction manager goes through project director and into general management.

References

1. The Business Round Table, *More Construction for the Money,* 200 Park Avenue, New York, NY 10166, 1983, reprinted 1986. [See Appendix A for list of Construction Industry Cost Effectiveness (CICE) Reports]
2. James J. O'Brien and Robert G. Zilly, *Contractor's Management Handbook,* McGraw-Hill, New York, 1991.
3. Peter Hall, *Great Planning Disasters,* Weidenfield and Nicholson, London, 1980.
4. Steven S. Ross, *Construction Disasters: Design Failures, Causes, and Prevention,* McGraw-Hill, New York, 1984. (An *Engineering News Record* book)
5. F. I. Herzberg, *Work and the Nature of Man,* World Publishing Company, 1966.
6. Donald S. Barrie, *Directions in Managing Construction,* John Wiley and Sons, New York, 1981.
7. Joseph P. Frein, ed., *Handbook of Construction Management and Organization,* 2d ed., Van Nostrand Reinhold, New York, 1980.

Introduction to Case Studies

A major purpose of this book is to serve as a textbook for training programs in construction management. To support that feature, a set of case study instructions is given at the end of each chapter. This will give the reader an opportunity to apply some of the ideas developed in that chapter.

Appendix C contains a selection of some typical case study projects. Included is a generic project format, which allows the reader to tailor a project to his or her specific area of interest. Use of the original project selection throughout the book will result in a complete construction project execution file as the case studies evolve.

In a group-training environment, use a team approach to bring more minds to bear on each aspect of the case study. There are no set solutions to the case studies, which encourage readers to develop new approaches to handling a given situation. Critiqueing each group's presentation solution by the others is also an excellent character-building device.

A goal of the case study approach is to stimulate the flow of your creative juices, to get you to develop innovative solutions to some everyday construction situations. If you are fortunate enough to be in a group, you can try some of your ideas out on your peers before using them in actual combat.

Please refer to Appendix C to make your project selections for use with the instructions listed at the end of each chapter. Set up a project file or notebook as a record of the solutions you develop as we proceed through the project life cycle. That will give you a complete project profile for future reference.

Case Study Instructions

1. Assume that your company has not previously operated in a strong construction management mode. Your top management has asked you to formulate and present a plan for installing an effective construction management system in the firm. Develop an outline of the main features of your recommended plan for the installation of such a system.

2. Prepare a job description for your assignment as construction manager on the project that you selected from the list in Appendix C.

3. Based on the general project goals discussed in this chapter, prepare a list of specific project goals for the goal-oriented groups involved in your selected project.

4. Use the chart in Table 1.2 to classify your project as to size and complexity. Prepare a construction execution master plan for presentation to your management. Include your recommendations for labor stance, subcontracting, staffing, scheduling system, and standard company operating procedures for use on the project. Base them on the expected company resources required versus those available.

2

Bids, Proposals, and Contracts

The bidding, proposing, and contracting process plays a key role in total construction project management. This process is the lifeblood of the engineering-construction industry. Until we have reached an agreement or signed a contract, no steps are taken to construct the facility. Entire books[1,9] are devoted to the subject we have to cover in this one chapter. Our specific interest here, however, is to discuss the construction project manager's role in the overall process.

The term *contractor selection process* refers to a system for selecting the contractor and negotiating the contract. The contract itself sets the ground rules and apportions the risks for executing the construction work. The overall process is shown schematically in Fig. 2.1.

The whole process has a significant effect on meeting the owner's and contractor's project goals. Another often-overlooked element in this process is the effect of the contract format on project cost. The selection budget itself is a cost factor, as well as the project cost effects of the resulting contract. These points are discussed in more detail in *Contractual Arrangements,* CICE Report A-7.[2]

The contracting process can conveniently be divided into two major phases: contract formation and contract administration.[9] Contract formation consists of taking and analyzing the proposals and signing a contract. Contract administration consists of working with the contract until project completion and satisfying all the terms and conditions.

The contract itself sets out the scope of work and responsibility for the execution of the work. It also assigns and limits the risk sharing

Development phase	Contracting phase	Execution phase
Activities	**Activities**	**Activities**
Project planning	Contracting plan	Detailed engineering
Market development	Contractor screening	Procurement
Process planning	Selection of bidders	Construction
Cost estimating	Invitation for proposals	
Basic design	Contractor's proposals	
	Bid review	
	Contract award	
. . . By owner	. . . By owner and contractor	. . . By contractor

Figure 2.1 The contracting process. (*Reproduced with permission from* Proceedings of 10th Annual Seminar/Symposium, *Project Management Institute, October 1978, p. II-L.1.*)

for the contracting parties.[2,9] Constructing any facility involves plenty of risk for all parties concerned, be they owner, design firm, prime contractor, subcontractor, or material supplier.

Our continuing drift toward a litigious society has significantly affected the construction contracting environment. Fifty years ago, contracts were often signed, filed, and forgotten until project completion. Not so today! Lawyers have turned this area into one of their more lucrative business opportunities. The legal staff will be well represented to protect the interest of both parties in developing the construction contract. That is why it is so important that construction management people be represented to endorse the technical and administrative aspects of the final document for both the owner and the contractor.

The construction manager is responsible for the administration of the contract after it has been signed. Often the CM is named in the contract as the prime contact for interaction between the contractor and the owner. That certainly makes it vital that the CM be thoroughly knowledgeable about every facet of the contract.

The best possible arrangement is for the contractor's CM to participate in estimating, preparing the proposal, and closing the contract for the project. Sometimes such arrangements are not feasible, and the CM inherits a contract executed by others. He or she may even come into the project after the project has started. In that case, the CM must become completely familiar with the content of the contract immediately. That includes those situations that require the CM to put out some ongoing fires the moment he or she hits the ground!

Construction Execution Approach

The construction bidding, proposing, and contracting process is initiated by the owner who has a project to build. That means there are about as many approaches to contracting as there are owners. Each owner-project combination has some unique features that need to be covered in the contract. Even the so-called standard contract formats developed by the American Institute of Architects (AIA) and the Association of General Contractors (AGC) have blanks to be filled in and appending attachments to adapt them to the specific contract. These formats have been discussed at length in other books, so I won't go into detail on that subject here.[1,3,10] Here we are concentrating on the underlying management concepts of construction contracting, which also applies to the standard contract formats.

There are at least four basic approaches to construction contracting, depending on how the owner wishes to do the work. Each one of these approaches is subject to further variations introduced by the type of contract used.

The owner can use the following contracting approaches:

1. Single construction contractor using a self-perform and subcontract.

2. Design-build (turnkey) by single contractor or joint venture team

3. Owner lets design-procure to A/E and third-party build contracts.

4. Owner lets construction management contract.

These four basic contracting strategies are shown graphically in Fig. 2.2, along with several alternatives. The traditional case (upper left) is the one used for projects involving public funds. The other five alternatives are used to varying degrees in the private sector. Even the basic plans shown can be cut and pasted to give virtually unlimited contracting possibilities. The owner's project team must consider the advantages and disadvantages of each approach, and match the most favorable plan to the environment and goals of each project.[4]

The single-contract, self-perform mode means that the contractor performs the bulk of the construction work with its own forces.

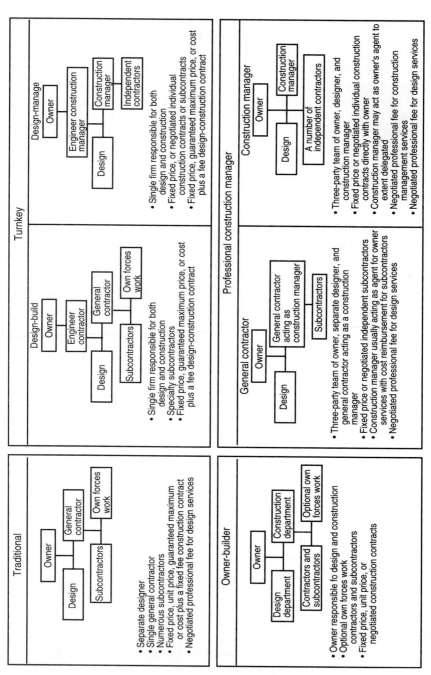

Figure 2.2 Contracting strategies. (*Reproduced with permission from Professional Construction Management by Donald S. Barrie and Boyd C. Paulson, McGraw-Hill, New York, 1978.*)

Certain specialty trades such as roofing, ceramic tile, insulation, HVAC, electrical, and so forth are generally subcontracted to others. This mode covers most of the projects where the design has already been completed by a design firm prior to bidding the construction services to detailed plans and specifications.

In a design-build mode, a single firm or joint venture provides the engineering, procurement, and construction (EPC) services under a single contract. The construction can be handled in either the self-perform or the construction management format. In this mode, CMs may also find themselves in the role of project manager (director) or reporting to a project director. In either case, the CM is directly responsible for administration of the construction portion of the contract.

The design-build mode is popular with industrial project owners, because it allows for fast-track scheduling while still controlling costs. This arrangement allows the design and procurement to start together, and construction to start when the design is about 30 percent complete. Projects can come on line six to nine months earlier using this contracting mode. The cost basis is usually cost plus a fixed fee (CPFF), based on a detailed project estimate at the 60 percent design phase. The owner's contract administration is somewhat more complex and costly to operate in the CPFF mode.

Another advantage of the design-build mode is the availability of construction technology input through constructibility reviews early in the design phase. These can substantially reduce the project construction cost by incorporating suggested construction improvements as the design phase develops. We will discuss constructibility reviews in later chapters.

The design/procure–third-party construct mode is often used when the owner wants to employ a specialty design firm and an open-shop construction contractor. The third-party constructor can be brought on-board early enough in the project to perform the constructibility input during the design phase. This mode also allows for fast-tracking the design, procurement, and construction efforts to expedite early project completion. The procurement function is usually split, with the design firm buying the engineered equipment and the constructor buying the bulk materials. Because the engineer and the constructor have prime contracts with the owner, it is the owner's responsibility to coordinate between the designer and constructor to control quality, schedule, and budgets.

The construction management mode means that the construction work will be entirely subcontracted to other contractors under the direction of the construction management firm. This means that the prime contractor does not direct-hire any craft labor. The CM coordinates the schedule, quality control, and work execution through the subcontractor's site superintendents. The CM uses a staff of project-control people, subcontract administrators, and craft specialists to attain the project goals of quality, schedule, and budget. The size of the staff is dependent on the size of the project. For example, a single-family home builder building several homes may be well enough organized to perform all the CM functions personally. A superproject may require a CM field staff of 40 to 50 people to manage the job.

A variation on this option is for the owner to act as its own general contractor by letting and coordinating the subcontracts. Most owners do not carry a construction management staff qualified to operate in this mode, so they use a qualified construction management firm to act as their agent. In either case the contractual obligations are directly between the owner and the subcontractors.

The selection of the contracting mode is largely a function of the owner's project needs. The single-contractor, self-perform mode is used by government owners who are required by law to operate in a fixed-price construction contract mode. The design work is fully completed prior to construction bidding, which normally promotes the highly competitive pricing environment required for public works. The fixed-price contract format minimizes the owner's contract administration and field cost-control efforts, which suits the makeup of most public owners' organizations. The owner usually makes the general contractor responsible for coordinating the total field effort, even when separate contracts for mechanical and electrical work are let by the owner.

The construction management mode has come into increasing use in recent years. It gives the owner the option of hiring a firm to manage the construction activities as a temporary extension of the owner's organization. Construction management services are offered by both construction and design firms. In either case, the people performing the field operations must be experienced in the administration and control of construction contracts.

The construction management method also allows the project to be run on a fast-track basis to shorten the schedule. As various construction packages become available from the design group, lump-sum bids are taken and awarded by the managing firm. The firm can also assist in reducing project cost by supplying constructibility input during the design phase. The managing firm furnishes the overall coordination of the subcontractors to ensure that the owner's goals of quality, schedule, and cost are being met. These three advantages of fast-track schedule, constructibility input, and field coordination can more than offset the fixed fee paid to the managing firm.

It's vital to project success that the owner select a contract execution plan during the conceptual phase, for this affects project cost and schedule. This in turn can affect the feasibility and the granting of approval to proceed with the project. The contracting plan can be revised if necessary as the conceptual phase develops.

Owner-contractor operating mode

Once the owner decides to go outside for help on a project, it introduces another party into the project proceedings. This complicates matters slightly. We must now have an arm's-length agreement between the owner and the outside party, which brings a proposal and contractor selection process into the picture.

The owner's decision to enter a proposal phase brings a number of managers into the action. The first one is the owner's project manager, who heads up the owner's contractor selection team. The owner's request for proposal (RFP) will in turn activate several contractors' project or construction managers, who are responsible for much of the proposal effort. Remember, one of the precepts of good management is for project and construction managers to be responsible for their projects from initial proposal to project closeout. The project architect or engineer, usually employed by the owner, can also be involved as a noncontracting party.

The Contractor Selection Process

The process for contractor selection that I am going to discuss here will apply across the board, from major prime contractors to minor subcontractors. Naturally, the process is less detailed when it is used for proposals on a smaller project with a smaller work scope. I hope that you will remember to use the KISS Principle when you find yourself engaging in this activity!

The selection procedure discussed here comes from a paper given by Bill Emmons at the 10th Annual Seminar/Symposium of the Project Management Institute in Anaheim, California, in 1978.[5] I recommend that you read that paper if you become responsible for contractor or major subcontractor selection. The goal of the contractor selection process is to arrive at a mutually agreeable contract with an excellent supplier of the desired services. That offers the best chance of meeting all the owner's project goals.

Figure 2.1 shows the part of the contracting cycle in which we are now working. The contracting phase fits neatly between the project development and execution phases. The bottom line on this graphic shows the party with the primary responsibility for the major activity taking place in the block above.

The owner's input

The owner must do a thorough job in the project development phase to allow the subsequent contracting and execution phases to proceed smoothly and efficiently. The design and construction phases are critical, because they involve the largest commitment of human and financial resources. Poor definition during project conception often brings delay and indecision, which are costly and goal-wrenching in the phases that follow.

Contractor selection

The contractor selection process must follow the contracting plan discussed earlier in the chapter. Changing the plan in mid-selection will likely cost money and valuable time in the strategic planning schedule.

Successful contractor selection can best be accomplished through an open and honest approach by all parties. Using a list of prequalified contractors can further improve the owner's selection process. There is no point in working with contractors with whom you are not willing to enter into a contract.

That point becomes more complicated when public bidding of government work is required by law. The selection team in public bidding must be especially careful to closely examine the financial and technical capabilities of all contractors who request bids. This must be done before bidding, because it is difficult to eliminate an unqualified low bidder after the bids have been opened.

In private work it is not absolutely essential to give the work to the lowest bidder. In fact, negotiated contracts without any bidding process are possible in the private sector. The negotiated contract approach doesn't exclude the selection process, but it does simplify it.

Selecting a contractor for larger projects requires a competent team, one that represents a broad spectrum of technical and business skills. A project or construction manager usually heads up the selection team to coordinate the other departmental inputs to the selection process. In addition to the technical people on the team, input from tax, risk management, procurement, legal, and accounting specialists is also necessary. The selection team should participate for the duration of the selection process to minimize the adverse effects of repeated learning curves.

The owner's contracting plan is the foundation of the selection process. A well-conceived and management-approved plan is necessary at this point to avoid the need to revise it later. Management should tailor the contracting plan toward the best arrangement to meet the owner's goals for this project. In the private sector, the contracting plan should not have to conform to long-standing corporate policy if a different arrangement offers advantages.

It's wise to make a complete analysis and evaluation of all contracting alternatives and project factors to arrive at a sound project strategy. Major items to consider are:

- Project needs
- Requirements of the project execution plan
- Key schedule milestone dates
- The scope of contractors' services
- Possible contracting alternatives
- Local project conditions
- Contracting market conditions

The owner is looking for the contractor who will perform best by delivering a quality facility on time at the lowest overall cost. In striking the ideal owner-contractor match, the owner should select a contractor from a list of firms that really want the job. Factors affecting a contractor's desires are: present workload, prestige, repeat business, market position, profitability, and the like. Incentives are necessary on both sides of the contracting equation to get outstanding performance and results. It is the merging of the owner's goals with those of the contractor that leads to a successful contract award.

Contractor selection is much like a miniproject; it needs a project management approach. This means that a total project management approach to the planning, organizing, and controlling of the selection effort is necessary. The bar chart in Fig. 2.3 shows a typical schedule for major activities in the selection process. You will have to add the applicable time scale, and any other activities that suit your particular project. In many cases, setting the proposal opening date controls the timing of the preceding activities. Be sure to allow enough time for the contractors to prepare sound bids and proposals.

The contracting basis used causes the time scale for the proposal to vary widely. A cost plus fixed fee (CPFF) based proposal requires less time to prepare than a lump-sum (LS) proposal. Fixed-price proposals require preparation of detailed cost estimates by the contractor and bidding documents by the owner. Documentation for CPFF proposals is much less complicated and time-consuming for both parties.

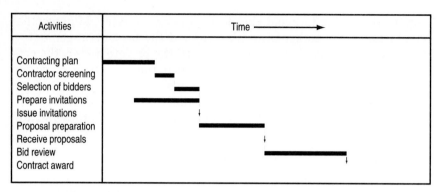

Activities	Time ————————▶
Contracting plan	
Contractor screening	
Selection of bidders	
Prepare invitations	
Issue invitations	
Proposal preparation	
Receive proposals	
Bid review	
Contract award	

Figure 2.3 Sequence of contracting activities. (*Reproduced with permission from* Proceedings of 10th Annual Seminar/Symposium, *Project Management Institute, October 1978, p. II-L.2.*)

Contractor screening

The contractor screening process again depends upon the type of owner involved. In government projects, the owner must publicly advertise the project to open the work to all interested firms. The projects also are listed in various builder's exchanges, which list all construction projects as a service to the contracting industry. In this case, the owner must ensure that the interested firms are qualified to bid the work. The problem here is keeping the list of bidders to a reasonable length. Government owners normally include the taking of public bids in the A&E's scope of services as part of the preparation of the bidding documents. The owner's project manager works with the A&E to see that the owner's bidding goals are being met.

In private work, the owner can restrict the list of bidders to only those firms deemed qualified or desirable to perform the work. The responsibility then rests with the contractor's business development group to get their firm onto the preferred bidders list.

The screening process for private work begins by issuing of a request for a statement of interest from a prequalified list of about 10 to 20 firms. For smaller projects, the screening may even be a telephone canvass of five or six firms to determine their present workload and degree of interest. The goal of the screening activity is to arrive at a bidders list of four to six firms that are sure to respond with viable proposals.

The shortened bidders list approach has several goals. The first is reducing the selection team's work in reviewing the detailed proposals. Second, the approach helps to control expenses for proposal preparation. Contractor proposals for medium and large projects often require investing of several hundred thousand dollars. Since there can be only one winner, the unsuccessful firms consume a total of five times several hundred thousand dollars. The proposing firms are profit-making entities, so they must charge as overhead the money spent on unsuccessful proposals. All owners eventually pay for the unsuccessful proposal expenses as part of the contractor's overhead when they employ contractors. Therefore, owners are wasting their own money when they request an unnecessarily high number of proposals.

Selecting a short list of eager, qualified, and competitive contractors should assure owners of a good, representative selection of proposals. Some major points to consider in the screening process are these:

- Screen as wide an area as needed to satisfy your needs.
- Use a written screening document: letter, fax, or mailgram.
- Prequalify the screening list.
- Use a simple screening request document.

Although concise, the screening document should cover all the critical points needed to select a good bidders list of quality firms. The screening process really requests miniproposals to get some preliminary facts from the contractors. Some of the key points to cover are:

- A brief project description
- A statement of the scope of services
- Key dates for proposals, contract award, project start, and desired completion date
- Tentative project plan and schedule
- A statement of contractor's interest
- Location of the work
- Request for status of contractor's existing work load
- List of contractor's current technical personnel capability
- Pertinent experience in area of project site
- Any other project-pertinent factors worthy of evaluation

The screening analysis requires a thorough review of the positive replies to the screening document. A major item to check is the contractor's capability to handle your project within the required time frame. Since you have approached only those firms capable of handling your work, your main concern is with their ability to staff your project properly. For this, the selection team needs to study the contractor's personnel loading curves, submitted with the screening reply. Figure 2.4 provides a typical format indicating the contractor's current workload data.

The examples show construction personnel loading curves for two prospective contractors. Contractor A's curve shows the current back-

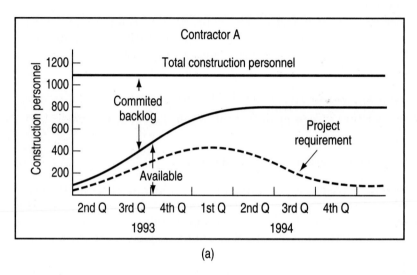

(a)

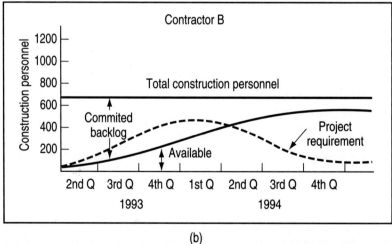

(b)

Figure 2.4 Contractor personnel loading curves. (*Reproduced with permission from* Proceedings of 10th Annual Seminar/Symposium, *Project Management Institute, October 1978, p. II-L.3.*)

log tapering off in time to add your expected project's workload with some capacity to spare. The curve for Contractor B shows that a backlog will not permit the commitment of enough personnel from the contractor's pool to staff your project to meet the schedule. The only alternatives for Contractor B are to increase staff or subcontract some of the work. Those options must be investigated thoroughly before selecting that contractor.

At this stage you are making only a preliminary evaluation of workload, and will do a more thorough investigation later on. Also at this stage, contractors will try to show themselves in the most favorable light, so the personnel loading curves will be somewhat optimistic. Furthermore, the contractor may book some additional work during the proposal selection process. Check the remaining responses for other pertinent factors such as: experience, location, dedication, interest, project execution plan, and worthwhile suggestions. Scoring these individual factors on a weighted scale of 1 to 5 points gives a total score for each proposal. The short list for proposals consists of the five or six highest scores.

Preparing the RFP

As shown on the proposal schedule in Fig. 2.2, the work for preparing the RFP documents proceeds concurrently with the screening activity. Thus there should be no delay in issuing the RFP as soon as the proposal slate is ready.

At a minimum, the RFP documents should include the following major sections:

- Proposal instructions
- Form of proposal
- Scope of work and services
- Pro forma contract
- Coordination procedure and job standards

The size and complexity of RFP documents can range from a few pages to several books, depending on type of contract and size of project. A small air-conditioning job, for example, might require several pages, while a grassroots paper mill might need several volumes.

The proposal instructions should include the following subjects to ensure that uniform proposals are received, permitting a meaningful comparison. Receiving proposals in a variety of formats makes comparison difficult, and can even lead to a poor selection and unmet project goals. Important areas to include in the instructions to contractors are:

- *General project requirements*—project description, proposal document index, scope of work, site location and description, etc.
- *Special project conditions*—owner contacts, site visitation, proposal meetings, etc.
- *Technical proposal requirements*—as needed for contractor's technical input on design-build projects.
- *Project execution proposal requirements*—could include a table of contents of key information that the owner desires (i.e., corporate organization chart, experience record, project execution plan, project organization chart, key personnel assignments, current workload charts, preliminary schedule, description of project-control functions, and any other pertinent data desired).

The proposal form is at the heart of the contractor's commercial proposal. It requests all of the contractor's business terms, such as pay scales, fringe benefits, overheads, fees, travel expense policy, heavy equipment charges, and computer charges. This section also asks contractors to state any exceptions to the terms of the proposal agreement, the pro forma contract, and any other terms of the proposal invitation.

The pro forma contract is a draft of the contract that the owner plans to execute with the successful firm. The number of exceptions taken to your pro forma contract will be directly proportional to the volume of the document. Standard short-form contracts, such as those developed by the American Institute of Architects (AIA) and the National Society of Professional Engineers (NSPE), can offer some particular advantages in this regard. At the other end of the scale are the long-form contracts generated by large corporate legal staffs and the federal government. The long-form contracts often generate more problems than they solve. Most legal people don't recognize the benefits of the KISS Principle.

The coordination procedure includes the basic standards and practices that owners expect contractors to follow in executing the project. This section may include the owner's design practices and standards, local government requirements, safety regulations, machinery standards, procurement procedures, mechanical catalogs, operating manuals, accounting standards, and acceptance of the work. This section is probably the largest volume producer in the proposal documents, depending on how much of this material an owner normally uses. Owners with limited standards can select the standards needed from a menu provided by the contractor and still have a good project.

Most of the material contained in the RFP is an outline for the execution of the project. Design it to get the contractors more deeply involved in thinking about your project. One goal of the RFP is to get contractors' creative juices flowing to offer any unique construction technology and to get management input to improve project cost or schedule. Good contractors are eager to display their competence and improve their standing with the owner, as a way of improving their chances of winning the contract.

Evaluation of the proposals

The purpose of having strict requirements in the RFP documents is to generate consistent proposals leading to easy and accurate comparison on a common basis. Nonconforming proposals are subject to rejection or revision, which adversely reflects on the errant proposer. The three crucial sections of the proposal request requiring special attention are:

- The business proposal
- The technical proposal
- The project execution plan

Assign these sections to the separate specialty people on the proposal team for their expert input. Reviewing the business and technical portions separately avoids any influence of the commercial terms on the technical portion of the proposal. Finally, combine the commercial and technical evaluations to select the best overall proposal. The selection team then presents their findings to upper management for approval.

The commercial evaluation focuses on the total cost of the various contractors' services. If the proposals are on a lump-sum basis, the commercial evaluation is quite simple and involves comparing a few lump-sum cost figures. The technical and performance evaluations are then more critical to ensuring that the offerers are pricing the full scope of services. That is necessary to ensure that the owner's project expectations will be satisfactorily met.

For a CPFF proposal, the commercial evaluation is more complicated. In this case the evaluation team must reconstruct the total cost of the services from the estimated project labor-hours, multiplied by the average hourly rates of each proposal. Since the labor-hour estimate may vary for each proposal, evaluators must compare the proposals against a standard set of labor-hours. The selection team then multiplies the estimated standard labor-hours by each offerer's total personnel costs, to arrive at the expected total cost for services. This puts the expected labor costs for all proposals on a common-cost basis. Remember, this type of cost evaluation is only approximate, so it doesn't carry the same weight in an evaluation as a lump-sum proposed price.

Before making the final selection, the selection team should visit the contractors' offices for an on-site examination of the firm's proposed personnel, systems, equipment, and facilities. If at all possible, you should plan to visit at least one of each short-listed contractor's construction sites to observe their field organizations in action.

Schedule your office and site visits well in advance, so the contractor's top management and key project personnel are present for personal interviews. The appraisal team should study the organization from top to bottom to get a feel for their ability to meet your project needs. Especially look for top-management commitment, capable personnel, modern facilities and equipment, workable systems and procedures, job history, and reference list. Now is the time to recheck the workload curves to see that they are still factual. Review each contractor's work in progress, as well as any outstanding proposals besides your own. Also, now is a good time to pursue the details of any technical proposal features that may need clarification or further discussions with the contractor's top technical people.

If the bidder is offering a design-build consortium, you should visit both the design and constructor's offices. Give special attention to how they propose to coordinate the design, procurement, and build interfaces.

The contractor will supply much of the information you seek as part of a dog-and-pony show presented on your behalf. Be polite as you enjoy this show, but don't accept it as the final answer to your questions. Remember, the main purpose of the presentation, from the contractor's point of view, is to downplay any weaknesses in its proposal!

With the final office visits completed and evaluated, you are ready to make the final selection. The three main areas to rate are the technical, commercial, and project execution proposals. To get the final rating, combine the scores from the three major areas with the results of the office and site visits. You are then in a position to make a recommendation to top management for approval.

It is good practice to rate the three top contenders in one-two-three order of preference before starting final negotiations with number one. Then if you can't reach agreement with number one, you can drop back to the number-two contractor, and so on. It's usually not a problem to reach an agreement with your prime candidate, if all matters were fairly presented by all parties. Using a well-organized selection process such as that described above should ensure a good selection and a happy project!

The final work of the selection team is to tell the unsuccessful candidates of the final decision as soon as possible. You may expect that the unsuccessful offerers will approach you for a debriefing as to why their proposal was not accepted. If the request is made in the spirit of wanting to improve their performance on the next proposal, a fair and open discussion of nonconfidential matters is in order. This provides the contractor with at least a partial payback for its efforts and investment on your behalf.[2]

Developing Construction Proposals

The contractor's project and construction management people meanwhile are performing their functions on the other side of the proposal preparation process. However, having a good understanding of the selection process from the owner's viewpoint is often more valuable than the how-to of this section.

Many project and construction managers regard proposal assignments as fill-in work to be done only when they are not assigned to an active project. This is indeed a shortsighted attitude toward a function that is the lifeblood[1,6] of their organizations. After all, it is successful proposals that make everyone's job in the company possible.

The construction manager who gives business development and top management a strong performance when it comes to selling jobs makes a valuable contribution to their companies and themselves. A truly professional construction manager cannot afford to take proposal assignments lightly!

The proposal goals

Managers of construction proposals have the goal of turning out winning endeavors on a tight budget within the time allotted. As with owner project managers, contractor CMs must treat the proposal work as a miniproject using the same planning, organizing, and controlling functions. Miniproject schedules allow no room for false starts, missed targets, or mistakes along the way.

The contents of a construction proposal are highly variable, depending on the type of contracting strategy used. Most will require some sort of bid price or cost estimate as part of the commercial portion.

The detailed cost estimating falls into the estimating department's domain, subject to review and approval by the CM and top management. We will go into more detail on the various forms of estimating in later chapters.

In fixed-price bidding for publicly funded work, having a competitive price is very important to gaining consideration for your proposal. However, it is also very important to have all the boilerplate items in the RFQ covered. Does the proposal address the RFP requirements for affirmative-action plans, small business set-asides, subcontractor qualifications, progress evaluation and reporting systems, and the plethora of other government regulations involved in public projects? Failure to address any one of these areas gives the owner an opportunity to reject an otherwise winning bid. Also, the construction estimates must be based on the construction-execution plan as developed by the CM and upper management. All of which adds up to the conclusion that the right price is at least half the battle when it comes to winning a project.

In private work, where having the lowest price is somewhat less important, developing a powerful project-execution approach for the proposal is even more compelling. In the private sector, owners look much harder at contractor performance, quality, and schedule than at price alone.

The CM's role in proposal development

Responsibility for proposal generation varies from contractor to contractor, depending on company policy. The business development (BD) group is usually responsible for the actual production and delivery of the proposal documents. BD also coordinates the live presentations to, and visitations from, the prospective client.

If a turnkey contract approach is proposed, a major technical part of the proposal will be devoted to the design and procurement services involved. In that case, a project manager (director) may be assigned to coordinate the technical input for the proposal. The PM normally delegates the development of the construction portion of the proposal effort to the CM.

The construction manager is responsible for developing a dynamic construction-project-execution proposal. The CM must be certain that the construction technology approach to the work is sound, and is presented in such a way as to match the goals of the prospective client and the requirements of the RFP. In developing the construction plan, the CM marshals the best company resources available, i.e., uses the expertise of the appropriate company department heads. The task of editing and polishing the specialist inputs to create a unified proposal format also falls to the construction manager. This means that the CM must be an excellent communicator.

The CM, with the approval of top management, also develops the construction organization chart and key personnel assignments for the proposal. The organization must present a can-do project execution team when the client selection committee makes its office visit. The CM should also prepare and rehearse the proposed project team for any live presentation to the client. The training program and rehearsal culminate in a dry run before a devil's-advocate tribunal. The tribunal should include top management, so the presentation had better be right! Make any final adjustments to it based on the comments received during the dry run.

The construction site survey

A key proposal function of the CM is to lead the construction site inspection team and to oversee the preparation of the site survey report. The amount of effort put into the site inspection is a function of the project's size, complexity, and location. The input can vary from one person on a small project to a dozen or more on a superproject. The cost in the latter case can be considerable, but the benefits to the proposal, and to project initiation after a successful proposal, can also be considerable. Our discussion will be based on the requirements for a large project; smaller jobs can be scaled back to suit their size requirements.

If the construction site is located in an area where your firm works regularly, you already know more about the local conditions than your outside competitors. However, a site visit is still necessary to get the information pertaining to this specific project, and to make a quick review of local conditions to confirm that your existing data is still valid. If the site is distant, the site survey should be set to coincide quite closely with any contractor orientation meetings scheduled by the owner or the architect-engineer.

An effective site survey requires a sound plan, good organization, and constant control by the survey team leader. Make sure that all the participants have thoroughly reviewed the bidding documents, plans, specifications, and project data before they visit the site. Team members should develop lists of questions and points that need clarification if they are to properly bid the job.

The team should consist of anyone who has a major contribution to make to the site survey for that type of project. At a minimum the project estimator, a labor relations specialist, a field engineer, a risk manager, and a personnel specialist should be considered.

The CM should hold an organizational meeting, with all team members present, to ensure that everyone understands the goals of the proposal and the site survey. Each team member's specific assignment should be discussed and coordinated to eliminate possible overlaps and confusion during the inspection trip. A professional approach to the survey can make a very favorable impression on the owner and on local agencies, and this will enhance the contractor's image immensely.

A site inspection checklist for all types of projects is too voluminous to include here, but the following points should be investigated at a minimum:

1. Project location
   a. Remote or developed area
   b. Research climatology and history of natural disasters
   c. Site access roads, laydown areas, and traffic problems
   d. Available transportation systems
   e. Soil conditions and site survey maps
   f. Location of neighbors
   g. Existing facility, ongoing operations
   h. Offsite warehouse facilities
   i. Local legal and licensing requirements
   j. Applicable local codes and regulations
   k. Availability of construction staff housing
   l. Local economic climate and cost of living
   m. Availability of local sources for construction supplies and services

2. Labor survey
   a. Project labor posture—union versus open shop
   b. Availability and quality of local craft labor
   c. Local labor practices and history
   d. Availability of local-hire office and service staff
   e. Quality and availability of subcontractors
   f. Prevailing wage rates, fringes, and other costs
   g. Area labor productivity
   h. Competing projects in the area (ongoing and future)
   i. Local craft-training posture

3. Site development
   a. Green-field site or existing plant
   b. Demolition work required
   c. Evaluation of site development and drainage design
   d. Construction waste disposal—solid, liquid, fumes, etc.
   e. Applicable government regulations

4. Temporary site facilities
   a. Availability of temporary construction utilities
   b. Temporary offices; trailers, portable, other
   c. Construction warehouses; existing, temporary, other
   d. Heavy equipment rental and services
   e. Communication systems; voice, data, other
   f. Site safety and security considerations—client requirements
   g. Local services available; banking, health, police, etc.

5. Public relations
   a. Check on how project is being received in community
   b. Contact involved local government agencies
   c. Collect local newspaper articles relating to project
   d. Check any antiproject activist activity anticipated
   e. Evaluate any potential environmental problem areas
   f. Visit possible ongoing projects in the area to assess potential local problem areas

6. Client relations
   a. Establish rapport with client organization
   b. Obtain any necessary missing project information
   c. Discuss client's project goals for the construction program and facility
   d. Review your planned project execution approach with client to eliminate any potential conflicts
   e. Check out client's facility labor posture

The CM should ensure that the team members have the necessary tools and data to carry out an effective survey. Dictating devices, video and still cameras, bidding data, measuring devices, portable builder's level, and any other labor-saving devices are a must to make the best use of the time and effort spent. Every effort should be made to do the job right the first time, eliminating the need for repeat trips later.

The format of the final site survey report should be presented in an organized and timely manner for use by the proposal team. This means that it should be clearly written, and published as soon after the trip as possible. A poorly prepared report will waste the considerable investment made in the survey trip, as well as ensure the failure of the proposal effort. Again, the site survey should be treated as a miniproject!

Reviewing the pro forma contract

The construction manager should take the lead in reviewing and responding to any administrative problem areas in the pro forma contract presented in the RFP. The legal people will handle some of this activity, as it affects the liabilities and other risks assumed by the company. The construction manager is responsible for reviewing the day-to-day operating requirements called for in the contract. The specific areas of interest in the contract for the PM are:

1. Scope of work

2. Payment terms

3. Change order procedure

4. Quality control

5. Guarantees and warranties

6. Schedule requirements

7. Incentive clauses

8. Suspension and termination

Your key role on the proposal team puts you in a position to come up with the "hooks" that will make the selection committee pick your proposal. This is an opportunity for creative thinking, so don't pass it up. You must reduce the potential impact of any negative proposal factors—human resource shortages, poor office or site location, adverse costs, etc.—that might hurt your chances of winning the proposal. This work is more exciting if one approaches the proposal as a contest of wits in trying to beat out the competition. You are probably spending good money on computer games that are not half as much fun as this real-life challenge!

Let's assume that you have been good enough at your craft to win the proposal. You developed the most creative problem solutions, organized the strongest project team, orchestrated an outstanding dog-and-pony show, and made the fewest mistakes. On that happy note, you are now ready to play a key role on the contract negotiation team. This requires a solid background in the area of construction contracting, which just happens to be our next subject.

Construction Contracting

The basic format of construction contracting has not changed much in the last 30 years. The fixed-price approach has been with us for centuries, and the construction management and incentive-type reimbursable contracts blossomed in the 1950s. This stability in the contracting area should come as no surprise, because the legal basis for contracting in the United States is founded on centuries-old English Common Law. I don't recommend that you get involved in the convolutions of the legal aspects of contracts with your friendly lawyer. The discussion is likely to make you want to switch to some other business! We will concentrate our efforts on those aspects of contract negotiation and administration that will allow the CM to keep the company out of court!

The CMs on both sides of the construction contract equation play influential roles in the negotiation and acceptance of the contract document. These are the people held responsible for fulfilling the terms of the document and making it work. If all goes well, and it often does, the lawyers and top management involved may never have to look at the contract again.

The reason for the contract is to clearly and fairly set forth the responsibilities of each party involved in the project before the work begins. It also spells out the remedies one party may seek from the other for failure to perform or for defaulting on any of the terms. When such a situation occurs, the contract becomes the primary document for settling any later arbitration or litigation claims.

Contracts for construction of capital projects have assumed more importance in recent years, because we have become a litigious society. Years ago, the executed contract sometimes arrived closer to the end of the project than the beginning. The trend toward a tightening up of contracting procedure has not been all bad, for it has forced the project participants to finish their homework *before* starting the project.

Due to the complexity of capital projects, the contract cannot cover all possible situations that are likely to arise. This means that there must be a good deal of give-and-take, as well as reasonableness on both sides. Having a cooperative approach to contract administration wins at least half the battle, with the other half being won by having a sound and equitable contract document to start with.

This discussion of contracts is from the standpoint of the CM, not from the legal side. It covers what the parties need to know to arrive at a sound contract through a fair and equitable distribution of the risks involved.[2] An approach that allows all parties to meet their project goals usually works out best.

Here are some of the major areas of contractual risk that must be considered when selecting a construction contracting strategy:

- Degree of project definition
- Possibility of ongoing design changes
- Possibility that the facility will not meet its performance goals
- Potential escalation of labor and material costs
- Unknown labor productivity in project locale

- Difficult climatic conditions
- Inadequate or unqualified labor supply
- Occurrence of labor strikes or civil unrest
- Occurrence of force majeure or natural disaster
- Possibility that bids will exceed the estimate
- Unfeasible strategic completion date
- A process design that has not been adequately proven
- Inadequate licensing package
- Casualty loss not covered by insurance
- Liability to third parties

The owner and the contractor must evaluate the applicable items on this list to determine which party will assume which risks. Some of the owner's risks can readily be covered by insurance and bonds, that are available at a fixed cost. Clauses to that effect must be included in the contract.

The owner eventually assumes the lion's share of the risks in selecting the basic design of the facility, developing the conceptual design, and selecting the site for the work. The contractor's risks occur in the construction management areas of labor supply, productivity, schedule, and local site conditions. All of the latter risks are assumed by the contractor at a protected price. To put it another way, the owner has to pay the contractor's contingency to guard against the potential risk. However, in a buyers' market, the contingency may well be a bargain.

Contract formats

The number of contract formats is virtually infinite. There are literally no two formats exactly alike, just as there are no two projects that are exactly alike. We can, however, classify the basic contract formats as listed in Table 2.1. Brief descriptions of the advantages and disadvantages and typical applications are also given. Although this chart first appeared about 20 years ago in *Chemical Engineering* magazine, I have never been able to find a more comprehensive summary of contracts in any later literature.

The contract format used on a given project depends on many factors. Some of these are: project definition, owner preferences, public laws (government contracts), current market conditions, project location, project financing, schedule, assumption of risks, and scope of work. Factors affecting format will vary over time and from job to job.

TABLE 2.1a Types of Contracts

Primary advantages	Primary disadvantages	Typical applications	Comments
	Cost-Plus		
	PD: Minimal (Scope of work does not have to be clearly defined.)		
1. Eliminates detailed scope definition and proposal preparation time. 2. Eliminates costly extra negotiations if many changes are contemplated. 3. Allows client complete flexibility to supervise design and/or construction.	1. Client must exercise tight cost control over project expenditures. 2. Project cost is usually not optimized.	1. Major revamping of existing facilities. 2. Development projects where technology is not well defined. 3. Confidential projects where minimum industry exposure is desired. 4. Projects where minimum time schedule is critical.	Cost-plus contracts should be used only when client has sufficient engineering staff to supervise work.
	Cost-Plus, with guaranteed maximum PD: General specifications and preliminary layout drawings		
1. Maximum price is established without preparation of detailed design drawings. 2. Client retains option to approve all major project decisions. 3. All savings under maximum price remain with client.	1. Contractor has little incentive to reduce cost. 2. Contractor's fee and contingency are relatively higher than for other fixed-price contracts, because price is fixed on preliminary design data. 3. Client must exercise tight cost control over project expenditures.	When client desires fast time schedule, with a guaranteed limit on maximum project cost.	
	Cost-Plus, with guaranteed maximum and incentive PD: General specifications and preliminary layout drawings		
1. Maximum price is established without preparation of detailed design drawings. 2. Client retains option to approve all major project decisions. 3. Contractor has incentive to improve performance, since he shares in savings.	Contractor's fee and contingency are relatively higher than for other fixed-price contracts, because price is fixed on preliminary design data.	When client desires fast time schedule, with a guaranteed limit on maximum cost and assurance that the contractor will be motivated to try for cost savings.	Incentive may be provided to optimize features other than capital cost, e.g., operating cost.

PD = Project Definition

TABLE 2.1b Types of Contracts (continued)

Primary advantages	Primary disadvantages	Typical applications	Comments
Cost-Plus, with guaranteed maximum and provision for escalation PD: general specifications and preliminary layout drawings			
1. Maximum price is established without preparation of detailed design drawings. 2. Client retains option to approve all major project decisions. 3. Protects contractor against inflationary periods.	1. Contractor has little incentive to reduce cost. 2. Contractor's fee and contingency are relatively higher than for other fixed-price contracts, because price is fixed on preliminary design data. 3. Client must exercise tight cost control over project expenditures.	1. Project involving financing in semi-industrialized countries. 2. Projects requiring long time schedules.	1. Escalation cost-reimbursement terms should be based on recognized industrial index. 2. Escalation clause should be negotiated prior to contract signing.
Bonus/penalty, time, and completion PD: Variable, depending on other aspects of contract			
1. Extreme pressure is exerted on contractor to complete project ahead of schedule. 2. Under carefully controlled conditions, will result in minimum design and construction time.	1. Defining the cause for delays during project execution may involve considerable discussion and disagreement between client and contractor. 2. Application of penalty under certain conditions may result in considerable loss to contractor. 3. Pressure for early completion may result in lower quality of work.	Usually applied to lump-sum contracts when completion of project is absolute necessity to client in order to fulfill customer commitments.	1. Project execution should be carefully documented to minimize disagreements on reasons for delay. 2. The power to apply penalties should not be used lightly; maximum penalty should not exceed total expected contractor profit.
Bonus/penalty, operation, and performance PD: Variable, depending on other aspects of contract			
Directs contractor's peak performance toward area of particular importance to client.	1. Application of penalty under certain conditions may result in considerable loss to contractor. 2. Difficult to obtain exact operating conditions needed to verify performance guarantee.	When client desires maximum production of a particular by-product in a new process plant to meet market requirements.	Power to apply penalties should not be used lightly.

TABLE 2.1c Types of Contracts (Continued)

Lump sum, based on definitive specifications
PD: General specifications, design, drawings, and layout—all complete

1. Usually results in maximum construction efficiency. 2. Detailed project definition assures client of desired quality.	1. Seperate design and construction contracts increase overall project schedule. 2. Noncompetitive design may result in use of overly conservative design basis. 3. Responsibility is divided between designer and constructor.	When a client solicits construction bids on a distinctive building designed by an architectural firm or when a federal government bureau solicits construction bids on a project designed by an outside firm.	Clients are cautioned against use of this type of contract if project is not well defined.

Lump sum, based on preliminary specifications
PD: Complete general specifications, preliminary layout, and well-defined design

1. Competitive engineering design often results in cost-reducing features 2. Reduces overall project time by overlapping design and construction. 3. Single-party responsibility leads to efficient project execution. 4. Allows contractor to increase profit by superior performance.	1. Contractor's proposal cost is high. 2. Fixed price is based on preliminary drawings. 3. Contract and proposal require careful and lengthy client review.	1. Turnkey contract to design and construct fertilizer plant. 2. Turnkey contract to design and construct foreign power generation plant.	1. Bids should be solicited only from contractors experienced in particular field. 2. Client should review project team proposed by contractor.

Unit price contracts, flat rate
PD: Scope of work well defined qualitatively, with approximate quantity known

1. Construction work can commence without knowing exact quantities involved. 2. Reimbursement terms are clearly defined.	1. Large quantity estimate errors may result in client's paying unnecessarily high costs or contract extra. 2. Extensive client field supervision is required to measure installed quantities.	1. Gas transmission piping project. 2. Highway building. 3. Insulation work in process plants.	Contractor should define the methods of field measurement before the contract is awarded

TABLE 2.1d Types of Contracts (Continued)

Primary advantages	Primary disadvantages	Typical applications	Comments
	Unit price contracts, sliding rate PD: Scope of work well defined qualitatively		
1. Construction work can commence without knowing quantity requirements. 2. Reimbursement terms are clearly defined.	Extensive client field supervision is required to measure installed quantities.	1. Gas transmission piping project. 2. Highway building. 3. Insulation work in process plants.	Contractor should clearly define the methods of field measurement before the contract is awarded.
	Convertible contracts PD: Variable, depends on type of contract conversion		
1. Design work can commence without delay of soliciting competitive bids. 2. Construction price is fixed at time of contract conversion, when project is reasonably well defined. 3. Overall design and construction schedule is minimum, with reasonable cost.	1. Design may not be optimum. 2. Difficult to obtain competitive bids, since other contractors are reluctant to bid against contractor who performed design work.	1. When client has confidential project requiring a balance of minimum project time with reasonable cost. 2. When client selects particular contractor based on superior past performance.	Contractors selected on this basis should be well known to client.
	Time and materials PD: General scope of project		
1. Client may exercise close control over contractor's execution methods. 2. Contractor is assured a reasonable profit. 3. Reimbursement terms are clearly defined.	1. Project cost may not be minimized. 2. Extensive client supervision is required.	Management engineering services supplied by consulting engineering firm.	Eliminates lengthy scope definition and proposal preparation time.

SOURCE: Reproduced with permission from "A Fresh Look at Engineering and Construction Contracts," *Chemical Engineering*, Sept. 11, 1967, pp. 220–221.

That makes it difficult to create, at any given time, to one standard contract format for all types of projects.

Lately, there does seem to be a trend in the industrial/commercial project area toward shorter contract formats. Even the legal people have come around to the opinion that a concise contract format makes for a more effective agreement. Many state and local government bodies, along with some commercial developers, have embraced the shorter association formats. An example of a construction version is AIA Document A201, "General Conditions of the Contract for Construction." Sidney Levy gives it a detailed review in *Project Management in Construction*.[3] Up-to-date copies of these printed contract forms are available directly from the various professional associations or through legal document dealers.

Let's review some of the key considerations involved in format selection by discussing the formats listed in Table 2.1. The 12 formats shown break down into 4 major categories as follows:

- Cost-reimbursable contracts (Cost-plus)

  Cost-plus

  Cost-plus with guaranteed maximum

  Cost-plus with guaranteed maximum and incentive

  Cost-plus with guaranteed maximum and provision for escalation

  Time and materials

- Bonus-penalty contracts

  Based on time and completion

  Based on operation and performance

- Lump-sum contracts

  Based on definitive specifications

  Based on preliminary specifications

- Unit-price contracts

  Flat rate

  Unit price, sliding rate

Another category is a convertible contract format, which changes any of the above reimbursable contracts into a lump-sum agreement. Conversion can occur at any previously agreed-upon point in the work. For example, you could start with a reimbursable-type contract when the project scope is not very firm, then convert to a lump-sum basis after you have finished some of the design. The parties can then negotiate a lump-sum price, including any contingency or incentives, based on the refined scope of work.

Selecting the Contract Form

The selection of the contract format is usually made by the owner when formulating the contracting plan. Occasionally a contractor may sell the owner on a different contract format, as when there is an unusually favorable owner-contractor fit on certain types of projects.

As we refer to Table 2.1, let's look at some factors that lead to matching the best contract form to the specific project environment. The number-one factor to consider is the degree of project definition (PD) available at the time of contractor selection. The first law of contract format selection is: *The better the project definition, the easier it is to get reasonable fixed-price bids.*

Owner advantages and disadvantages in format selection

We said earlier that most public agencies split their contracts between the design and construction phases. Thus, when they are ready to commit about 85 percent of the total project cost for construction, the project scope is completely defined. The work is fully defined by owner-approved plans, specifications, and contract documents. Such detailed project definition allows a slate of bidders to develop fixed prices with minimal contingencies. If a good pattern of pricing, in line with the engineer's estimate, occurs, the owner then has a good handle on its costs. This method offers definite cost-control advantages to an owner spending public funds.

A major sacrifice that the owner makes in getting the cost-control advantage is prolonging the project schedule. It usually takes about 25 to 30 percent longer to get from concept to facility use using the split-contract mode. The bar charts in Figs. 4.1 and 4.2 illustrate the potential timesaving resulting from a CPFF contract format over a fixed-price format. The freedom to make an early start on procurement and construction during the design phase results in the shorter schedule. Many commercial and industrial project developers can't afford the luxury of a delayed schedule in the fast-paced world of international competition.

Obviously, the best available alternative to shorten the completion date is to use a cost plus fixed fee (CPFF) contract format, allowing the design, procurement, and construction activities to proceed simultaneously right after the conceptual phase. A fringe benefit is the additional time saved by the early ordering of any long-delivery equipment and materials that are on the critical path. Most government agencies do not have the personnel to control CPFF work, or the willpower to keep the project scope under control during the development of the design. The history of cost overruns has led to the passage of laws requiring lump-sum bids on most government work, and often that doesn't even work.

The CPFF approach was pioneered by the international oil companies during the 1950s and 1960s, when they were building their huge refinery and petrochemical complexes around the world. That in turn led to the development of the engineering construction giants who served that market segment. The need for speedy completion of those superprojects offset the relatively minor cost of the added owner supervision of the CPFF contract requirements. That added cost was more than offset by the savings in contractor contingencies that would have been included in a fixed-price bidding approach. This approach is referred to as *fast-tracking the schedule*. The faster schedule delivers the facility earlier and offers an early entry into the marketplace, with potential for earlier return on investment (ROI).

With effective supervision of the CPFF contract, the difference in contractor supervisory cost versus the fixed-price method is minor. Usually it is more than offset by the savings in contingencies, inflation, and earlier ROI. The extra expense results from the added owner's personnel required to control project costs. Lump-sum contracts are self-policing on both the owner and the contractor sides. Both are working against the agreed lump-sum price for the work that controls both changes and costs. However, the owner can expect change-order requests from the lump-sum contractor at the slightest variation in the contract documents.

Owners can get the best of both worlds by converting from a CPFF to a fixed-price basis after finishing the project definition. At that time both parties can agree on a reliable, open-book cost estimate, including contingency, and agree on a fair not-to-exceed price for the work. The contract then switches into a guaranteed-maximum format with the contractor assuming primary responsibility for controlling the final cost of the work. I firmly believe in the philosophy that construction contractors perform better when they have a financial incentive. Linking the incentive to a meeting of project goals ensures a good return for the owner. With a well-designed incentive contract, an owner can inspire a superior performance from contractor teams and improve the odds of meeting or exceeding the project goals. It is difficult to include quality in the incentive contract without sacrificing the overall quality of the facility. The incentive method usually is applied only to the goals of cost and schedule.

In the incentive method, the owner and the contractor agree on the estimated cost of the project. They add a contingency or upset amount to the targeted cost to arrive at an *upset price*. This upset amount will be shared by the contractor and the owner on the basis of the savings actually made at the end of the project. This pot can be split (usually 60 percent owner, 40 percent contractor) based on the contractor meeting or bettering the agreed-upon targets of cost and schedule. Industrial owners have used this arrangement for many years on critical projects with tight schedules. It actually offers a double-carrot approach instead of the carrot-and-stick approach of the bonus-penalty contract format.

If a bonus-penalty contract format is felt to be necessary, remember that it works best only when one contractor has sole responsibility for the work. Otherwise there will be constant claims for time extensions, resulting from delays caused by the various parties responsible for the work, including the owner. This format is no longer common in industrial and commercial practice.

However, government contracts still use the penalty part in the form of liquidated damages, particularly on construction work. Liquidated damages claim to indemnify the owner against losses incurred by dint of not having the facility ready on time. Actually they work as fines set against the contractors, forcing attention to the schedule and completion of the work. These damage clauses are difficult to enforce. When the project does get into serious trouble, they usually do more harm than good.

Before leaving the subject of reimbursable and lump-sum formats, I should warn against using them simultaneously on one project with the same contractor. To do so runs the risk of mixing two different forms of payment, which can result in the owner paying twice for the same work. Some minor work, such as extras, can be done on a time and material (T&M) basis on a lump-sum job with careful controls.

One can also consider a T&M basis for small amounts of work, when it is not convenient to use another form of contract. But open-ended T&M work orders are anathema to construction managers. Avoid them whenever possible, because they are fraught with surprises, particularly when accumulated charges arrive near the end of what you thought was a successful project. T&M work should always have an authorized, not-to-exceed value that triggers an evaluation of the work to date when that dollar value is billed out.

Unit-price contracts are acceptable when it is not convenient to use one of the other methods. The key to controlling the cost is to get a unit price that gives the contractor an incentive to perform efficiently. The main disadvantage to this format is the added cost needed to count the work units. It is a good idea to have prearranged unit prices in the contract or subcontract for use when pricing additions to and deletions from the work. Unit prices are convenient for relatively standard types of work such as steel erection, masonry, paving, excavation, and the like. They are more difficult to use for complex trades such as HVAC, plumbing, process piping, instrumentation, and the like.

Contractor advantages and disadvantages in format selection

The contractor, as a seller of construction services, starts from an inherently weaker position than the owner in the contracting process. There are always plenty of competitors offering similar services of equal quality at a reasonable price. That makes it a buyer's market just about 99 percent of the time. That in turn usually places the contractor in the position of having to accept more risk than the owner. However, owners must exercise some restraint when using the buyer's power, to guard against poisoning the contractual arrangement and thereby placing the project goals in jeopardy.

Those contractors that can successfully manage the increased risk of lump-sum work stay in business, those that can't fall by the wayside or take only reimbursable work. In general, this means that any contract format that reduces the contractor's exposure to risk is advantageous to the contractor. Looking down the list of Fixed Price vs. Reimbursable columns in Appendix B of BRT Report A-7,[2] one readily sees that the owner assumes most of the risk for costs on reimbursable work and the contractor most in fixed-price work. This readily explains why owners spend much more time and care in selecting reimbursable contractors. It also explains why having some sort of performance incentive plan in reimbursable contracts is so vital to giving a financial incentive to the contractor.

Although the contractor sheds most of the financial risks with a reimbursable contract format, that doesn't mean that the CM can forget about cost control. Every reimbursable project must have a budget to set the owner's financial goals. The construction management team must make every effort to meet or better that budget, even without an incentive clause in the contract. The project should be run like a fixed-price job by both the owner and contractor project teams, if they are to have any chance of meeting the financial goals. Often it is too easy for the owner to fall into the trap of making costly changes during the construction phase. Each change must go through an approval process and be thoroughly documented to maintain a current budget and projected cost-to-completion for the project.

The same cost-control thinking must be applied to craft-labor productivity and field indirect costs if they are to meet their established budgets. The only real difference between the fixed-price and reimbursable modes of operation is that the reimbursable work is done with open-book cost estimates and budgets. In fixed-price work, the only financial records of the project open to the owner are the lump-sum price and the contractors' payment schedules.

Contract formats with liquidated damages and penalty clauses are considered disadvantageous by contractors. Such clauses are usually introduced by owners to ensure facility performance and completion on schedule in the fixed-price contracting arena. The contractor who accepts penalty clauses will normally have some sort of contingency, bonding, or insurance cost in the price to cover those costs, should they be invoked. If protection is not included, the contractor's risk as to profitability on the project is quite high.

I recently encountered an extreme case of owner liquidated-damages protection on a large flue gas wet scrubbing system being built by an electric utility company. The scrubbing process was designed to produce wallboard-grade gypsum as a by-product. The bidding documents included a liquidated-damages clause making the turnkey contractor liable for the cost of 30 years' production of gypsum if the specified grade of product could not be produced. A rough estimate of the value of 30 years' production at today's market price is over $750 million, and the contract did not even include a contractor's limit of liability clause! That little item was included on top of some normal penalties for not meeting schedule, cost of power lost to down time, and an extremely rigid start-up schedule with penalties. The accumulation of all the penalties, without limit of liability, in a fixed-price contract could easily wipe out any contractor in the world. Because we were not "lucky" enough to win the contract, I was not privy to the final contract negotiations.

No company owner, architect-engineer, or contractor can accept unlimited financial liability in such a complex undertaking as the design and construction of a capital project. Much of the financial risk can be offset by buying suitable insurance coverage. Owners must remember, however, that eventually they pay for all insurance premiums, either directly as a project cost or in the contractor's overhead. Setting an acceptable limit of liability for the contractor can reduce the overall project cost.

A general risk area encountered in construction contracting is the owner or engineer's use of exculpatory clauses. These are catchall clauses that absolve the owner or engineer from all responsibility for anything that can possibly go wrong on the project. I like to refer to them as "hunting licenses," because they give permission to go after the contractor's hide if things go wrong. Generally these clauses are unenforceable in most states, but they do put the contractor's legal defense at a disadvantage. Owner's exculpatory clauses must be flagged by the construction manager during the contract review, so that the legal people can try to get them changed.

The Role of the CM in Contract Negotiations

The first time the proposed contract format surfaces is in the pro forma contract issued with the bidding documents. As the chief administrator of the final construction contract, the project or construction manager should become thoroughly familiar with its contents. Any potentially injurious or ambiguous areas should be flagged for swift resolution by your legal and top management people.

The degree of negotiability of the nebulous points depends on the type of contract format used by the owner. Use of the AIA-type standard formats in public works bidding usually leaves very little room for negotiation. The contract bids are often offered by the owner on a "take-it-or-leave-it" basis.[6] As mentioned earlier, construction contracting generally takes place in a buyers' market.

Contracts for private sector work are often more negotiable. If you can base your requested change on giving the owner improved project results, an acceptable compromise can be worked out. As the owner proceeds through the negotiation process, the contractor's bargaining position improves, because the owner doesn't want to go back to square one with another contractor.

As CMs, we often think of contracts as being strictly the domain of the legal department. Nothing could be further from the truth when we look at just how the contract comes into being and how it is handled during project execution. The legal people are responsible for developing the format and properly protecting the company against possible legal actions and uninsured casualty losses. Most legal people have only a limited view of the technical and management aspects of construction project execution.

Therefore, construction managers are responsible for developing or organizing the technical and commercial input for the contract. If the contractor prepares the contract, the owner's project or construction manager has the responsibility of reviewing the technical content for the legal and top-management people. With that sort of responsibility resting on their shoulders, neither the owner nor the contractor CM can afford to be cavalier when it comes to their knowledge of contracting!

It is good business for both the owner and contractor managers to know the project contract in intimate detail. If they did not partici-pate in the negotiation of the document as I recommended earlier, they must become conversant with it immediately upon arriving on the project. They are often named as contract administrators for their respective firms, so they had better know what the contract says.

To facilitate discussion of the most important areas of concern to the construction manager, I have sorted the relevant clauses into technical and commercial terms. They will not appear in the contract in that order, but the grouping should help to organize our discussion. The CM has primary responsibility for developing and defining the technical terms, and secondary responsibility for refining and coordi-nating the input for the commercial area.

Contract Technical Terms

The technical terms of the contracting procedure that we shall dis-cuss are:

1. The scope of work
2. Performance guarantees
3. Engineering warranties
4. Equipment guarantees
5. Schedule guarantees
6. Responsibilities of owner and contractor
7. Site safety and security

The scope of work

This is the most critical section of any contract, because it defines in specific terms the services performed or the work to be done. Defining the scope is especially critical in lump-sum contracts, where there is very little room for error. If scope definition is incomplete, contract and financial problems are sure to occur.

The scope section directly specifies the boundaries of the work. This includes buildings, offsite units, access roads, battery limits, remote units, and interconnections. There may be portions of the work, or even specific equipment, furnished by the owner or third parties. Defining those interfaces is important if misunderstanding between the parties is to be avoided. The problem is defining the scope in a direct and concise manner, without writing volumes to do it.

One way to define a complex scope with minimum wording is by reference to other detailed documents. A good example of that type of definition is where the A&E has completed the design and prepared bidding documents. In a less well-defined setting, you should include any existing documentation in the contract by reference. The documents can be drawings, studies, design manuals, and any other descriptive material prepared during the early stages of the project. The only criterion for using such documentation is that it should be a reasonably accurate representation of the work.

Another important item of scope definition is the scope of services to be performed by the contractor. In a design-build format this can include the design, procurement, construction, and start-up of the project under a single contract. Such an all-inclusive approach introduces a plethora of third-party interfaces not present in a simple construction services contract. A scope of services that broad will probably require a long-form contract to cover the legal implications introduced by the design, procurement, and construction activities.

A sound definition of the scope of work on any project is so critical that you should have other key project personnel review your final efforts to help ensure completeness and clarity. Another approach is to place yourself in the role of a project outsider with only limited knowledge of the work and then see whether you can clearly understand the scope definition that you have written.

Performance guarantees

The contractor's management must understand the complexities involved in performance guarantees. Failure to meet a performance guarantee can be expensive in make-good work. The CM must be careful to follow the process requirements of the process design closely, to eliminate any claim of faulty construction by the owner when guarantee problems surface. That means a thorough plant checkout against the process design, and the use of a carefully planned plant commissioning and start-up procedure.

The performance guarantee spells out exactly how the facility will perform as to products, quality, utilities, on-stream time, etc. Engineering/construction contractors rarely give a plant performance guarantee unless the contractor is furnishing basic process design as part of its services.

The term *performance guarantee* means that the contractor will modify the plant at its expense until the specified performance conditions have been met. Even this type of guarantee has a limit of liability or a specific penalty payment. This limit of liability comes into play in cases when the actual performance did not meet the guarantee in all respects. The construction contractor is usually kept on the site to handle any plant modifications, should they be required.

Engineering warranties

Design firms give this type of warranty to cover their detailed design work. Use of the term *warranty* rather than *guarantee* means that the design firm corrects any errors in its plans and specifications at its own cost. Although the design contractor furnishes corrected design documents at its cost, it usually does not accept responsibility for associated construction costs caused by the errors. On a split design-and-build contract format, the owner usually accepts responsibility for the costs of correcting construction work caused by design errors.

Equipment guarantees

The contractor passes these guarantees through to the owner on any equipment and materials bought by the contractor. They are the usual one-year materials and workmanship guarantees required of the equipment vendors. The contractor merely serves as the clearinghouse for owner complaints if equipment problems arise during the contractor's one-year facility warranty. All make-good costs under these guarantees are borne by the vendor. They are never underwritten by the contractor should the vendor fail to perform for any reason.

Schedule guarantees

Earlier, we talked about ensuring schedule performance through incentives or bonus-penalty clauses in a variety of contract formats. In the absence of incentive or penalty clauses, the contract should include a "time is of the essence" or a "best efforts to complete" clause. This type of clause does not create any financial obligations for either party. However, it does put contractors on notice as to the strategic completion date for the project. Thus contractors cannot plead ignorance when the end date arrives and the project is not finished. If they wish to protect their performance image, CMs must be careful to document schedule delays, even though there are no financial penalties involved.

If schedule guarantees are involved in the contract, the CM must ensure that the schedule is not impossible to meet under any circumstances. If the CMs are overridden by their managements and forced to accept the impossible schedule, they should go on record describing the problem areas. A statement to the effect that they will give their best effort to make the impossible date should be included.

Responsibilities of owner and contractor

The responsibilities of owners and contractors are set forth in two sections of the contract that define any obligations beyond those defined in the detailed scope of work. For example, they may involve the owner supplying by certain dates the project design basis, property surveys, government approvals, owner-furnished equipment, or project financing. Because any of these things can affect job progress, be sure to clearly define them in the contract and coordinate them into the overall project schedule.

An additional contractor responsibility might include the owner's request to furnish some additional items that are too minor to rate inclusion in the scope of work. These could be special reports, help with financing, certain types of schedules, project cost estimates, special inspections, subsurface investigations, etc. Many of these items can entail real added project costs and therefore must be included in the budget.

All of the above technical terms require strong input from the construction manager and the project technical staff. The respective project and/or construction managers are often the only technical representatives present at the contract negotiation. They must be ready to spot problem areas on either side of the table and bring the necessary technical expertise to bear on them.

Site safety and security

There must be a contractual requirement requiring the contractor to conform with owner's and governmental regulations pertaining to safety in the field. This clause should set the minimum standards for the contractor's site safety program, particularly when working in an existing operating facility. Ineffective contractor safety programs can be expensive to the owner in the form of insurance claims and high workers' compensation rates. (See Chapter 10 for details on safety requirements.)

The contractor's security liabilities should also be covered in the contract. This includes defining when the transfer of ownership of the materials and the completed facility will actually take place. Security can cost a lot of money when unforeseen site security problems arise. For example, I was the owner's representative on a school for the mentally retarded being built in a high-security-risk area. My boss was so against fences on that type of facility that he wouldn't even let us put up a temporary security fence during construction. This presented special security problems, because the school's presence was not being well received by the community. We had to pay for costly night watchman services to protect against nightly vandalism.

Contract Commercial Terms

The commercial terms of the contract that we will discuss are:

1. The type of contract and fee structure
2. Payment terms, including retention and the fee
3. Limits of liability
4. Project change notices
5. Cost-escalator clauses
6. Suspension and termination of the work
7. Force majeure
8. Insurance and risk management
9. Applicable law
10. Arbitration
11. Subcontracts

Type of contract, and the fee

Selecting the type of contract and the fee structure should occur during the formulation of the original contracting plan. Changes to the plan may occur as a result of additional ideas developed during the proposal and negotiating stages. In any case, there must be a clear statement as to the fee basis and payment terms in the final agreement.

When doing lump-sum work, make a clear statement that all costs, overheads, and profits for performing the scope of work are in the contract price. List any unit rates for optional additional services by the contractor in this section. These might include rates for inspection services, costs for soils investigations, and laboratory services. It is sometimes convenient to handle these costs as reimbursable items in an otherwise lump-sum contract.

A cost-plus contract should define in detail the allowable reimbursable costs such as overheads, labor rates, out-of-pocket costs, equipment rates, and the like. Make these documents attachments to the contract rather than part of the text. Also, put the fee statement in this section. Some owners use percentage fees, but I don't recom-

mend them. A percentage fee actually gives the contractor an incentive to increase project costs. To avoid that situation, convert the percentage fee to a fixed fee of a specific amount. The only grounds for
changing the fixed fee is a major change in project scope—upward or
downward.

State clearly any incentive or penalty terms to the fee structure in
this section. Spell out the formula for calculating the incentives or
penalties to everyone's satisfaction. Set up the percentage split now,
even though the parties may not yet know the exact amount of the
upset price.

On cost-plus work, the owner reserves the right to audit the contractor's books covering their project. The owner may also want to
approve the contractor's accounting system for the project. On lump-
sum work, however, owners give up the right to audit in return for
the fixed price.

Payment terms

Payment terms are important to both parties, since they define how
the money flows from the owner to the contractor. The owner wants
to pay the contractor on the basis of project progress, so there must be
enough work in place to justify the payment. In the U.S., we don't
usually pay contractors in advance. In some foreign countries that is
not the case, and advance payments may be necessary to get the project started.

Contractors, on the other hand, do not want to finance projects for
owners, so their interest in regular payments is very keen.
Contractors usually invoice monthly, for payment within 15 to 30
days. In the project design phase, most contractors will finance their
own services for 30 days. In the procurement and construction phases, however, the size of the invoices requires faster payment to avoid
the high cost of financing the work.

For example, on an $80 million reimbursable project, the monthly
billing for a peak month might be 6 percent of the project cost, or $4.8
million. The monthly interest earned on this amount at 8 percent
amounts to $32,000 per month, which is a lot of money for anybody to
be losing. Over a 10-month period in the life of the contract, this
amounts to $320,000. Therefore, contractors request the use of a zero-
balance bank account for processing project invoices. That way, the
owner bears the project financing costs. The contractor forecasts the
project expenses for the labor, material, and out-of-pocket expenses for
the coming month. As we will learn later, that number is readily available from the project cashflow curve. The owner then transfers the
projected amount of money into the zero-balance bank account, and

the contractor writes checks against that account for the approved project invoices, running the balance down to zero each month. If the account does not run down to zero in a given month, any balance gets deducted from the following month's cash-transfer request.

On lump-sum work, the contractor bills the owner monthly or biweekly for costs of work accomplished within that period. Usually the owner, engineer, and contractor will agree to the percent completion of the work before the billing date. That way there will be no surprises to delay the payment. A key point in setting up the payment terms is to develop a simple and reliable system ensuring a smooth flow of money. Delays can upset the teamwork of owner and contractor necessary to execute a successful project. It is up to the respective project and construction managers to make the payment system work smoothly. Each party to the contract has a vested interest in setting up a simple and effective payment system at the outset.

Payment terms for international contracts introduce some additional requirements, because of problems inherent in collecting money across borders. Also, many international projects have marginal financing. I have also known instances where governments prohibited the exporting of cash for making progress payments except under very strict controls.

On international projects, contractors must insist on having the owner set up an irrevocable letter of credit at a bank in the contractor's country. That way, contractors are sure that the funds to pay their costs are already in the country. They submit their invoices against the letter of credit as work proceeds. Since the letter of credit is irrevocable, the owner cannot withdraw any funds from the account for any other purpose without the approval of the contractor.

Although the respective accounting departments handle the mechanics of preparing and paying the project invoices, the PMs and CMs are responsible for reviewing and approving them for submission and payment. Their reviews should ensure that the invoices conform to the terms of the contract. Thus, payment approval should require only a routine arithmetical check by the owner's accountant.

On many lump-sum contracts, the owner will withhold a certain percentage of each invoice to build up a retention fund against the contractor's payments. The purpose of the retention fund is to build up a sort of performance bond to maintain the contractor's interest to satisfy the terms of the contract. The amount withheld can range from 5 to 15 percent, with the average being 10 percent. Contractors consider retentions to be financially onerous, because they lose the interest on the funds held by the owner. In the state of Ohio, for example, contractors have pushed through a law that stops the withholding of retention funds at 50 percent completion of the work. The owner holds the retained funds in an escrow account, with the interest accruing to the contractor. The owner releases the funds on acceptance of the work. This law applies only to government work within the state and does not affect private sector work.

Another contractor device to avoid the cost of retainage is the substituting of a performance bond for the retainage near the end of the job. The performance bond protects the owner's interest in having the contractor meet the terms of the contract. However, it costs the contractor only a fraction of the interest lost on the retained funds. This satisfies the needs of both parties until there has been acceptance of the work by the owner.

On lump-sum work, where retainage is most onerous, contractors try to load the front end of the payment schedule in order to generate more cashflow early in the project to offset the retainage. Most owner's project managers have caught on to this ploy by now and will take steps to inhibit the practice.

Limits of liability

Liability clauses place a limit on any liabilities assumed by either party to the contract. Either party may request a limit-of-liability clause to cover any costs not covered by insurance. The main one usually comes from the party offering the guarantees and warranties on the project.

Agreeing on liability limits can be a sticking point. Contractors prefer to set the limit at something less than their profit or fee on the project. That will allow them to earn something for the part of the job that was done right. Owners, however, want to set the liability limit high enough to keep the contractor's attention, should problems arise on the project. This area is more critical to the legal people, who usually work out a settlement without a lot of input from the construction manager.

Project change notices

The one thing certain on any project is that there will be changes occurring along the way—sometimes even before the signing of the contract! Every contract must have a specific procedure covering the process for handling changes to the work. The most frequent type of changes met in constructing capital projects are changes in the original scope of work or those that arise from unexpected conditions in the field.

Changes in scope or design, and therefore price, are more critical on lump-sum contracts. The contractor has proposed to do a certain amount of work for a fixed amount of money. If the scope changes, the contract price also must change. On cost-plus contracts, scope changes automatically go into the owner's cost account. Although contractors do not absorb the added cost, both parties must still record the changes on CPFF jobs to control the original project budget. Also, the scope may change so drastically over the life of the contract that adjustments to the fixed fee are required. We will be discussing the control of change orders in relation to cost control in later chapters.

The contractual procedures for handling project changes should be as simple and direct as possible. It's best to require only the signatures of both project or construction managers without higher approval. Added levels of approval always delay the processing of the changes, and thus delay the project. Individual company policies usually place a dollar limitation on the project manager's authority to approve changes. Chapter 13 in *Managing Construction Projects*[9] presents a thorough discussion of change order procedures, including forms. The proposed system as used in a larger project is quite detailed, so you may wish to scale it down to suit your type of project.

A clear formula for pricing changes is essential to reduce the bickering that normally results when change orders arise. Allowable percentages for overhead and profit, reimbursable costs, out-of-pocket expenses, etc. should be part of the contract. This protects the interests of both parties and eases the processing of change orders.

Project changes come in all shapes and sizes, in addition to changes in scope. Extra costs can arise from unknown construction conditions, delays due to force majeure, suspension of the work, or some other extra cost.[9] Claims of this type are hard to define and much harder to resolve, because the added costs usually do not add value to the owner's facility.

Delay claims also are difficult to calculate, since they involve such vague subjects as loss of efficiency, cost escalation, and extension of overhead and supervision charges. Reaching agreement between owner and contractor on the validity and value of this type of claim can be very involved. Resolution often comes via a settlement just before going to court or arbitration. As the CM representing your company, you must carefully review the contract language covering these matters. More importantly, as CM you must document the project to prepare or defend such claims when the situation requires it.

Cost escalator clauses

Cost escalator clauses are not common in contracts for capital projects in the U.S. This is because most owners want contractors to include their estimated escalation in the original proposal. Escalation was a serious factor, even in the U.S., during periods of high inflation following the world energy crisis of 1973–74. The moderate inflation rates of recent years have calmed things down again. However, one is still likely to find escalation clauses in places where high inflation is a way of life.

Getting a good, simple, mutually agreeable formula for equal protection of both parties against the effects of inflation is not easy. A simple price-adjustment formula involving design, craft labor, and material costs appears in *Applied Cost Engineering* by Clark and Lorenzoni.[8] It offers a possible model of a simple escalation formula for adaptation to your particular project needs.

The various cost indices must be those applicable to the area of the work. They should be agreed upon and stated in the escalation terms of the contract, along with the formula. A key factor in escalator clauses is that they sometimes result in broad swings in dollar value when there are large swings in the indices. I recommend trying a few sample calculations with the formula. This will test its validity for a range of probable inflation scenarios over the life of the project. In that way you can tune out any really wide swings in the cost of the escalation, avoiding catastrophic costs to either party.

Suspension and termination clauses

Suspension and termination clauses don't come into play very often on well-conceived and well-financed projects. Yet their inclusion in the contract is critical, because projects can be suspended or canceled for various reasons. We always discuss these two clauses together, because a termination is really a permanent suspension.

Both the owner and the contractor need to protect their interests when suspension or termination of the work becomes necessary. Largely, the problem areas requiring protection revolve around payments during the suspension, disposition of owner and contractor personnel, ownership of the finished work, and extra costs.

Contractors also want contractual rights to suspend the work for nonpayment, or other default, on the part of the owner. In extreme cases, contractors will also want to reserve their right to terminate the contract if the owner continues to default on payments.

Likewise, owners reserve their right to suspend or terminate work with or without default on the part of the contractor. This right includes a clause to cover possible bankruptcy of the contractor. Owners must have the right to complete the work in a timely manner with other contractors.

An important point to consider in this area is the proof necessary when claiming the other party to be in contractual default. Usually an official notice is required, along with reasonable time allowed to correct the alleged default before invoking suspension or termination of the contract.

Suspension and termination notices must include reasonable time limits before they take effect, since they can have a devastating affect on the project organization. Both parties must plan the temporary disposition of their project forces during and after the suspension. Owners do not like to lose the advantage of keeping their own and the contractor's trained people on the project during the suspension. On the other hand, they often cannot afford to keep nonproductive people on the project.

If the contractor must move personnel to other projects, there is no guarantee that they will be available to return. Assigning new personnel to the project, with the associated learning curve, reduces efficiency and costs money. On a lump-sum project, the contractor will certainly submit a delay claim to recover the added costs.

Suspension or termination notices must provide for timely notice for the event to take place. The minimum length of notice is directly proportional to the size of the project and the human and physical resources involved. On smaller jobs, the notice period can be relatively short; say, one to two weeks. On larger jobs, 30 days is more reasonable. If the termination is more organized, owners may grant contractors a longer period to properly mothball the work in progress. An organized closeout procedure serves to protect the investment of both parties and reduces losses.

Suspension and termination clauses must also address the final payment of the contractor's costs to date and payment of fees and incentives. Resolution of the payment terms permits the owner to gain full title to the work in progress according to the contract.

Force majeure

A force majeure clause is an old standard in all construction contracts; it gives contractors certain rights to excused delays owing to matters beyond their control. The conditions that are beyond their control are the so-called "acts of God." They consist of such things as fires, earthquakes, tornadoes, hurricanes, riots, and civil commotion.

Some other conditions, such as strikes, work slowdowns, and labor shortages, which I call "acts of man," are often included in this clause.

One cannot argue much about the acts of God, because they are beyond the control of the contracting parties. The acts of man, however, are open to discussion. The owner may want to place some limits on the latter points, which puts more pressure on the contractor to overcome any adverse effects on project progress that the delays may have had. In areas of labor shortages and labor unrest, however, the contractor may have to insist on those factors being included under the heading of Force Majeure.

Force majeure excuses the contractor from performing to the project schedule if the delay is caused by any of the conditions listed in the clause. Causes can vary from minor delays due to labor problems to virtual demolition of the project by a major natural disaster. Fortunately for all of us, the natural disasters do not occur very often, but the contract must consider that chance.

A more insidious cause for delay can occur during unusual periods of unsettled weather, which seem to be happening more and more these days. The CM must analyze daily construction diaries to note any patterns of excessive loss of work days, resulting in schedule delays, due to inclement weather. If the number varies significantly from the norm, a good case for force majeure can be made.

Insurance, bonding, and risk management

Insurance clauses are there to protect the owner by insuring against all potential casualty losses that can occur on a project, mainly in the construction phase. The contractor is required by law to keep certain insurance in effect to protect its employees. Both parties require liability insurance to protect against third-party actions.

The normal hazards of the construction activity involved in the contract forces the owner to have comprehensive insurance coverage. That includes setting minimum insurance limits for all contractors and subcontractors as protection from claims made by the contractors and damaged third parties. In addition, the owner will need to have a construction all-risks insurance policy, one that covers the replacement cost of the construction works and materials on the project. Sometimes the owner will request that the contractor take out the policy, then reimburse the contractor for the premiums. Present-day practice includes taking out a supplementary umbrella policy covering all risks not covered by any of the other insurance in effect. So far this has been the ultimate in owner and contractor protection. The contract requires that insurance certificates be acquired prior to the start of work on the site.

Most government contracts require the contractors and subcontractors to have bid, performance, and payment bonds. These are available from a bonding company, and their cost depends on the contractor's financial strength and prior performance in those areas. The bid bond assures the owner that the contractor will sign a contract for the bid price or forfeit the bond to the owner.

The performance bond is a warranty by the bonding company that the contractor will finish the work or the bonding company will take over, hiring other contractors to finish the work. The original contractor's failure is usually caused by bankruptcy or some other unforeseen disaster.

The payment bond is protection of the owner against the contractor not paying his bills. This bond protects the owner against supplier or mechanics' liens when subcontractors are not paid. All these bonds are readily priced and added to the bid price, so they wind up being paid by the owner.

In larger owner and contractor companies, a risk management department handles bonding and insurance. The CM acts as the enforcer, keeping the proper insurance and bonds in force and initiating claims when losses occur. The CM also advises the risk management people of any unusual risk situations in which coverage is not enough or the insured value has changed. CMs working in firms without risk management departments must play a stronger role in that area when working with the firm's administrative staff.

Applicable law

The lawyers insert the applicable law clause in the contract, to set up the legal jurisdiction governing the particular project. The owner usually calls for the applicable law to be of the state or the country of its incorporation. In the U.S. this doesn't make much difference, because most states are fairly uniform when it comes to contract law. On foreign projects, however, it can become a problem if something goes wrong with the project and legal action becomes necessary.

The thought of entering legal action in a foreign court in some third world country does not sit well with the legal people. To get around this problem, international contracts often call for settlement of contract disputes by international arbitration. This avoids lawsuits in unfriendly courts. There is really nothing CMs can do about this clause except to be aware of its presence. They should take particular pains to document the project to suit the applicable legal jurisdiction.

Arbitration

The settling of contract disputes through arbitration has become increasingly common now that the courts have become so expensive and overcrowded. This clause merely states that both parties agree to submit contractual disputes to arbitration. This can be done without giving up any legal rights to sue if they feel that the occasion warrants it.

The arbitration machinery is already in place, and is covered by federal and state laws in most states. The procedure starts by contacting an arbitration service like the American Arbitration Association (AAA). A neutral group such as the AAA sets up a tribunal of three members from existing panels of experienced arbitrators. This panel hears the dispute and renders a judgment, much like a court but without the expense. Each side selects an arbitrator from the panel, and the arbitration board appoints the third member. The proceedings are much less formal and time-consuming than going to court. You can, in fact, present your own case before the tribunal without legal counsel if you so desire.

The operative word in any arbitration clause is *binding*. When you agree to binding arbitration, both parties agree in advance to accept the findings of the tribunal. If the basis is nonbinding, you are still free to sue in a court of law.

Information on U.S. arbitration procedures is available from the American Arbitration Association, 140 West 51st Street, New York, NY 10020. For information on international arbitration services, you can contact one of the international tribunals in Paris or Geneva.

Subcontracts

Most contracts for capital projects have a clause stating the owner's policy on subcontracting parts of the work. Normally, construction contracts permit the subcontracting of various specialty trades, with the owner reserving the right to approve the subcontractors. Design-build contracts may not include the right to subcontract design work. Occasionally, plans call for outside design-specialist help through subcontracting. That is a decision the owner must make when drawing up the prime contract.

Construction subcontracts fall within the purview of the construction manager, since execution for part of the construction passes over to others. Subcontracting design services usually risks dilution of project control, so approach it with care.

Usually a purchasing group handles the commercial aspects of subcontracting. The most critical factor in subcontracting is making certain that all key prime contract requirements extend to the subcontractors, especially in the area of change orders and claims for extras. That will ensure that the prime contractor does not get stuck with paying for missing items and change orders.

CMs are more likely to find themselves reviewing the subcontracting procedure rather than developing the details of the subcontracts. That role is quite different from their deep involvement in the prime contracting procedure. The reviewing duties are especially critical when subcontracting in the design area!

Contract Administration

We have spent a lot of time talking about arriving at a mutually satisfactory contract document, but that is only half the battle. CMs also have prime responsibility for overseeing the contract for the life of the construction project. A very large project (a superproject or above) may rate a contract administrator[9] to ensure accountability of all contract provisions. Most of us, however, will never experience the luxury of having a contract administrator. Therefore, we must work at becoming more contract-oriented without becoming lawyers in the process.

As a foundation to good owner-contractor relations, it is best for both parties to be proactive in conforming to the contract requirements. A good way to remind yourself of key contractual requirements is to set up a contract "tickle file" to automatically remind you of key notice dates and performance features required by the contract. Today there are software programs, such as "Expedition" from Primavera Systems, Inc., that permit control of all document and contractual submittals by personal computer. (See Chapter 13.)

CMs must document the project according to the terms and conditions of the contract, either manually or on a PC. Keep in mind that the contract sets up only the minimum documentation required. Professional CMs on both sides of the contract must document their projects thoroughly to prepare or defend any contract claims that are likely to arise.[9]

Treat the contract documents as company confidential material. Limit the distribution to those members of the project team who have a need to know. Keep a copy available in the CM's file, for ready access by the CM and the top field staff. Other members of the field team should get their information from the Field Procedure Manual, which summarizes the contractual information in everyday language. We will discuss the Field Procedure Manual in detail later in the book.

More detailed procedures for contract administration and legal factors can be found in Chapters 16 and 20 of the *Contractor's Management Handbook*.[6] This and any of the other references given at the end of the chapter are recommended as valuable additions to your management library. If you don't yet have such a library, now is a good time to start one!

Letters of intent

The one remaining contractual matter to cover is the letter of intent. There are times when owners want to start work immediately, but it will take several months to work out the contractual details. Issuing the contractor a letter of intent authorizing the work to proceed on a limited basis, pending the contract-signing, can solve the problem. Owners should remember that issuing a letter of intent weakens their position at the negotiating table, so they should use them only when relatively minor contractual points remain to be settled.

Letters of intent should spell out, in only one or two pages, the limited basis for starting the work. Referring to the RFP and the proposal to clarify the desired payment terms, scope, budgetary limits, etc. for the temporary authorization is normal. Figure 2.5 shows a typical letter of intent.

It is not advisable for either party to continue working under a letter of intent for longer than 90 days. Never use a letter of intent as an excuse for not completing the contract negotiations and getting the project on a fully committed footing.

Notice to proceed

Government contracts often include a clause that work is not to start without an official notification from the owner. That means that signing the contract does not authorize the contractor to commit funds on behalf of the owner. Any money spent before the notice to proceed is at the risk of the contractor. The notice to proceed is usually a brief statement that initiates the work, allowing commitment of funds and starting the schedule clock.

Summary

In this chapter we have touched on some of the major points in the proposal and contracting environment, as seen from the construction manager's viewpoint. This is an important area that often gets neglected as people develop a career in construction management. Construction firms live or die on successful proposals and contract execution. Your management will be looking for strength in this area when it is selecting CMs for assignment to key projects and promotion.

There is not much formal literature on these subjects, and most of what does exist is written by sales or legal people. Those people do not usually consider the subject from the project or construction manager's viewpoint. If you do study contracts and construction law in more detail, I hope that you will not get too worried about the strictly

ACE DEVELOPMENT COMPANY
200 Casino Avenue
Las Vegas, Nevada 11122

October 29, 1993

Mr. Sam Smith, President
XYZ Construction, Inc.
1330 Winner Street
Los Angeles, CA 33355

Subject: Four Aces Casino Project

Dear Mr. Smith:

This letter will confirm our intent to negotiate a contract with
your firm to furnish procurement and construction management
services in connection with the subject project within the next
60 days. The scope of work and the contract format will conform
to that described in our RFP dated August 2, 1993 and your proposal dated September 15, 1993.

Since our scheduled completion date is very critical to us, we
are authorizing you to proceed with procurement and field mobilization immediately. We are herewith authorizing you to spend up
to $100,000 for procurement and mobilization services prior to
the execution of our contract.

Our project manager will expect your construction team to meet in
our offices to kick the project off on November 10, 1993. Meanwhile, we will be sending you our proposed contract draft within
the next several days so that we may resolve the outstanding
differences as soon as possible.

We are looking forward to working with your people on this important project and hope that by working together as a team, we
can successfully complete a quality project on time and within
budget.

Sincerely yours,

Mr. James Jones, Vice President
Ace Development Company

Figure 2.5 Sample letter of intent.

legal aspects. That could lead you to become so entangled with the
legal risks that you never get the work done—let alone complete it as
specified, and within budget!

References

1. P. D. V. Marsh, *Contracting for Engineering and Construction Projects,* 3d ed.,
   Gower Publishing Company, Ltd., Aldershot, U.K., 1988.
2. The Business Round Table, "Contractual Arrangements" (CICE Report A-7), 200
   Park Ave., New York, NY 10166, 1986.

3. Sidney M. Levy, *Project Management in Construction,* McGraw-Hill, New York, 1987.
4. Donald S. Barrie, *Directions in Managing Construction,* John Wiley and Sons, New York, 1981.
5. "Project Management Starts Before Contract Award," Proceedings of the 10th Annual Seminar/Symposium, Project Management Institute, Anaheim, CA, October 1978, pp. II-L.1 to II-II-L.10.
6. James J. O'Brien and Robert G. Zilly, *Contractor's Management Handbook,* 2d ed., McGraw-Hill, New York, 1991.
7. Donald S. Barrie and Boyd C. Paulson, *Professional Construction Management,* McGraw-Hill, New York, 1978.
8. Forest D. Clark and A. B. Lorenzoni, *Applied Cost Engineering,* Marcel Dekker, New York, 1985.
9. Robert D. Gilbreath, *Managing Construction Contracts,* John Wiley and Sons, New York, 1983.
10. Joseph P. Frein, ed., *Handbook of Construction Management and Organization* 2d ed., Van Nostrand Reinhold, New York, 1980.

Case Study Instructions

1. As the owner's project or construction manager, prepare a proposed contracting plan for the execution of your selected project.

2. Assume that your management has decided to have your project handled by an outside firm. Prepare your plan for the request for proposals (RFP) and the selection of the prime contractor for your project. Again, the basis for the RFP should be in accordance with your contracting plan.

3. As a contractor's project or construction manager, prepare a plan for the preparation of the proposal requested in (2). Give special attention to any factors enhancing your firm's position to win the award, even though your backlog of work is quite full at the moment.

4. Draft the main points of the pro forma contract, which you will want to include in the above request for proposal. Base this draft on the contracting plan developed under (1).

5. Now put on your contractor's hat. List the points in the pro forma contract prepared in (4) that you want to take exception to or negotiate.

3

Project Planning and Initiation

We are ready to explore a total construction project management approach to executing capital projects. Planning is the first step of our total project management philosophy for planning, organizing, and controlling the execution of capital projects.

First, we will discuss the methods used to prepare a project execution plan (or master plan), then move on in Chapter 4 to scheduling detailed execution of the work. In later chapters we will cover the financial and project resources plans for the project.

I have deliberately separated project planning from project scheduling, to stress the point that these are two separate and distinct functions. The project and construction managers and their key staff members prepare the master plan, the scheduling people put the plan on the time schedule.

Another administrative consideration that has to be settled is that of selecting the sort of construction projects we are going to discuss. Up to this point we have been talking about three general types of construction projects, namely, design-build, construct only, and construction management modes.

The design-build mode involves a major personnel and technology input on the design side that is not under the direct control of the construction manager. Rather than include the details of how to execute the design portion of this mode, I refer the reader to my previous book, *Total Engineering Project Management,* also published by McGraw-Hill.[1] We will assume for the purposes of this book that the design-build project is led by an owner or contractor project manager who coordinates the project through a design manager and a construction manager. Here we will primarily concern ourselves with the work handled by the construction manager.

The initiation phase of construction projects is so critical, and so intertwined with the planning effort, that I have decided to combine both activities in this chapter. We will discuss planning first.

Planning Definitions

There are several good definitions of planning that I have used over the years. I am going to list all three of them, because each delivers a slightly different message:

- Planning is a bridge between the experience of the past and the proposed action that produces a favorable result in the future.

- Planning is a precaution by which we can reduce undesirable effects or unexpected happenings and thereby eliminate confusion, waste, and loss of efficiency.

- Planning is the prior determining and specifying of the factors, forces, effects, and relationships necessary to reach the desired goals.

The first definition reminds us to make use of our prior experience, often gained from past mistakes, to avoid repeating them in our present endeavor. It also says that we should not reinvent the wheel on each new project. The second definition cites the advantages of increased productivity by planning the unexpected and undesirable happenings out of existence before starting to work. The third one stresses making a conscious effort to find and control the variables in a capital project. We must do that before starting work if we are to meet our project goals. It also indicates the need for an organizational phase if we are to execute the plan.

All of the definitions point to the obvious conclusion that the first move on any project assignment is to do the necessary planning. Furthermore, that applies to every one of your activities throughout the project!

Planning Philosophy

Planning must be done logically, thoroughly, and honestly if you are to have a chance to succeed. The previous experience of many years has honed the basic planning logic for capital projects to a fine art. There is no point in trying to develop a new planning logic for each new project. Also, the owner has already set up the basic project-execution format in the contracting plan.

After selecting the normal planning logic for your project, you should examine the work for exceptional features affecting the normal logic of your plan. Look for any special problem areas that may be different this time around, such as unusual client requirements, an out-of-the-way location, or potential internal or external delaying factors that are likely to affect the normal logic of your plan.

Work these potential problem areas over in detail to reduce their negative effects on the master plan and later the schedule. You may leave the details of the schedule to your planning specialists, but you must set the basic scheduling logic for them.

Earlier I spent a lot of time telling you to opt for simplicity in project management. Now I want to stress being thorough in the area of planning. Each aspect of your plan calls for individual scrutiny. Discuss it with your staff to get all of the expert input available. Small details passed over in the early stages can rise up to smite you mightily when least expected later in the job. Enough things will go wrong by accident, so there is no need to increase the chances of that by overlooking problems in the planning stage.

Honesty in planning is also very important. Remember, you are the one responsible for carrying out the project plan. You will be the one finally held accountable for its success or failure. *You* may blame a late project on your planning people, but your management will not.

It is best to maintain a delicate balance between optimism and pessimism during this early planning exercise. A project plan without some float or contingency for unforeseen events is a trap. It will spring shut on you eventually, unless you have an inordinate amount of good luck on your side. Many times it is hard to find any float in the construction schedule, because it has all been used up in the prior conceptual and design phases. Often it falls to the CM to deliver the bad news as to a slippage of the strategic end date to the owner and designer who were responsible for burning up the float.

On the other hand, a fat plan is wasteful of time and money. Parkinson's Law tells us that "the work always expands to fill the time allowed." I have always felt that the highest project efficiency results from being slightly understaffed. Working on a little tighter-than-average schedule, coupled with suitable incentive programs, also helps.

Good construction managers must learn to work under pressure; if they don't, they will not survive. It turns out that we are lucky in this regard, because there are always plenty of people around to sharpen the schedule.

Types of Planning

Several types of planning are involved in any capital project. Let us define the types, and see how the construction manager fits into the overall planning process. The three major types of planning are:

1. Strategic planning, which involves the high-level selection of the project objectives

2. Operational planning, which involves the detailed planning required to meet the strategic objectives

3. Scheduling, which puts the detailed operational plan on a time scale set by the strategic objectives

Does the construction team do the strategic planning? No, that is done by the owner's corporate planners. They decide what project to build and what the completion date has to be to meet the owner's project goals. The project development phase involves a great deal of strategic planning. The development phase requires the input of market analysis, financing planning, project feasibility, and so on. Those areas need thorough study before the project can get the green light. Usually the strategic completion date is fixed before making the decision to proceed without allowing more time on the end date. That uses up the schedule float on the front end and results in a tight schedule for the remaining work.

The construction team formulates the master construction execution plan within the guidelines called for in the strategic and contracting plans. This chapter will explain how to make an operational plan for a typical project. The job of putting the plan onto a time schedule falls to the project schedulers. That points up the difference between the terms *planning* and *scheduling,* which so often are mistakenly thought to be interchangeable.

Operational Planning Questions

Operational planning usually raises some interesting questions for resolution in the construction master planning phase:

- Will the operational plan meet the strategic planning target date?
- Are sufficient construction resources and services available within the company to meet the project objectives?
- What is the impact of the new project on the existing workload?
- Where will we get the resources to handle any overload?
- What company policies may prevent the plan from meeting the target date?

- Are unusually long-delivery equipment or materials involved?
- Are the project concepts and design firmly established and ready to start the construction?
- Is the original contracting plan still valid?
- Will it be more economical to use a fast-track scheduling approach?

We must answer those and any other pertinent questions in preparing the construction master plan before the detailed scheduling can start. Preparing a detailed construction schedule before a logical construction execution plan has been formulated is a waste of time and money.

The Construction Master Plan

The master plan must address how we will plan, organize, and control the major work activities to meet our goals of finishing the work on time, within budget, and as specified.

A major consideration in formulating the master plan is the contracting plan, which helps us to answer a lot of questions: Are we going union or open shop? Are we going with a construction management format or self-perform? How much work is subcontracted?

Many other questions are *not* answered by the contracting plan: Do we need more resources for design or only for construction? What are the owner's policies or legal requirements in this regard? What government and social restraints come into play in the execution of this project? How do the contractual requirements affect the master plan?

We must unravel those questions and a host of others during the development of the project execution plan. Some of these are answerable immediately; others must wait for information that develops as the project progresses.

Project-execution plans are subject to review and evaluation as the work progresses. Minor variations are common, but you should consider major changes with extreme caution. Changes to the plan often bring on a great deal of trauma and should not be made lightly. Because all parts of the plan are so closely interdependent, a change in one major part can affect the interaction of all of the other parts.

When we have completed the master plan and gotten it approved, we can start thinking about more detailed operational plans:

- The time plan (schedule)
- The money plan (budget)
- A project resources plan (people, materials, and services)

The time plan, which results in the project schedule, will be the subject of the next chapter. There we will address the activities that we will schedule, the scheduling methods available to us, and how to select the best system to suit conditions.

The construction manager's role

The construction manager is the prime mover in preparing the construction master plan. The CM accomplishes the master planning through the judicious use of the complete construction team's many talents. The CM gets input from the key people in the owner and designer organizations, as well as company management and service groups. The CM may even find the need for planning input from major subcontractors who have not yet been signed on.

Lead people from subcontractors and from the service departments such as estimating and cost control, scheduling, personnel, accounting, safety, and engineering are assigned to the project. They have the responsibility for planning the portions of the work normally performed by them. It's the CM's duty to coordinate the contributing groups and to see that they have a sound basis for their part of the master plan.

The CM organizes and chairs a series of planning meetings early in the project. The various contributors to the plan can interact with one another to resolve any operational differences that may arise. The differences not readily resolved by the meeting participants must be resolved by the CM.

The master planning effort should include early establishment and dissemination of the specific project goals. That allows the master plan contributors to formulate their individual plans to meet the overall project goals. The CM also is responsible for handling all the missing information and decisions required to complete the master planning effort on time.

After receiving all parts of the plan, the CM integrates them into a written master plan for presentation to company management for approval prior to submitting it to the owner. The CM makes any changes resulting from the final approval stage and issues the plan as part of the Field Procedure Manual.

Project Execution Formats for Capital Projects

Unfortunately, there is no single project execution format that fits all the various types of projects shown in Table 1.1. As one moves from left to right in Table 1.1, the construction activities are quite different. CMs generally spend their working lives in one or two of the special areas without making a broad jump across the lines. That should not be a problem, however, because the management techniques are common to all, with minor variations due to technology differences. When reading Fig. 3.1, the reader may have to adapt the blocks to suit the particular nature of that construction area. I hope readers will have some interest in how the different groups of the capital projects industry execute their projects. One never knows when crossing the line into another type of expertise may become necessary. Also, there is a good possibility that CMs will find a new and valuable technique to use in their operating environment.

The involvement of manufacturing processes creates the difference between process and nonprocess projects. A chemical or mechanical process weaves its way through the entire design, procurement, construction, and project activities necessary to complete a process facility. Nonprocess projects tend to be more people-oriented, so aesthetics and creature comforts are stressed in the design.

Another way to state the difference is that manufacturing facilities are basically process-design-driven, whereas nonmanufacturing projects are architectural or civil-design-driven. Architectural and civil works projects involve satisfying such human and infrastructural needs as schools, office buildings, hospitals, highways, dams, and public transportation systems. In process projects, the civil and architectural input serves to house and support the needs of the processing operations.

The nuances of the processes involved do make for subtle differences in the construction management techniques used. Although CMs use the same management techniques in different technical environments, they rarely seem to cross over the line to handle projects outside their fields of specialization.

A typical construction project format

The development of any project follows the major phases shown on the project life-cycle curve shown in Fig. 1.3. Our discussion model envisions the owner using its central engineering group or an outside design firm to design the facility, and a construction contractor to build it. The scope of this book is to examine the management of the construction effort for the project.

The scope of the major work activities required to execute a typical construction project is shown graphically in Fig. 3.1. Each of the major activities shown breaks down into several individual work activities, so that a large project can have as many as 2000 to 3000 activities to schedule. If the construction team is to meet its goals, it

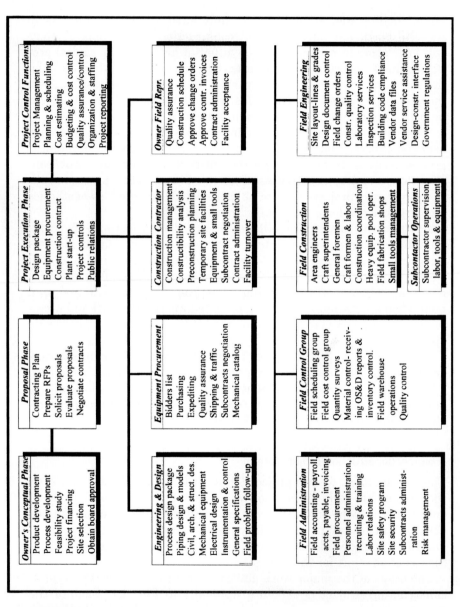

Figure 3.1 Major work activities for a process construction project.

must successfully plan, organize, and control their work activities so that they are performed in proper sequence and on time.

A brief description of the major project operations should give readers a common understanding of the actual work performed by each group in a large capital project team. Many of the activities must happen before the construction phase starts. Starting with the owner's conceptual phase and working through the project-execution phase to the final turnover, we will encounter the following major areas of project execution:

1. Owner's conceptual phase
2. Proposal phase
3. Project design phase
   a. Engineering
   b. Procurement (major equipment)
   c. Project control functions
   d. Construction input
4. Procurement
5. Construction
6. Facility start-up and turnover

The top row of blocks in Fig. 3.1 primarily represents owner functions. The second row introduces engineering and design, equipment procurement, and construction work, all of which are normally contractor functions. The third layer shows the detailed field activities performed by the construction contractor.

Owner's conceptual phase

The owner usually assigns a project manager to direct and coordinate the conceptual phase of a proposed new facility. The major activities involved in this might be:

- Product development
- Process development or process license
- Marketing surveys
- Setting project scope and design basis
- Capital cost estimating
- Project financing
- Economic feasibility studies
- Board approval of the project

The owner's marketing or R&D group develops a new product or an improved process for making an existing product. If that work offers a chance for increasing the owner's profit, there is an incentive to pursue the venture. An organized and well-thought-out plan to build a new (or revamped) facility to manufacture the product or provide a service should result. But first, a proven process for making the product must be developed and tested by using experimental equipment in the laboratory or a computer model. Alternatively, the owner may choose to buy a process license from an outside firm specializing in the process. In that case the R&D steps are not required if the owner buys a license.

While the R&D work is progressing, the sales department makes the marketing survey to determine the projected market volume, probable selling price, market entry costs, and other expenses involved in bringing the product or service to market. The plans to finance the project are proceeding concurrently with the above technical work to make available the data necessary to perform a feasibility study. The feasibility study is the culmination of all the work done during the initial part of the conceptual phase. The conceptual group presents the study to the board of directors, for preliminary approval to proceed with further development of the project.

The initial strategic planning results from this first phase of the work. The strategic project completion date is established along with the basic project financial plan in which the project budget plays a major role. Good planning dictates that you take account of some contingencies in the financial and time plans to cover the usual errors and omissions that will show up as the plan progresses. Construction input to the feasibility study should begin during this preliminary phase. The construction group makes contributions in the area of scheduling, cost estimating, and other constructibility matters.

Since we are still very low on the overall project life-cycle curve, the financial risks at this point are still controllable. The owner's project team makes further evaluation of the project as the detailed development work evolves along the life-cycle curve. The final "go/no-go" decision occurs when the detailed design is about 30 percent complete and commitments for major equipment purchases are ready to proceed.

The elapsed time for this part of the project can run from several months for a simple project to three years or more for a complex one. Since the amount of work in that early stage may not warrant a full-time PM, he or she may handle several conceptual projects simultaneously. The conceptual PM rarely carries the project through the execution phase to completion. The specialized skills developed in handling the diverse conceptual phase activities are so different from those required to execute the detailed project execution phase that not many people can make the transition.

When the project baton passes between these two phases, it is extremely important that the project definition and the feasibility of making and marketing the product at a profit are well proven and documented. Many a project has started down the road to disaster because of incomplete or sloppy workmanship during that critical phase. Some of the worst disasters that I have experienced have occurred because the owner tried to skip the conceptual phase altogether![3]

Proposal phase

Once the board has given its approval to proceed with the project, the owner is ready to enter the proposal phase and select a contractor. That phase involves the following activities:

- Preparing a contracting plan
- Prequalifying contractor slate
- Preparing a request for proposal (RFP)
- Receiving and analyzing the proposals
- Selecting the best proposal
- Negotiating a contract

These activities were described in detail in Chapter 2, so it's not necessary to repeat them here. Suffice it to say that the owner still has not committed a great deal of money to reach this point.

Project execution phase

At this point, the owner's project manager is ready to execute the project in accordance with the owner's contracting plan. The main decision affecting our discussion is whether the project is to be let on a turnkey or separate design-and-construct basis. Details of the design activities are not included here because of space.

The project life-cycle curve starts to slope upward with the initiation of the project execution phase. That's the result of significant commitments of project financial resources to the following work activities:

- Engineering design phase
- Equipment procurement activities
- Early construction activities
- Project control functions

The engineering design phase covers those activities required to generate the plans and specifications for the procurement of the equipment and materials and the construction of the facility (Fig. 3.1). The various technical disciplines involved are:

- Process design
- Mechanical design
- Civil, architectural, and structural design
- Piping design
- Electrical design
- Instrumentation design
- General specifications
- Construction input

If the construction manager is responsible for the total turnkey project, he or she will probably delegate the design phase to an engineering design manager or a project engineer. For a more detailed description of the duties of the design supervisor, see my previous book, *Total Engineering Project Management.*[4]

Construction input

When possible, construction people should be brought on board during the detailed design phase to provide construction input to the detail design team. Bringing this expertise to the design area doesn't have to involve a large number of personnel hours. It can be one general construction expert assisted by specialists on an as-needed basis.[5]

Procurement activities

The activities of the procurement group are closely interwoven with those of the engineering and construction groups. Procurement is basically responsible for getting the materials and equipment from the design stage to the construction site as specified and on time to meet the construction schedule. A key early output from the procurement group is the project procurement plan, which plays a key role in making the project master plan. The procurement plan is also a major part of the overall project materials management plan.

The owner is responsible for assigning the project procurement functions as part of the contracting plan. In a split design-and-construct approach, the design firm usually buys the engineered equipment and the construction contractor buys the bulk materials. On an EPC turnkey approach, the contractor usually buys the engineered equipment in the home office and the bulk materials in the field office.

Project and construction managers sometimes overlook the procurement activities because they get too wrapped up in the design or construction activities. No matter how well the design and construction teams perform, you cannot build the facility on schedule without the materials and equipment. I will stress that point about procurement several times in this volume, to emphasize the importance of procurement in effective project and construction management.

Project control functions

A team of control specialists performs the necessary project control functions to ensure that the project goals relating to budget, schedule, and quality are effectively met. Cost engineers will be monitoring all project commitments and expenditures to see that they conform to the budget and cashflow projections. A monthly project cost report presents the data to the project team. The scheduling group monitors the project schedule on a regular basis and reports any drift off target at least monthly. The project team should reevaluate the schedule periodically, and rework it whenever 25 percent or more of the target dates have been missed.

The various groups responsible for each type of material closely monitor and report the status for their particular area on a regular basis. Material and project quality are checked regularly by the responsible functions in engineering, procurement, and construction.

The contractor's project and/or construction manager coordinates the control functions and issues monthly progress reports to the managements of the owner and the contractor. Any off-target items are highlighted and discussed, along with recommended solutions for any problems.

The onset of the project-execution phase triggers a significant commitment of major human, physical, and financial resources to the project, as we start to climb the life-cycle curve. Design costs, for example, can run from 8 to 12 percent of the total plant cost. The placement of purchase orders for process equipment and materials involves an even greater commitment of funds. This is a good time for the prudent owner to look at the project financial plan to see if the return on investment is still valid. Using the expanded design database to make a capital-cost estimate with an expected accuracy of plus or minus 15 percent would be a good investment. That would be a big improvement over the plus or minus 40 percent estimate used in the original feasibility study.

The owner's field representative

The function of the owner's field representative (OFR) can fall into several categories, depending on the size and complexity of the project. On large turnkey EPC projects, the owner usually assigns a resident engineer to monitor the performance of the construction work. On a really large project, the OFR has a staff of quality and control specialists to assist in monitoring the contractor's field activities. In that situation, the owner's field team and the contractor's construction management team work very closely on a day-to-day basis.

In a split design-and-construct format, the owner often retains the A&E firm to monitor the construction contractor's performance. On large projects, that arrangement usually involves a resident A&E field representative, who calls in design office specialists as required to interpret the design and perform quality-control inspections.

In any case, the owner's field representative performs the functions shown in the OFR block in Fig. 3.1.

The Role of the Construction Contractor

Construction activities involve the largest single commitment of resources on the project, about 50 to 60 percent of the total project budget. Before starting any field work, the construction team performs preconstruction activities and supplies construction input to the master planning effort. This preliminary work usually conforms to the contracting philosophy and the field labor survey made during the proposal phase.

The construction group's primary mission is to construct the facility to a point of mechanical or substantial completion ready for start-up. Field schedules and cost-control procedures are set up to ensure that the project budget and schedule are met in accordance with the project-execution plan and the terms of the contract.

The construction team is also responsible for final system-testing, plant-commissioning, and clearing the punch lists resulting from the owner-designer final inspection team. The construction group supports the start-up team with maintenance services until the owner's plant forces can take over. After plant acceptance, the construction team closes out the field operations.

Preconstruction activities

The scope and timing of the preconstruction activities depends on the owner's desires and the contracting plan. If possible, the owner should bring the construction contractor on board during design, so that preconstruction activities begin with constructibility analyses.[5]

Essential construction input to the overall schedule early in the project planning and scheduling phase is also possible. At about 40 percent completion of the engineering, field operations can start when the design-build mode is used. This involves installing the temporary construction facilities and utilities, as well as planning the field site layout and contractor laydown areas.

Getting early construction input during the planning and design phases can save the owner time and money, which more than makes up for the small extra cost involved. When the contractor bids on completed plans and specifications, much valuable time and construction input is lost. The construction contractor is virtually starting from ground zero when the contract is finally awarded. Any possible cost savings from early construction input are long past. The procurement program starts late, and the contractor commits only minimal costs during the proposal phase. Nothing tangible moves in the construction area until the notice to proceed has been issued.

Most private sector owners have long since decided that they can't afford that type of delayed contract execution in today's competitive marketplace. Therefore, our discussion here is conducted on a full-scope EPC contracting basis.

The construction contractor has the responsibility for planning, organizing, and controlling the total construction-related activities for the project. Those activities are divided between the home office and field operations. The construction contractor appoints the construction manager, who is made responsible for all field operations for the project.

Because the construction manager is often not available for assignment until shortly before the opening of field operations, much of the preconstruction input is done by home office construction people. If possible, the preconstruction input should be done by the same people who performed the site survey during the proposal phase.

Constructibility analysis

The constructibility analysis consists of reviewing the design documents during the preliminary design phase. Suggestions are made to simplify the design so as to reduce construction costs where possible. The installation of heavy equipment is studied, to reduce installation costs and to plan the heavy-lifting equipment required. The design is reviewed for the application of the best construction technology to minimize field costs, such as earth-moving, concrete forms, lifting lugs, site accessibility, and the like.

Input to preconstruction planning also consists of making the preliminary construction schedule for the overall project planning effort.

Construction input for the schedule logic and elapsed times for key construction activities in the CPM schedule is vital for development of a rational project schedule. That input also includes scheduling the arrival of the long-delivery equipment and materials that will suit the field schedule.

The preconstruction input also must consider the contracting plan, and how the strategic plan for the project will affect available human, physical, and financial resources for the project. How well this preliminary work is done has a profound impact on the CM's performance when the baton is passed to start the field activities.

Major Field Activities

The major areas delegated by the CM to the key players on the field team are shown in the boxes located in the bottom row of Fig. 3.1. The activities carried out in these areas are the cutting edge of the construction project management and execution.

The Field Construction block is where the physical construction work actually gets done. This group is led by a field superintendent who reports directly to the CM. The superintendent is assisted in executing the construction work by an organization of area engineers, craft superintendents, general foremen, and subcontractor supervisors. All the other field groups perform their duties to support the Field Construction operations.

The field superintendent, in conjunction with the CM, coordinates the field construction activities to staff the project, mobilize the tools and equipment, order materials, direct the subcontractors, monitor the field productivity, and perform any other functions required to prosecute the construction work. The superintendent works closely with the various staff support groups to maximize their input to promote the success of the construction operations.

The Field Engineering group is the technical arm of the field operation. Their work starts with setting the lines and grades and surveying monuments for laying out the facility on the site. They also are the first unit on the site to supervise the installation of temporary facilities and supervise the site development subcontractor.

The field engineer receives and distributes the technical documentation for the field organization. They manage the design-construction interface to ensure that the project is built according to the design documents. All design clarifications, design and field changes, change orders, as-built drawings, vendor assistance contacts, and the like must pass through the field engineering office.

The field engineer is also responsible for quality control for the construction operations. This includes all quality-control testing services, laboratory reports, radiography services, etc. The field engineer also maintains copies of the applicable codes and government regulations and interprets their application to the project.

In a process facility, the field engineer's office maintains the documentation for final testing and acceptance of the process systems as they are completed by the construction group. The complete field technical documentation file is turned over to the owner as a record of the quality control for the facility. The field engineer also supports the field start-up team and participates in the facility acceptance procedures and closes out the field files at the end of the job.

The Field Control Group is a staff group that reports directly to the CM. Its main function is to monitor the field schedule, oversee the field cost-control system, and control the material purchased and received at the site. The scheduling group is responsible for keeping the field schedule current as to task-planning and for reporting progress to date. They run the weekly field scheduling meeting.

The cost-control group monitors the project cost commitments, expenditures, indirect costs, and labor productivity, to ensure adherence to the field budget. These factors are summarized and reported monthly in the field cost report. Off-target items are discussed with the CM for enforcement of corrective action.

The material controls group is responsible for implementing and operating the construction materials management plan. Its function includes field procurement, receiving, and warehousing all construction supplies and materials.

The field warehousing facilities are set up to receive, store, and control the construction materials and equipment delivered to the site. That includes all construction tools and equipment, as well as materials bought in the field. The warehousing function is a major part of the project materials management program. An important warehousing planning decision is whether new project buildings can be made ready for use for warehousing, or whether temporary structures will be needed.

The Field Administration function at the site is an important staff function required to support the construction activity. The accounting function handles the field payrolls, accounts payable, project invoicing, and local banking to maintain cashflow on the project. It is important to plan the computerization and data links with the home office if the field accounting operations are to run smoothly in the field office.

An effective field personnel group is critical to field operations to ensure a reliable source of local personnel and craft labor in accordance with the project labor posture. This group also handles the site labor relations, which can make or break the performance of the entire field operation. Establishing a sound and effective site labor agreement is a key master-planning input for this group. Planning, organizing, and controlling the site safety program is an important function of the administrative group. The CM plays a key role in setting up the policies and practices to be followed by the site safety group to plan and enforce an effective site safety and health program.

The CM is responsible for setting policies for site security in accordance with the owner's requirements and prudent risk management. The site security plan should address personnel access, control of physical loss, and contacts with local fire and police departments.

The administrative group handles the business portion of the subcontractors function. It processes the paperwork for payments, change orders, and subcontract administration.

The risk management function of the administrative group oversees those functions relating to site insurance requirements. These people ensure that insurance policies are maintained in force, process insurance claims, and assess risks at the site.

Facility start-up activities

Although the facility start-up is the last activity on the project, it has to be considered in the project master plan. The start-up plan establishes the order for putting the operating units into service. That in turn sets up the strategic date for mechanical completion of the operating units. This agenda for completion of the operating units must be considered in the master plan and the detailed schedule for the project.

The amount of construction participation in the startup must also be considered in the scope of services along with the money and resource plans for the overall project. Even though the services are not to be performed until very late in the project, the CM must not overlook them during the planning phase.

Construction Project Initiation

Getting a construction project off on a good footing is vital to project success. If the project initiation is slow or flawed, the adverse effects will be felt throughout the project which will most likely result in unmet project goals.[6] Formulation of the construction master plan and project initiation usually occur simultaneously. We must not allow them to get in each other's way!

Some project initiation activities actually occur during the proposal effort, which was discussed in Chapter 2. To make possible the preparation of the original cost estimate, we had to make some assumptions based on the project scope, our original site survey, and our estimate of field indirect costs. If we are awarded the contract and are ready to start the project, these data must be reviewed and reconfirmed with the current project plan before they are used. The project initiation activities occur in the home office between the dates of contract award and opening the field office. This is an extremely busy time for the CM, when most of the project planning and organizing is done. Sidney Levy devotes a whole chapter to this critical time in his book *Project Management in Construction.*[7]

The best way to describe the project initiation procedure is to develop a Project Initiation Checklist. The following items must be considered as a minimum approach in kicking off a construction project.

A project initiation checklist

1 Become completely conversant with the contract terms and conditions, especially performance, scope, cost, and schedule.

2. Review the contracting plan, and update it if necessary. Establish a project priority list.

3. Study available design documents to become knowledgeable on the technical aspects of the work.

4. Prepare a project master plan for approval by management.

5. Finalize the labor posture for the project, and develop a local site agreement if required.

6. Become knowledgeable on the project cost estimate, budget, and schedule, if you didn't participate in their development.

7. Develop and issue a field procedure manual (FPM), including a statement of the project goals.

8. Prepare a project organization and staffing plan, in accordance with the project master plan, and include it in the FPM.

9. Bring key staff members on board as required for project initiation.

10. Initiate the project material control plan, including procurement activities, especially if your scope includes all procurement. Concentrate on long-delivery items.

11. Review subcontract proposals and let subcontracts, especially those connected with site development or demolition.

12. Make a site layout for installing the temporary site facilities, laydown areas, and utilities, and start arrangements for their installation.

13. Establish rapport with the client's organization.

14. Organize and chair internal and client project kickoff meetings, per your duties and responsibilities.

15. Establish policies and procedures for expense accounts, and field assignment allowances for supervisory staff.

16. Keep your and the client's (if required) managements informed of your project initiation plans.

17. Start schedule and cost-control activities. You are spending time and money!

18. Set up contract "tickler file" for early warning on contractual obligations.

19. Ensure that project risk analysis and bonding has been done, and that required insurance coverage and bonds are in place.

20. Establish job files, and transfer them to field office when ready.

21. Develop a project safety program applicable to the specific project.

22. Perform (or review) the heavy-lift requirements for the project. Ensure that the constructibility reviews have been done, and that the recommendations have been incorporated into the design.

23. Initiate arrangements for mobilizing the heavy equipment, items of special construction technology, and the small tools required for the project.

24. Establish the interface between the applicable design and construction organizations.

25. Establish the administrative procedures required by the contract and/or company policies.

If you were fortunate enough to be assigned as CM during the proposal phase, your project-initiation duties will be much easier. However, if you are assigned to the project only after contract award, you will have a lot of ground to make up in a hurry. An even worse situation occurs when you are assigned to take over a project that has already been started by someone else. In that case you will have to learn the job and handle the ongoing problems at the same time. In all these cases remember the eleventh commandment: *Know thine contract as thyself.*

Executing the project-initiation procedure depends on the size and complexity of your project. On small projects you will be doing most of the listed activities yourself. On larger or more complex projects, you will have to delegate some of the duties to available staff specialists. They may be home office staff people or certain field staff already assigned to the project.

Item 9 on the checklist calls for bringing key field staff onto the project as required. Care must be exercised here bringing people on board too early, and adversely affecting indirect budget. Make sure that the people brought on early have continuing work to keep them productively busy until the field opens.

The FPM is an important document to have early in the project, to assist in indoctrinating your new people as they come on board. We will be discussing this document in detail in Chapter 7 so I won't go into detail now. The best approach to getting at least a preliminary issue of the FPM is to model it from a previous job that is similar to yours. You don't have time to reinvent the wheel during this critical phase of the job.

Starting the procurement program is key to getting started early. That is especially true if you are bidding to completed plans and specs on a fixed-price basis. You need to get the long-delivery items to the field as early as possible, to avoid losing the schedule and incurring penalties. Early procurement is also necessary to preclude the vendors from raising prices before you can get the orders placed.

Item 22 on the checklist indicates that constructibility analyses were made during design, and that preconstruction services can take place only when the constructor is brought on board early in the design phase. That service usually is not possible on fixed-price separate design and build projects.

Because there are a lot of things happening at the same time during this hectic period, it's absolutely essential that you make a priority list for handling the key issues first. Once you get a priority list started, it's a good practice to keep it going throughout the job!

Summary

In this chapter we have discussed the construction manager's role in the overall project planning and project initiation effort. Sound planning is the cornerstone of effective construction management. Planning the broad range of activities involved with any type of capital project is essential if you are to meet your project goals. Nobody is ever lucky enough to have a successful project without a well-conceived project plan!

Concurrent with the master planning, the CM must get the project started on a solid footing. Project initiation is an art that must be learned early if you want to become a successful construction manager.

References

1. Arthur E. Kerridge and Charles H. Vervalen, *Engineering and Construction Project Management,* Gulf Publishing Company, Houston, TX, 1986.

2. William Peña, *Problem Seeking—An Architectural Programming Primer,* 3d ed., AIA Press, Washington, DC, 1987.
3. Peter Hall, *Great Planning Disasters,* Weidenfeld and Nicholson, London and New York, 1980.
4. George J. Ritz, *Total Engineering Project Management,* McGraw-Hill, New York, 1990.
5. The Business Roundtable, CICE Report B-1, *Integrating Construction Resources and Technology into Engineering,* 200 Park Ave. New York, NY 10166, 1982.
6. Robert D. Gilbreath, *Winning at Project Management,* John Wiley and Sons, New York, 1986.
7. Sidney M. Levy, *Project Management in Construction,* McGraw-Hill, New York, 1987.

Case Study Instructions

1. As the owner's or contractor's project or construction manager, prepare a detailed outline of the proposed project-execution plan based on the use of a turnkey prime contractor to perform the design, procurement, and construction for your selected project. Prepare an alternative plan using a third-party constructor with design and equipment procurement done by the design firm.

2. As the owner's or design contractor's project manager, prepare a detailed outline of the project execution plan based on a typical split contract with a public bid-opening procedure. The construction contractor will also perform the procurement. Now put on your construction project manager's hat and do the same thing for the construction phase.

3. Make an evaluation of the overall time you feel it will take to execute your selected project by using the three different execution plans. Make a comparison chart showing the key milestone dates for the three cases. Were you able to meet the strategic planning completion date in all three cases?

4. As the CM on your selected project, make a list of the specific project initiation activities you feel are necessary to properly kick off construction on your project. Make a priority list of the critical activities.

4

Construction Scheduling

In the preceding chapter we discussed the essential project master-planning effort, which lasts a relatively short time at the beginning of the project. In this chapter we will discuss the scheduling effort as it specifically relates to project construction activities. Scheduling activities are continuous throughout the project life cycle.

In all but the smallest construction organizations, a scheduling person or department is available to assist the construction manager in preparing a detailed construction schedule. The CM must remember, however, that the responsibility for making and monitoring the construction schedule is only *delegated* to the project scheduler. The CM is ultimately responsible for the project's schedule performance.

CMs must make a major input to the construction schedule by way of the master plan discussed in the previous chapter. They must also review the schedule product in detail to check the scheduler's logical approach and the elapsed times used for the major work activities. The critical analysis should start right after the scheduler has made a first pass and debugged the schedule. The CM must be satisfied with the construction schedule before publishing it for approval by other departments and the client.

Scheduling Approach

Again we face the problem of discussing a scheduling approach that is applicable to the differing types of contracting plans introduced in Chapter 2. In the design-build or turnkey approach, the contractor is responsible for scheduling the complete project, including the design, procurement, and construction. If the reader is responsible for scheduling the complete project, please refer to Chapter 4 of *Total Engineering Project Management*[1] which describes that process. Here we will concentrate on scheduling the construct-only approach, to avoid confusing our construction-oriented efforts in this chapter.

There are two other commonly used contracting options to cover; namely the construct-only based on compete design, and the third-party constructor working with the owner's design firm. The only basic difference between these two options is the handling of the design-construction interface, which is relatively straightforward.

Also, the problem of differences in process versus nonprocess types of projects does not assert itself in our scheduling discussions. Experienced CM practitioners in each project arena know the specific technologies that require special scheduling attention. Any major differences will be clarified in the general discussion.

The Owner's Schedule

Two scheduling areas that will have to be discussed are from the owner's and contractor's side. The owner's schedule is a macro schedule stressing the strategic planning goals. The contractor's schedule is a fully detailed operational schedule covering all the construction work activities and finishing within the owner's strategic end date. Both schedules must consider the critical design-procurement-construction interfaces.

Owner's schedule for separate design-bid contracts

Owner project management organizations come in all shapes and sizes. Large corporations handling larger projects normally have a good-sized scheduling department, complete with computers and software tools to serve their project managers. Medium-sized owners may have a scheduler, or they may delegate overall project scheduling to their PMs, who use PCs and software to do their own scheduling. Small owner organizations may have the project leader schedule manually or not at all. In the latter two cases, the owner often delegates the detailed operational scheduling to the design and construction contractors.

Let's first consider a public owner who must use the option of completing a separate design and public advertising for construction bids. This overall macro type of schedule is normally made before the project is started and is updated during the conceptual phase.

The owner's major project execution activity milestones, to be scheduled after project conception and approval, are as follows:

1. Let design contract.

2. Complete design contract.

3. Approve design documentation.

4. Advertise for construction bids.

5. Open and evaluate construction bids.

6. Compare bids with engineer's estimate.

7. Redesign by A&E (if necessary) to meet budgeted costs.

8. Rebid new design.

9. Open and evaluate revised construction bids.

10. Selection and approval of contractor(s).

11. Negotiate contract(s) (including construction schedule).

12. Give contractor(s) notice to proceed.

13. Monitor field schedule(s) performance.

14. Monitor schedule changes.

15. Final inspection and facility acceptance.

16. Final payment and release of retainage.

17. Close out construction contract(s).

These 17 activities can easily be scheduled on a manual bar chart, as shown in Fig. 4.1. They also can be run on a PC using one of the simpler critical path method (CPM) scheduling software programs readily available on the market.[3,4]

The first thing that strikes one about the owner's bar-chart schedule is that all the activities occur in series with each activity serving as a predecessor to the one following. In short, each of the major activities is on the critical path. There is no alternate path to shorten the schedule, especially if items 7, 8, and 9, requiring redesigning and rebidding the job, unexpectedly kick in. If no extra float for items 7, 8, and 9 were allowed, the start of construction would be delayed by at least 6 to 8 weeks.

That emphasizes the importance of owners including nominal float in their overall project activities if the schedule has any chance of being successfully met. Even when owners go into the detail required on an overall project schedule, they may fail to consider all the activities involved or be overly optimistic about the strategic end date. The planned float for each activity is not always strictly held in each operation. When early operations run out of float, it is conveniently *borrowed* from ensuing activities, which is the only way to keep the strategic end date. This means that the construction, which is the last and largest

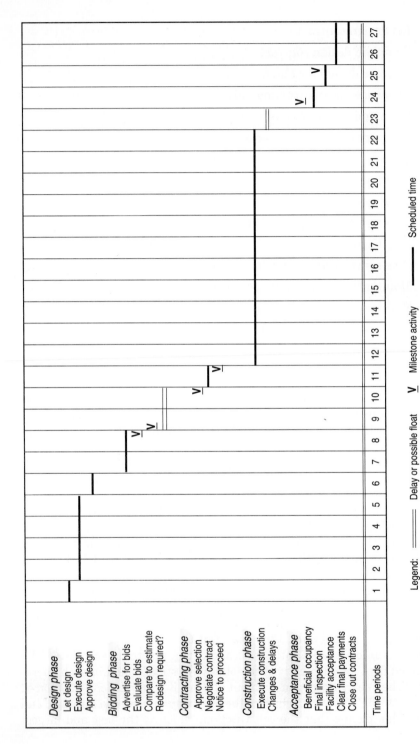

Figure 4.1 Owner's bar chart schedule—separate design construct, A&E project.

Legend: ══ Delay or possible float   V̱ Milestone activity   —— Scheduled time

Design phase
Let design
Execute design
Approve design

Bidding phase
Advertise for bids
Evaluate bids
Compare to estimate
Redesign required?

Contracting phase
Approve selection
Negotiate contract
Notice to proceed

Construction phase
Execute construction
Changes & delays

Acceptance phase
Beneficial occupancy
Final inspection
Facility acceptance
Clear final payments
Close out contracts

Time periods   1   2   3   4   5   6   7   8   9   10   11   12   13   14   15   16   17   18   19   20   21   22   23   24   25   26   27 operation in creating the facility, is usually short of schedule float. That inevitably leads constructors to accept an overly tight construction schedule in an attempt to keep an unrealistic strategic end date.

Prior to getting the constructor on board, the owner has only its and the A&E'S personal experience as to how long the construction should logically take. The constructor's first opinion as to elapsed construction time first surfaces in the original construction bid. In their effort to avoid losing the bid due to schedule problems, many constructors' managements will agree to an overly optimistic construction schedule, despite the CM's and schedulers protestations to the contrary. This happens even when the contract calls for financial penalties for being late!

Owners should be very careful not to place contractors in the position of accepting unrealistic schedules, hoping for some magic on the contractors' part to pull it off. Any penalties, even if collectible, will not offset the problems of poor quality, possible litigation, and lost time that inevitably result from forcing an unlikely schedule on the constructor. Owners should make sure that the selected contractor has done a logical and thorough job on the schedule, and should set a mutually agreeable completion date viable for both parties.

Readers also should note that there were no preconstruction activities involved prior to bidding the construction work. The first preconstruction activity by constructors is the preparation of their construction bids and schedules. A rare exception to that situation occurs when an owner hires a separate construction consultant during the design phase of a particularly complex project.

Owner's schedule for third-party constructor contract

Now let's consider a private owner's schedule using a design, procure, and third-party constructor contracting combination for a complex project heavy in piping and engineered equipment. The process design has not been finalized, so the owner is proceeding on a cost plus fixed-fee basis for both major contracts.

With this contracting arrangement, the owner selects the construction contractor shortly after letting the design-procurement contract. Thus the constructor's input is available early in the project planning, scheduling, and design phases of the project. This contracting arrangement closely resembles the design-build contract approach with some minor variations.

With this contracting arrangement, the owner's list of major project execution activities is as follows:

1. Let design-procure contract.

2. Develop and approve conceptual design.

3. Start detail design with design-procure contractor.

4. Let third-party construction contract.

5. Start preconstruction activities.

6. Start engineered equipment procurement.

7. Start procurement of construction bulk materials and subcontracts.

8. At 30 to 40 percent design completion, open field operations.

9. Complete design-procure contractor's scope.

10. Design-procure contractor's field inspection and follow-up.

11. Constructor's mechanical completion.

12. Constructor's final system testing and commissioning.

13. Owner and design contractor's final plant checkout.

14. Plant start-up.

15. Test runs and acceptance.

16. Close out contracts.

These 16 activities can easily be scheduled with a manual bar chart or a time-scaled logic diagram, produced on a PC with CPM software. Figure 4.2 shows a typical manual bar chart.

With that selected contracting plan we get all the schedule-improving advantages of fast-tracking, with the associated disadvantage of the owner assuming most of the cost-performance risk of two cost-plus prime contractors. Even that risk can be lowered by converting to a guaranteed maximum plus incentive contract, when good project definition is available around the 40 to 50 percent design point.

The biggest advantages of this contracting plan are the many paths of parallel major activity on the schedule, plus the early preconstruction technology and scheduling inputs from the constructor. These advantages and disadvantages are typically found on a fast-track project, but they do allow the earliest possible strategic completion date.

A study of Fig. 4.2 shows that controlling the project execution remains with the owner through the delegation of the major design, procurement, and construction activities to those who are most qualified to do them. The design and construction contractors make their own schedules in close cooperation with each other. This early contractor cooperation produces short and reliable schedules, maximizing the potential for project cost saving.

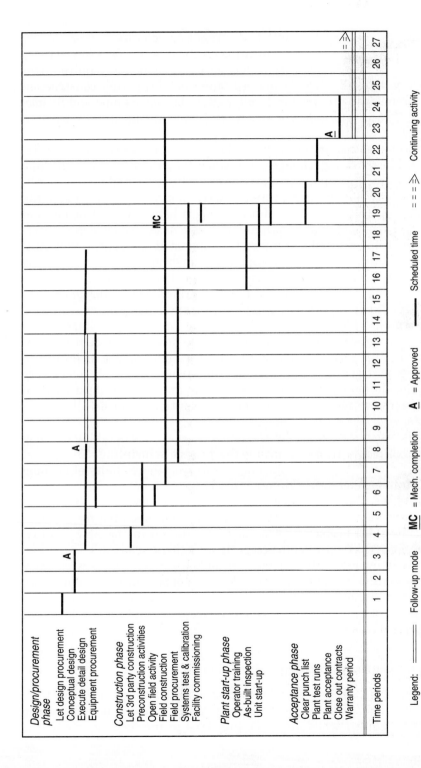

Figure 4.2 Owner's bar chart schedule—separate design, third-party construct process project.

Legend: ══ Follow-up mode   **MC** = Mech. completion   **A** = Approved   ─── Scheduled time   **A** = Approved   = = =⟩ Continuing activity

115

The designer can take effective action to accelerate delivery of any long-delivery items by concentrating on them early. That key activity is aided by having early construction input as to the required field delivery dates. Early bulk material takeoffs also permit construction procurement to order critical bulk materials such as pipe and fittings early. Those activities are especially important in process-oriented projects.

Preconstruction input in the design phase occurs early, thus avoiding necessary design changes later when they will upset the schedule. This contracting environment requires early, close, and formal coordination between the design and construction contractors under the owner's overall leadership. Mutual support between the designer and constructor ensures a smooth and effective design-construction interface, which is a schedule (and cost) improver. It is essential that owners maintain this mutually supportive environment throughout the project to final closeout to attain the most effective project execution. This arrangement also fosters the "integrated owner, designer, constructor team" approach, which is presently being advanced as a major project productivity enhancer by the construction industry.

The design-construction interface in the design-lump-sum bid contracting approach is perhaps the most common cause of failure to meet project goals. Without strong owner leadership, it becomes very easy for the contracting parties to fall into an adversarial rather than a cooperative mode. Such a scenario eventually harms all participants, *most especially the owner.*

This discussion points out the importance of owners taking a strong leadership role in coordinating the project scheduling. They must make an overall project schedule to check on the viability of keeping the strategic end date. Owners must keep track of the overall project float to see that each major project contributor uses only that which has been allotted to that organization. If float is shifted for any reason, the total project float must be evaluated in the new context. Don't force contractors into accepting unrealistic schedules.

Closely evaluate all construction contractors' schedules when they are presented for approval. Do make them prove any overly optimistic projections. Don't let them burn up their float early in the schedule. Do monitor their schedule reports carefully, and do check actual progress in the field.

Developing the Construction Schedule

Now let's discuss how a contractor prepares an operational construction schedule. This is the construction schedule a CM uses for the weekly detailed scheduling meetings held in the field, so it must be formatted in sufficient detail to meet that goal. These schedules have a much more detailed work breakdown structure than an owner's schedule.

Project execution and scheduling philosophy

Construction execution and scheduling philosophy are two items that greatly affect the preparation of the construction schedule. The construction execution philosophy is actually the construction master plan we discussed in Chapter 3. The plan lays down the basic ground rules for construction execution. It also answers such questions as: What is the construction scope? Will design be completed or concurrent? Who does procurement? Will we self-perform or subcontract construction? Each answer affects the selection of a suitable scheduling approach and format.

Scheduling philosophy refers to the selection of the scheduling system. For example: Will we use bar charts or CPM? How often will the schedule be cycled? How many activities are there? Do we have trained people for that type of schedule? What are the contractual requirements for progress reporting? These questions must be decided along with the contracting basis prior to the first scheduling meeting. It's the CM's responsibility to establish those two philosophies and to get management approval on them. Changing either of the philosophies later can be very expensive and disruptive of the project work.

As a major part of the project master plan, the construction schedule must also dovetail with the other major project activities such as design and procurement. The construction schedule should also reflect the start-up sequence of the various units making up the facility. Naturally, units that start up first must finish first.

The reliability of the construction schedule is a function of the degree of design completion available when the schedule is made. That is no problem in the case where design is completed before bidding the construction. In other project execution modes, design should be 30 to 50 percent complete if a reliable operational construction schedule is to be prepared.

On small projects, the construction milestone schedule may be the only one used. On larger projects, the construction schedule serves as the basis for making more detailed weekly work plans in the field for each major activity. In any case, the approved construction schedule is the fundamental working document used by the owner and the contractor to set the major milestones and monitor the actual construction progress.

Detailed field scheduling

Even a detailed CPM schedule with an extended work breakdown structure is not suitable for scheduling the day-to-day activities in the field. Detailed field planning is necessary to make efficient use of field manpower on the priority list of tasks required to meet the CPM milestone dates. This detailed planning also concentrates on those items of work that are on the critical path.

The field scheduling group is responsible to the CM for planning the construction activities each week. Field schedulers list the work activities to be accomplished in the following week. The list of activities is discussed in the weekly planning and scheduling meeting with the field superintendent, chief field engineer, area engineers, major subcontractors, and the CM. The meeting is chaired by the field scheduling engineer. Prior to the meeting, the scheduling engineers have checked to see that all the necessary labor, hardware, and materials are on hand to perform the scheduled work. After the meeting, the weekly task lists are published and distributed to all concerned with the work, so the individual supervisors can plan its execution.

Any work not completed in the past week is also discussed, along with the outlook for the next two to four weeks. The field schedulers must also check the longer-range plans by checking the critical CPM milestone dates four to six weeks ahead.

Proposal schedules

The construction schedule first comes under serious discussion when the bidding documents or the request for proposal (RFP) arrives at the contractor's office. The RFP or bidding documents contain a section devoted to the construction schedule. Usually the documents ask the contractors to develop a preliminary bar-chart schedule showing the major construction activities and their durations in their proposals.

The purpose of that request is to get the contractor to look at the strategic end date in general terms to see if it's still feasible. The owners use the contractor's preliminary bidding schedule to compare it with the schedule they have in mind. This is probably their first chance to get a constructor's professional opinion on how long the construction work will actually take.

Typically, constructors don't spend much time in developing construction schedules during the bidding process. After all, they are not assured of getting the work at this point, so why risk additional proposal or bidding costs? An exception to that rule occurs when the selection of the contractor may revolve around being able to meet a tight completion date. Then the schedule becomes a *selling tool* that may be well worth the extra bidding expense. In that case, the proposal schedule will undergo serious scrutiny during the bid evaluation, so it must stand up to a critical analysis.

In the proposal stage, the CM must evaluate the schedule in relation to the contracting plan and the construction technology available to improve the schedule. For example, there might be some corporate standards or philosophy affecting the contracting plan that could be changed for this project to improve the schedule. These suggestions have to be reviewed with contractor and owner managements to see if the suggested changes are workable. The same is true for suggesting construction technology improvements that will shorten the schedule. This is an excellent spot for the CMs to bring their creative thinking and know-how to the project!

An important point to remember with proposal schedules occurs when it comes time to sign a contract. CMs must make their feelings about the finally agreed-upon schedule known to their managements if they don't agree with what is going into the contract. Silence at this point will be taken as agreement by management.

Scheduling Systems

Now we are ready to discuss the various methods for scheduling capital construction projects that are available to us. This discussion will not be in enough detail to turn you into a trained scheduler, because we are looking at scheduling only from the viewpoint of the construction manager. We just want to make sure that you get a good understanding of the basic theory involved, so that you can use schedules effectively in planning and executing your construction projects.

Many of the examples of scheduling techniques in this section were taken from references 3 and 4 in the listing at the end of this chapter. Those volumes or one of the others listed there will be valuable additions to your construction management reference library. Attending a good seminar on construction scheduling is another excellent way to increase your basic scheduling knowledge, should you wish to learn more about it.

The two basic methods we will be discussing are bar charts and logic-diagram-based schedules. Both methods are used extensively, and sometimes interchangeably, in project and construction work. Each method has its advantages and disadvantages. Knowing when to select the correct method is half the battle in successfully making and controlling your project schedule.

Bar charts

When one analyzes the history of project scheduling, one sees that preplanned written schedules came into use on capital projects in the early twenties. The forerunner to the bar chart was developed by two industrial engineers, Frederick W. Taylor and Henry L. Gantt, for scheduling production operations during World War I. The name "Gantt chart" is still in use today to designate certain types of bar charts. It was sometime after World War I that bar-charting was adapted to the scheduling of construction projects.

There appears to have been little formal scheduling done for the large capital projects developed over the preceding centuries. Since no schedules have been found among the documents in the Egyptian tombs, we must assume that they were planned in the minds of the builders. The supervisors of the construction, however, had certain methods for improving schedule that we do not have available today. When things got behind schedule in the old days, the Pharaoh just sent out the army to round up more slaves to get the work back on schedule. Present-day construction managers can't fall back on that practice to get their jobs back on schedule, as we will discuss later under the subject of labor relations!

Bar charts are the simplest form of scheduling and have been in use the longest of any of the systems we have available. They offer the advantage of being cheap and simple to prepare; they are easy to read and update, and they are readily understood by anyone with a basic knowledge of the capital projects business. They are still in wide use today, even as a final product of the computerized CPM scheduling system.

The main disadvantage of the bar chart is its inability to show enough detail to cover all the activities on larger, complex projects. On large projects, the number of pages required to bar-chart the project becomes cumbersome, and interrelation of work activities becomes difficult to follow from page to page. It is very difficult to closely control the work from a large multipage bar-chart schedule.

Figure 4.3 illustrates those difficulties. In the bar charts seen there, many activities have been condensed into one line to get the complete chart on one page. Each of the macroactivity bars really needs to be broken down into a series of individual tasks to control the project in more detail. The bar-chart sample shows little more than the strategic dates for starting and finishing major portions of the overall project along with a few milestones.

This means that each bar would also take another series of bars to show major activities in that section of the work. For example, under construction, the Site Prep bar might have a number of major operations covering clearing and grubbing, rough grading, roads, drainage,

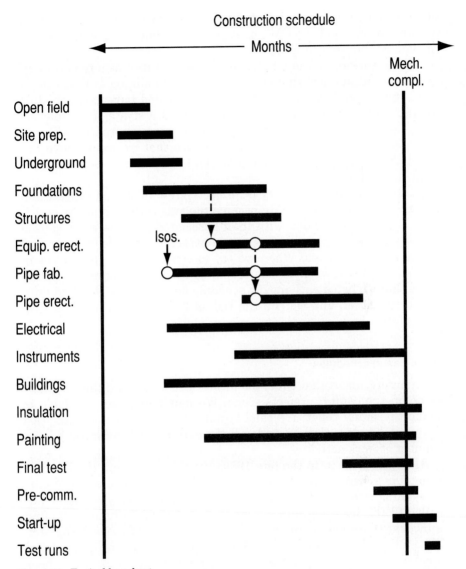

Figure 4.3 Typical bar chart.

utilities, and finished grading on a larger project. But showing the project in that sort of detail could easily take 15 to 20 pages of bar-chart schedules just to list all the major construction activities. That problem could be solved by using a manual or PC-generated spreadsheet to monitor and record progress on each site-development activity and by reporting only overall progress on the bar chart. That arrangement involves a lot of detailed record keeping, which soon absorbs too much of the CM's or supervisor's time to allow him or her to manage the work properly.

I recall working in the late forties and early fifties as a project engineer on a megaproject that cost only about $90 million at that time. I can't remember having seen even an integrated bar-chart schedule for the project. Since no one ever mentioned a strategic completion date, I presume we were using a seat-of-the-pants approach similar to that used on the pyramids! I am also sure that we would have finished that megaproject earlier if we had used a more sophisticated form of scheduling.

As the size and complexity of projects grew in the late fifties and sixties, finishing projects late became the rule rather than the exception. Late finishes, along with their associated cost overruns, caused increased pressure on owners and contractors to develop improved scheduling techniques. When we try to schedule a larger project in that sort of detail with bar charts, we quickly lose most of the advantages that we listed earlier. The schedule becomes unwieldy and difficult to interpret, and we run the risk of losing control of the project time plan.

Logic-based schedules

Fortunately, about that time, the network schedule and the computer came on the capital projects scene. We now had a tool available to make the many repetitive calculations for the early and late start dates, and a place to store and sort the data needed to control a large number of work activities.

As I remember it, in the late 1950s the U.S. Navy and the Du Pont Company concurrently developed two different logic-diagram-based scheduling systems at about the same time. The Navy's system was called PERT, for Program Evaluation and Review Technique. Its first successful application was on the Polaris Missile Program. Scheduling that first-of-a-kind, highly complex, submarine-launched weapons system program with bar charts alone would have been nearly impossible. There were 250 prime contractors and over 9000 subcontractors!

At about the same time, Du Pont first successfully used their Critical Path Method (CPM) of logic diagram scheduling on several new petrochemical plants. Later, Du Pont found it useful when doing critical plant turnarounds, which required a minimum of downtime to be economical.

Other owners and contractors lost no time in adapting the new scheduling methods to their projects in order to improve their timely completion performance. The CPM system was somewhat simpler than the PERT method, so it soon became the system favored for use on commercial and industrial capital projects. The KISS principle triumphed again! The basic logic-diagraming principles developed in the 1960s are still in use today. Most of the improvements have been in the areas of improved sorting and graphical output. In the 1980s, the development of the relatively low-cost PC made the use of the CPM system possible for even the smallest companies.

Shortly after the introduction of the PERT/CPM systems in the early sixties, the pendulum swung from simple bar charting to the side of overly detailed, computerized schedules. That didn't work out as well as the early success with the systems had seemed to indicate it would. If a little bit of CPM was good, more had to be better! Everyone promptly defied the KISS Principle and started to schedule in too much detail on each activity. The result was reams and reams of computer output that virtually inundated many untrained people. I observed several old-time construction managers happily tossing piles of computer printouts into the trash can after quizzically inspecting the first page for a few minutes. Their only recourse was to run the job by the seat of their pants as before, with the expected mixed results.

That was an outstanding example of installing a new and complex system without properly training the users. Most of the construction managers and field schedulers of that period were entrepreneurial craft people who had worked themselves up through the ranks. In many cases they were literally untrainable in the new technology of computerized CPM scheduling.

Fortunately, some of the users of the newly developed techniques remembered the KISS principle and developed some easy-to-use systems before the paper flood turned everybody off. Several good mainframe programs came onto the market, including McDonnell Automation's MSCS system, Metier's Artemis system, and IBM's PCS system, to name a few of the 100 or so programs offered.[3] As computer capacity and new software bloomed, the programs developed and improved rapidly over the next 20 years.

The rapid development of low-cost PC hardware and software has now virtually taken over the CPM capital projects scheduling market. Mainframe computers are now required only on the very largest and most complex projects, those whose logistics demands are too great for the memory capabilities of the PC.

Basic Network Diagraming

To understand how a logic network works, one must at least understand the very basic network diagraming techniques used in their development. I will briefly cover some of the basic practices here, but I strongly recommend that you pursue the subject more intensively in the reference books listed at the end of this chapter, and/or take a training course in CPM scheduling. CMs don't need to become full-fledged schedulers, but if they are to successfully deal with CPM schedules and schedulers, they must understand basic CPM philosophy and the terminology.

The two systems that I will address here are the arrow diagraming method (ADM) and the precedence diagraming method (PDM), both of which are applicable to the CPM system. Most of today's CPM schedules use the PDM method because of its greater flexibility in using lead and lag factors, which are discussed later.

Logic-diagram scheduling has its own terminology that has developed over the years. I have included a brief glossary of some of the more common logic-diagraming terminology in Fig. 4.4.

PERT method	Program evaluation research technique
CPM schedule	Schedule using the critical path method
PDM schedule	Schedule using the precedence diagraming method
Arrow diagram	CPM diagraming method using arrows
Logic diagram	Arrow diagram of complete project or section of a project
Time-scaled chart	Logic diagram with a time scale
Activity	Any significant item of work on a project
Activity list	List of work items for a project; also work breakdown structure
Activity duration	Elapsed time to perform an activity
Optimistic time	Earliest completion: shortest time
Pessimistic time	Latest completion: longest time
Realistic time	Normal completion: average time
Activity number	Number assigned to each activity
Early start date	Earliest date activity could start
Late start date	Latest date activity could start
Float	Measure of spare time on activity
Free float	Time by which activity can be changed without affecting next activity
Total float	Total free time on any activity or project
Negative float	Time a critical activity is late

Figure 4.4 Glossary of CPM terminology.

Arrow diagraming

In arrow diagraming, the basic unit is a work activity that occurs between two events or nodes. The events or nodes are numbered sequentially, and the activity is identified by the beginning and ending event numbers. Those numbers are also designated as the i-j number as shown in Fig. 4.5. The sample Activity of 3.5 (time duration) is designated as i-j number 1-2 .

Activities 0-1 and 1-2 occur in series, so 0-1 must be complete before 1-2 can start. Activities 2-3 and 2-5 run in parallel and start after 1-2 is completed, and so on.

Each activity has an elapsed time necessary to accomplish the work involved. The estimate of elapsed time for the activity must consider the scope of the activity and any historical data available from previous similar projects. If there are no historical elapsed-time data on the activity, the scheduler must use the best estimate (guess) of the value, based on input from the people most experienced in performing that activity. The value estimated for the sample arrow in Fig. 4.5 is shown below the activity arrow. The unit of elapsed-time value can be hours, days, weeks, or months, depending on the scale selected for the schedule. Most elapsed-time values on capital projects are done in weeks, but short, turnaround schedules are often done in hours.

The object of the arrow diagram is to determine the longest time path through the diagram, which is called the *critical path*. In Fig. 4.5, the longest time passes through 1-2, 2-5, 5-6, and 6-7, for a total of 22 days. Paths 1-2, 2-3, 3-4, and 4-7 total only 17 days, which leaves 5 days (22-17) of *slack time* or *float*. If the overall schedule between nodes 0 and 7 is to be shortened, improvement must be made on activities 1-2, 2-5, 5-6, or 6-7. The third path 0-8, 8-9, 9-10, 10-7 takes only 9 days, so it has float of 13 days.

In the PERT system, the elapsed time is calculated by assessing an optimistic time and a pessimistic time and then calculating an average time. The computer program for PERT automatically calculates the average time. Most present-day CPM programs for capital projects don't include the averaging exercise but merely accept the elapsed-time estimate input by the scheduler. Obviously, the proper selection of the elapsed time for the work activities is an important part in preparing the CPM schedule, because it directly affects the critical path and all the float times. That points up to the CM the importance of getting the most-educated guess possible from the most experienced specialists available on the project staff or subcontractors.

Off-the-cuff time estimates for whole projects are very difficult to make. However, breaking the work activities into manageable units greatly improves one's time-estimating accuracy. The elapsed-time

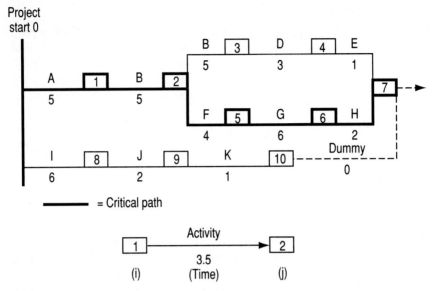

Figure 4.5   Basic arrow-diagraming terminology.

estimate is based on the unit quantities taken off to estimate the project cost. The project cost estimate contains a gold mine of information, such as numbers of masonry units, cubic yards of excavation and/or concrete, tons of structural steel, and the like. That information, combined with standard work units from the contractor's (or estimating services company's) files, is used to calculate the expected elapsed time for the activity. The number obtained is then shaded by any overriding job conditions such as complexity, weather, productivity rates, and so forth.

What makes the elapsed-time estimate for the CPM method work is the averaging out of offsetting pluses and minuses over many work activities. Hopefully we won't get all-pessimistic or all-optimistic inputs for our elapsed times. That could skew the schedule too far one way or the other. A detailed treatise on estimating event times, activity durations, and early and late start dates is given in O'Brien's book *CPM in Construction Management*.[3]

After running the first pass on the computer (or manually), it's a good idea to recheck the elapsed times of the critical and near-critical activities to ensure that the time estimates are reasonable. The key critical activities are the 20 percent or so that most directly affect the project completion date. You should also consider any abnormal factors—adverse seasonal conditions, conflicting workloads, critical facility usage, shortages of material and human resources, and transportation times—that could adversely affect an otherwise normal elapsed-time estimate. Keep in mind that field labor productivity reflects directly on elapsed-time estimates.

The construction of any network must start from a *work breakdown structure* (or *detailed activity list*) showing each activity to be scheduled. Sometimes it looks like an *activity tree,* as one major item is subdivided into its major parts. For example, we might have an activity called "Install foundations," which could be further subdivided into "excavation," "forming," "setting rebar," and "pouring concrete." The foundations could be further subdivided into "Buildings" (affecting structural-steel erection) and "Equipment" (affecting equipment delivery and erection). The decisions affecting the degree of work activity breakdown determine the length of the activity list and the complexity of the overall construction schedule.

Attention also must be given to the longest delivery items on the schedule to see which ones fall on the critical path. If there are several large field-erected items with varying completion dates, we might even break the most critical item into individual units to track the latest one. The same thinking also applies to time-critical subcontracts.

A comparison of the amount of detail covered in a CPM work breakdown structure and a comparable bar chart clearly shows the difference in the amount of detail covered by the two systems. Bar charts cannot even approach CPM schedules in activity numbers without becoming completely unworkable.

Contractor's Logic Diagram

Let's look at the basic layout of a simple contractor's logic diagram for site development, as shown in Fig. 4.6. This type of work commonly occurs on most construction projects for grassroots projects.

The major site-development activities in the project scope are as follows:

1. Clear site.
2. Survey site and layout buildings.
3. Rough-grade site.
4. Install underground utilities, consisting of natural gas line, water lines, sewer lines, and underground electrical service.
5. Provide site drainage and access road to rough asphalt stage.
6. Notes:

   a. The design shows that the utilities (except electric) are in the access road right-of-way, and the access road can't start until after the utilities are in.
   b. Elapsed times are in working days.

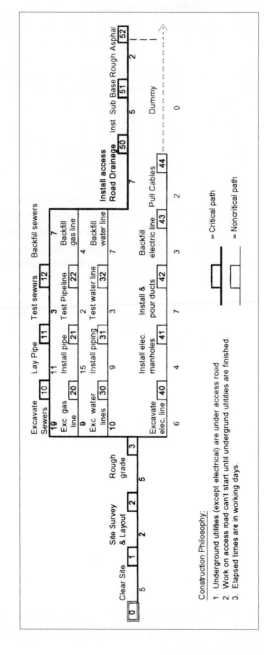

Figure 4.6 Sample contractor's ADM logic diagram.

The site work starts with the site-clearing, survey and layout, and rough-grading activities taking place in series because none of this work can proceed simultaneously. The utility work can all run in parallel because some is under the access road and some is not. The electrical work is entirely independent of the access road. The utility activities are broken down into the applicable subtasks of excavation, laying pipes or conduits, testing, and backfilling. Each of those activities is given a number in the 10, 20, 30, 40, and 50 series, to identify each category of work. The elapsed time in working days is shown under each activity.

Any restraint, known as a *dummy activity,* is shown as a dashed line with zero elapsed time. A dummy activity shows us that the node 53, Completion of Site Work, cannot be considered complete until the electric service has been installed. Because the electrical service is completely independent of the access road, it gets the benefit of the extra 14 days of float that the access road installation requires.

To analyze the site-development diagram for the critical path, we must add the elapsed times for the three parallel paths from (4) through (50) as follows:

	Path	Elapsed times	Total days	Float
(1)	4-10-11-12-50	19 + 11 + 3 + 7	= 40	0
(2)	4-20-21-22-50	9 + 15 + 2 + 4	= 30	10
(3)	4-30-31-32-50	10 + 9 + 3 + 7	= 29	11
(4)	4-40-41-42-43-44	6 + 4 + 7 + 3 + 2	= 22	18

Path (1) equals 40 days, which is the longest. That means the critical path for the site development passes through activities 1, 2, 3, 4, 10, 11, 12, 50, 51, and 52. If we are to improve the schedule for the site development, we must shorten the activities on the critical path. If we could shorten the path through the sewer lines by 10 days, we would have two critical paths of 30 days each. If we could shorten the sewer lines path by more than 10 days, the critical path would shift to the gas or water line paths.

Another way to describe the difference between paths (1), (2), (3), and (4) is to say that path (2) has 10 days of float, path (3) has 11 days, and (4) has 18 days. That also says that activity 4-20 can start up to 10 days later than activity 4-10 and still finish without extending the utility critical path of 40 days. Likewise, we could delay finishing any activity on the gas line by 10 days without delaying overall completion. Also, the total floats of 10, 11, and 18 days can be used to delay any of the activities on the noncritical paths.

Another form of logic diagraming notation is the *precedence diagraming method* (PDM), which grew out of the arrow diagraming method to overcome some of the ADM's faults. Most present-day CPM scheduling programs are PDM-based systems. Figure 4.7 shows the basic terminology for PDM diagraming. Three different notations are available to show a variety of start-to-finish relationships to fit any type of activity. This gives more flexibility to notation when building the logic diagram.

PDM shows the activity number in the block and not on the line, as shown in the sample precedence diagram in Fig. 4.8. Placing the activity in the block gives more area to describe the activity in more detail. The activity number also appears in the block as a simple code number instead of a dual-digit *i-j* number. As the elapsed time appears below the activity block, the activity connecting line is short, making the PDM diagram more compact.

Actually, the PDM diagram in Fig. 4.8 covers the same activities as the contractor's arrow diagram in Fig. 4.6. I think you will agree that the PDM diagram is cleaner and easier to read than the ADM diagram. The strongest advantage of the PDM system is its ability to show lead and lag factors when an activity can start before the preceding one has been fully completed. Figure 4.7 depicts the lead and lag notations that make PDM a more powerful scheduling tool.

Activity number 10, Excavate Sewers, in the PDM diagram in Fig. 4.8, illustrates the use of a lead factor. In that case we show that activity 11, Lay Pipe, can start before completion of the excavation. To show that potential for shortening the critical path on the ADM diagram, we would have had to break activity 3-10 into two arrows with an intermediate node. A much more detailed account of precedence networking appears in *Scheduling Construction Projects*[1] and *CPM in Construction Management*.[3]

Whichever the diagraming method, calculating float time and the early and late start dates for all activities in the logic diagram is the same. The calculation is quite simple, but too voluminous to cover in this discussion. Mr. O'Brien covers the principles very well in Chapter 2 of the *Scheduling Handbook*.[2] I recommend that you study that chapter to learn how to calculate a critical path manually if you do not already know how. It becomes apparent in studying the early and late start calculations that they are indeed manifold and repetitious. That makes the use of a computer program mandatory in all but the simplest of CPM schedules. Doing thousands of those calculations by hand is much too labor-intensive.

I have reduced the sample logic diagrams and discussion to their simplest forms to give nonscheduling readers an introduction to the

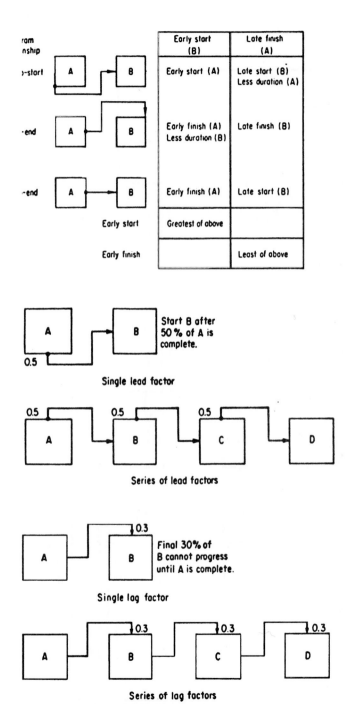

Figure 4.7   Basic PDM terminology.

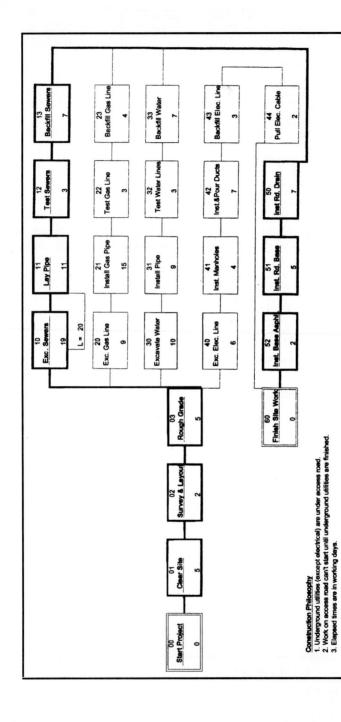

Figure 4.8 Sample contractor's PDM logic diagram.

Construction Philosophy
1. Underground utilities (except electrical) are under access road.
2. Work on access road can't start until underground utilities are finished.
3. Elapsed times are in working days.

workings of logic diagraming. My simple example is multiplied by up to several hundred times in a large project of several thousand activities. On larger projects, the CPM logic diagrams can become large enough to cover the walls of a typical construction management war room. The scheduling people like to have the complete diagram displayed so they can easily check the logic of the diagram in detail. The project managers may have need to refer to it occasionally to resolve particularly knotty problem areas. Usually, however, CMs work from the early and late start date, increasing float sort, milestone sorts, and bar-chart printouts for day-to-day control of the construction project.

Normally, fewer than 20 percent of project activities will fall on the critical path, so CPM is a management-by-exception technique. When more than 20 percent of the activities fall on the critical path, it is a sign that the schedule is in trouble and needs to be recycled. Having activities with negative float also is a sign that the schedule no longer is achievable in its present form.

Recycling the schedule should not be confused with the monthly progress evaluation. Recycling might be necessary over a three- or four-month period when your short-term goals are not being met. The recycling procedure addresses the question of how to get the work back on track without extending the strategic completion date.

Advantages and Disadvantages of CPM and Bar-Chart Methods

Evaluating the advantages and disadvantages of bar charts versus CPM allows us to select the most effective system for a given project. It also leads us to some simple rules that are applicable to the selection.

Advantages of CPM

The number-one advantage of the CPM system is its ability to handle many work activities on complex projects with ease. Let me introduce a word of caution on that point: Don't fall into the trap of using more activities than necessary, just because it is easy to do so. You risk getting your schedule bogged down in too much detail, which makes it harder to use and costs more money to operate. Remember, that's the same trap that almost killed the CPM system in its early days! One way to avoid the problem is to break out some of the less complicated scheduling areas and use bar charts for them. They could be offsite areas such as small office buildings, warehouses, tank farms, and roads. A blend of the two systems often results in a simpler and more effective overall project schedule.

Another outstanding advantage of CPM is the intangible benefit of forcing the project team to dissect the project into all of its working parts. This forces the early analysis of each work activity. The CPM logic diagram is actually a dry run on paper for all phases of the project. For that reason alone, it behooves the CM to play a major role in planning the CPM schedule and in checking the resulting logic diagram.

The actual scheduling phase, such as calculating the early and late start dates and the associated float, is best left to the scheduling technicians and the computer. It is usually necessary to run the first pass of the schedule several times, to test and debug the logic diagram before the final version is ready for review and approval.

The large menu of output sorts is another big advantage of a computerized CPM schedule. It allows the various interested members of the construction team to order the output sort best suited to their work. Most CPM programs will yield a sort menu as follows:

1. Total float per activity

2. Limited look-ahead sorts

3. Critical-path sort

4. Critical-equipment sort

5. Project-milestone sort

6. Bar-chart printout

7. Human resource leveling

Most CMs, for example, will find the sorts by total float and by milestone most valuable for their needs. The total-float sort starts with the low-float (most critical) work activities listed first for immediate attention. The less critical high-float items show up later on the list. By using the periodic look-ahead sorts, one can also home in on specific time periods. A 30-, 60-, or 90-day look-ahead sort will list only those critical items that will occur in the next 30, 60, or 90 days. These sorts are attractive for the field superintendent, area supervisors, and field schedulers.

Other members of the project team need other types of data sorts to make their work more productive. For example, material control people find the critical-item sort more convenient in tracking their required delivery dates than extracting those dates from a milestone chart. As the revised delivery data and actual progress are fed into the computer, revised printouts quickly reflect the delivery changes and their effect on the field schedule.

Field people usually find that the key-milestone-date sort better suits their needs. That section of the CPM schedule is the basic document that the field scheduler uses to make the detailed weekly work schedules in the field.

Most CPM scheduling software even delivers a bar-chart printout, which is most convenient for upper management and reporting purposes in reviewing project progress. Management people have neither the time nor the inclination to delve into the details of the logic diagram or early and late start dates. Simplified bar charts are usually included in the progress reports to give a graphic view of actual progress against the schedule. On smaller projects, a simple time-scaled bar chart might be used in the progress report.

The rapid turnaround of data by the computer also allows the project team to perform what-if exercises with the logic diagram. When scheduling problems arise, the project team can try alternative solutions by reworking elapsed times for problem activities. This generates new early and late start dates that can be shifted to improve the critical path. The computer calculates a new critical path in a matter of seconds, with immediate access to the new output data right there on the computer screen.

A CPM/computer system also simplifies recycling the schedule. Recycling becomes necessary whenever schedule deviations grow to a point at which some of the intermediate goals are in jeopardy. Recycling involves revising any target dates that may have slipped beyond repair, perhaps because a significant change in scope has occurred. Exercising some what-if options should allow you to obtain the scheduling-revision option best suited to keeping the project on schedule.

Human resources leveling

Another important advantage of the CPM/computer system is the ability to level peak personnel requirements, which occur during the project's design and construction phases. By taking advantage of the available float and rescheduling the start of noncritical activities, it's possible to shave personnel peaks. Leveling the personnel requirements leads to more effective use of the project's human resources.

This option is invaluable for smoothing out craft manpower peaks in key areas of the work. Judicious use of the early and late start dates can also keep subcontractors from getting in each other's way.

Disadvantages of CPM

There are only a few disadvantages to using the CPM method for project scheduling, and even these can be avoided with proper attention from the CM. However, overlooking any of the disadvantages can scuttle your efforts to control the project schedule!

It's extremely important that your key field people be trained in CPM techniques. That includes all levels from the design group to procurement, and through the construction management team. Remember the story about construction managers trashing the CPM/computer printouts in earlier days!

I don't recommend controlling a large project with only a newly trained crew, or using a new software system without running your old scheduling system in parallel, at least until the new system has been proved to work. If the new system breaks down for any reason, you will be without any means of controlling the end date on the project.

The cost of running a CPM schedule is likely to be higher than that of using bar charts, particularly on smaller projects. That was especially true of running the CPM schedule on a mainframe computer. In recent years the relatively modest cost of PC hardware, software, and training has enabled us to expense off that cost for computerized CPM scheduling on a medium-to-large project. It might take several small-sized projects to cover writing off a PC scheduling system.

The minimum hardware requirements to accomplish the CPM scheduling should consist of an IBM-compatible PC of 640K memory, an 80-megabyte hard disk, and a dot-matrix or laser printer. Fancy graphics require a plotter, but that is not essential to doing a decent job of schedule control. Any one of the 100 or so scheduling software programs that suits your type of project will do the job. The cost of the training could be the sleeper in the scheduling cost budget, depending on the experience and computer literacy of your project people. But regardless of the cost, training is the linchpin of the whole system, so don't ignore it.

The real savings in using a PC-computerized schedule is that it will generate a good deal more data than is possible with bar charts. That means the unit cost of the data is low. However if the data is not being used (or worse, is being improperly used!), you will not be getting your money's worth. It is up to the CM to see that the computerized schedule output is used in a cost-effective manner.

The cost effectiveness of using CPM

It is difficult to accurately quantify the cost effectiveness of using CPM scheduling systems on capital projects. First, there is no absolute measure of the time saved by using CPM versus bar charts. Second, the value of the time saved must be balanced against the value to the owner of having earlier access to the facility. Any comparisons of that nature have proved to be highly speculative and difficult to verify.

Most owners and contractors accept any additional cost of using CPM scheduling systems as a way to improve the odds of completing their projects on time. Those who do not believe that CPM saves money and ensures a project's earlier completion date can continue to use manual bar-charting with reasonable hope of success.

O'Brien's book, *CPM in Construction Management,*[3] contains a chapter on costs and some expected savings from using CPM. On average, the cost of applying a CPM system to a project is about 0.5 percent of the total facility cost. The major cost areas for using the system are schedulers' time, software cost, and computer time.

CMs must be aware of the type of scheduling system that is being proposed for their projects, so they can budget funds to cover the cost. Small projects can be done with a part-time scheduler; medium-size projects need at least one person full-time; and larger projects will require two or more schedulers to handle the workload. Include all computer costs (including necessary training) in the project budget. Be especially careful if a mainframe computer is used.

Advantages of bar-chart schedules

As I said earlier, bar charts are inexpensive to produce and are easily understood by people with a minimum of scheduling training. I heartily recommend them for small, less complex projects, as being suitable and cost-effective. The more *comprehensive* CPM system is often too complicated, and represents unnecessary overkill when used on small projects. There are some good PC-based CPM programs available for scheduling a series of small projects, which draw from the same resource pool.

The only thing that threatens the economic advantages of using bar charts on small projects has been the advent of PCs, along with less complicated scheduling software. It is easy to be tempted into the use of a PC, with the construction manager or engineer acting as the project scheduler. That can be all right if the designated scheduler does the scheduling work in his or her free time. If, however, the project leader gets so involved in running the CPM schedule that he or she lets the rest of the project direction go its own way, the project is doomed to failure.

Disadvantages of bar-chart schedules

As we showed in Fig. 4.3, bar charts have only a limited ability to show many detailed work activities and their associated interactions. They become bulky and unwieldy on larger projects with as few as 100 activities.

Bar charts cannot show clearly the interaction between early start and late finish dates of activities and the resulting float of noncritical activities. There is no clear identification of the critical path through the project that appears with the CPM system. Also, it's impossible to develop the wealth of scheduling detail with a bar chart like that developed and manipulated with the CPM system. With bar charts there are no concise information sorts as with the CPM.

Computer versus manual scheduling methods

The major factors in selecting computer over manual scheduling methods are project size and complexity. Small projects are best done manually, since good time control is possible at low cost. However, a complex plant turnaround project, with a relatively low budget but working three shifts on a tight schedule, definitely warrants a computerized CPM approach.

On larger projects using CPM, computer operation is a must if the many repetitive critical path calculations are to be performed in a short time. Manipulating and sorting the expanded database of project information is well worth the additional expense if the system is properly applied. The arrival of PCs and minicomputers, with their associated scheduling software, has brought the cost well within acceptable limits. The simpler operations of the PC-based systems has also reduced the cost of the necessary CPM training.

Scheduling System Selection

Our discussion of the advantages and disadvantages of the available scheduling systems should allow us to develop guidelines for selecting an effective scheduling system. The selection involves such factors as:

- Size and complexity of project
- Scope of services required
- Sophistication of user organizations (i.e., client, field organization, subcontractors, and so on)
- Available scheduling systems
- Scheduling budget
- Client preference
- Mixing schedule and cost

Size of project

We have already discussed this point earlier. The rule of thumb is bar charts and manual systems for small projects, and computerized CPM for medium-size and larger projects. The level of sophistication of the system tends to become greater as the projects become larger.

Complexity of project

Even small complex projects can make good use of computerized CPM schedules, if the fewer activities take place in a very short time span. An example is a plant turnaround worth $1 million or less, with only 10 days to do it. On the other hand, a $1 million project with a 12-month schedule might not warrant a computerized CPM schedule.

Scope of services

Full-scope design, procurement, and construction projects lend themselves to more complex scheduling methods, because of the extra interfaces among the many design, procurement, and construction activities. A project involving just one of these macroactivities could be effectively controlled with a less sophisticated and less costly system.

Sophistication of user organizations

The sophistication of user organizations is probably the most over-looked factor in selecting a scheduling system. Often, the need to produce a full-blown CPM schedule exists. However, one key project group may not be experienced enough in CPM to properly interpret their part in it. Assure yourself that the failure of that group to perform properly will not defeat the proposed scheduling method.

An example would be working with a client in a developing country. If the client's people lacked experience in the proposed scheduling system, they might not feel comfortable using it to track job progress. Also, they might not keep their contributions to the project on schedule. Another example would be an inexperienced construction force not being able to use the output of the CPM scheduling system, such as we discussed earlier.

The worst possible case would occur if the CM were not versed in the selected scheduling system! That underscores the need for present-day CMs to stay current on the latest CPM scheduling methods available in their companies and the marketplace. I definitely recommend that you success-oriented CMs do further, more in-depth study of logic-based scheduling than I have presented here.

If there is a shortage of CPM know-how in your organization, it is possible to hire a CPM consultant to handle your project scheduling. At least one member of your team, however, should have enough knowledge of the system and the project to monitor the consultant's work. That is the best way to ensure that the resulting schedule will be effective for your project.

Existing company systems available

The availability of company systems is important, because we want to use a system that has been in use within the organization, and that has been thoroughly tested on prior similar projects. Introducing a new system on a project often causes more problems than it solves. As an owner's project manager, you would do well to assure yourself that the contractor is proficient in the system before allowing its use on your project. It is also not a good idea to force the use of your corporate standard system on the contractor just because your organization is familiar with it. A much wiser course is to train your people in the use of the contractor's scheduling system.

Scheduling budget

If the field indirect cost budget does not allow for the expense of a sophisticated scheduling method, you are going to come up short of money. Most computerized CPM scheduling costs have a tendency to grow and overrun their budgets. A common problem is job stretchout, which increases the schedule cycles, which in turn runs up the scheduling personnel hours and computer time. A factual estimate of the total cost of the proposed scheduling system is needed if an effective system for the project is to be selected.

Client preferences

Owners who want computerized CPM schedules, and who are willing to pay for them, are entitled to have them. If the owner does not specify a preference for a scheduling system, some common ground for developing a cost-effective system will have to be found.

In recent years most federal government contracts have required adherence to a strict contractual standard, calling for use of CPM schedule-control and reporting. Careful attention must be given to investigating the latest scheduling requirements for any federal work on which you may be proposing.

Mixing Cost and Schedule Control

Lately, many software companies have been promoting programs that claim to control cost along with schedule.[3] That may work on very simple projects, where you can project costs closely into the schedule.

On larger projects it has never really been very successful. The budget numbers are broken down into different categories than the scheduled activities. Most cost-reporting systems seem to need the accuracy level required by good accounting practice. Combined schedule and cost-reporting systems have not yet been able to generate that type of accuracy. I would be wary of using a system that ties those two key areas of project control together, and thereby runs the risk of losing control of two very critical areas of project management.

References

1. Edward M. Willis, *Scheduling Construction Projects,* Prentice-Hall, New York, 1986.
2. James J. O'Brien, *Scheduling Handbook,* McGraw-Hill, New York, 1969.
3. James J. O'Brien, *CPM in Construction Management,* 4th ed., McGraw-Hill, New York, 1993.
4. Mike C. Tidwell, *Microcomputer Applications for Field Construction Projects,* McGraw-Hill, New York, 1992.
5. J. J. Moder, C. R. Phillips, and E. M. Davis, *Project Management with CPM, Pert, and Precedence Diagraming,* Van Nostrand Reinhold, New York, 1983.
6. Ira H. Krakow, *Managing Your Projects Using the IBM PC,* Brady Communications Co., Bowie, MD, 1984.

Case Study Instructions

1. The basic activities, with their elapsed times (in working weeks), required to construct a habitable frame house are listed below:

Clear site	2	Finish plumbing	4
Excavate basement	3	Interior painting	6
Rough walls & floors	9	Wallboard	8
Exterior siding	7	Exterior fixtures	2
Pour basement	6	Exterior paint	6
Roof	4	Finish flooring	5
Rough in plumbing	9	Roofing	4
Electrical	8	Landscaping	6

Arrange the work activities according to precedence requirements, and draw the arrow diagram.

- Calculate the early and late start dates.
- What is the earliest completion time?
- What is the critical path?

2. Assuming that you have access to a personal computer and an applicable construction scheduling software program, input the data in instruction 1 above, to check your manual work on that problem.

Practice the various sorts and options available in the program, including some what-if exercises to try to shorten the house completion date.

3. Evaluate your selected project in regard to the type of scheduling method you would recommend to your client or management. List the reasons for each of your recommendations.

4. Referring to your project master plan developed in your case study at the end of Chapter 3, assume some best-guess elapsed times for the major activities and long-delivery items for your project. Based on your assumptions and your knowledge of activity precedence, sketch out a logic diagram to indicate the anticipated critical path through the project.

Make a list of recommendations that would help to shorten the completion date, along with an explanation for each recommendation.

Chapter

5

Estimating, Budgeting, and Cost Control

The project money plan is the financial forecast for the project, and it sets the basis for the control of project costs and cashflow. Developing the money plan involves the functions of cost estimating, budgeting, cashflow, cost control, and project profitability. Taken together, these functions make up the field of cost engineering. The professional organization for cost engineers is the American Association of Cost Engineers (AACE), which promotes the development of cost engineering and publishes many excellent reference books on the subject.[1,2]

This chapter considers cost engineering from the construction manager's viewpoint. We don't expect CMs to become full-fledged cost engineers, but they had better know the ramifications of cost engineering as it applies to their projects. They will be dealing with cost engineers throughout their projects. In the early stages, they will be dealing with the project estimators. In the execution stage, CMs will be dealing with the field cost engineers in controlling project cost and issuing the monthly cost report. CMs should remember, however, that cost engineers only supply the CM with project cost data. It's the CM who bears the sole responsibility for using the data to ensure the financial success (or failure!) of the project.

Those of you who perhaps have made your own estimates on smaller projects will find yourselves at a definite advantage when you become a full-fledged construction manager. I strongly recommend that you add the applicable books listed in the reference section of this chapter to your library for use in formulating your project money plans. Of course, *reading* the books before consigning them to the library is also an excellent idea!

The preparation of the project financial plan cuts across virtually every line of communication on the project, as well as the total organization (see Fig. 1.2). Therefore the construction manager must be very tactful in handling the human relations aspect of developing the project money plan. A sound project financial plan also holds the key to reaching a major goal of the construction manager's creed of *finishing the project within budget.*

The Construction Cost Estimate

The construction cost estimate usually gives the first specific indication of the *total project cost* (TPC). The estimate of total project cost gives vital information to the owner, the project designer, and the construction contractor. The owner uses the latest TPC projection to verify the project's economic viability and cashflow needs. The designer uses the number to confirm the viability of its design and thereby meet the projected investment. The construction estimate also sets up the potential profit that the contractor hopes to realize on the project. The CM, of course, has a personal interest in the estimate, his or her desire to have a financially successful project. These are a few of the major goal-oriented project groups that focus their attention on the first detailed construction estimate.

The finally agreed-upon cost for the construction work, be it fixed-price, target price, or any construction pricing variation, sets the financial baseline for the construction cost. Usually the price starts as an estimated cost that is agreed to by the owner and contractor. The estimate used to reach the contract price is then converted to the project control budget for controlling the project costs. That's the most important reason for having a well-organized cost estimate based on the applicable code of accounts to facilitate converting the estimate to the budget.

Construction Cost Estimating

The foundation for the project money plan is a sound project cost estimate that is the *predicted* cost of executing the work. The construction cost estimate should be neither *optimistic* nor *pessimistic,* and to be truly effective it must be produced at a reasonable cost.

The decision to make a detailed construction cost estimate must not be taken lightly, because considerable cost is involved. As we mentioned in Chapter 2, submitting a project proposal without a detailed construction cost estimate can be costly. If the detailed construction estimate is included in the contractor's services on a design-build project, the cost is borne by the owner directly. If the construction estimate is for a competitive fixed-price bid, the cost is borne directly by the contractor. Actually the owner pays for the estimating cost indirectly, because contractors' estimating costs are recovered in the overhead charges on projects they do get.

In any event, before agreeing to submit a competitive fixed-price bid, the contractors involved usually evaluate their chances of success before committing the funds to prepare a bid. A preestimate meeting is held[5] to discuss the pros and cons of submitting a bid. Construction management will be there to provide input as to constructibility problems, applicable construction technology, labor climate, experience on prior similar projects, and the like. After weighing all the technical, marketing, and financial input, the firm's top management will make the decision whether to bid or not.

In defining a construction cost estimate as a *predicted* value, it is well to remember that the prediction is based on certain key factors existing at the time of the estimate. Some of those factors are:

- Project definition
- Contracting plan
- Construction schedule
- Construction technology used
- Labor productivity basis
- Estimating methods

The major factors listed should be clearly stated in the basis for the estimate, so that users of the estimate can make their own judgment as to the level of accuracy of the predicted cost. Confirming the estimate basis also sets the ground rules under which the estimate was made, should those ground rules happen to change during project execution. We will discuss the effect of the various basis-of-estimate factors as part of the subsequent discussion of the various types of estimates used in construction practice.

Detailed estimating of fully designed projects has long constituted the lion's share of construction estimating activity. However, construction input to preliminary or planning estimates does arise often enough to include it in our construction estimating discussion. The need for early construction-cost input occurs most often in the design-build contracting mode. In fully designed fixed-price bidding, the designer is at a disadvantage in getting preliminary construction bids without paying for them.

There is another good reason for including a discussion of all types of project estimates in a construction management book. CMs run into all sorts of statements of project cost during a project, so it is important for them to be able to evaluate how those numbers were developed. By the time the numbers reach the construction phase, they may have taken on more validity than they deserve.

Types of Cost Estimates

Cost estimates fall into four basic classes based on the purpose the estimate serves during the project's life cycle. The names of the estimates vary widely across and within the various types of disciplines dealing with capital project execution. Those that I have chosen for discussion are given in Table 5.1.

In addition to the differences in the names of the four classes of estimates, one can expect a good deal of variation in the expected accuracy figures. That is why I have shown a range of expected accuracies for each type of estimate except the definitive estimate. The exact percentage used depends on company policy and the degree of project definition available at the time of the estimate. For example, feasibility estimate accuracies could even run as high as plus or minus 50 percent, if only imprecise design data were available at the time.

Most construction estimates based on completed plans and specifications fall into the definitive class, because the design data is virtually complete, or at least well defined. With hardware and material quantities well defined, the major unknowns are the quantity and productivity of the construction labor. Most of that risk is passed off to subcontractors in today's lump-sum bidding environment.

Percentage of accuracy

The use of percent accuracy is perhaps the most misunderstood concept in the field of cost estimating. The figure is not a percent of contingency for the estimate. Rather it is an indication of the probability

TABLE 5.1   Names of the Types of Estimates

Type of Estimate	Purpose	Accuracy (%)
1. Feasibility	Determine project feasibility.	± 25 to 30%
2. Appropriation	Obtain project funding	± 15 to 25%
3. Capital cost or budget	Project control budget.	± 10 to 15%
4. Definitive	Final cost prediction.	± 5%

of overrunning or underrunning the estimated cost figure. The estimate should already include a contingency allowance to cover unexpected errors of omission and commission in making the estimate.

Because the various estimates are made as the degree of completion of the design work increases, the percentage of accuracy becomes progressively more refined. The increase in accuracy is due to the improvement in project definition and to improved pricing data that become available to the estimators. The lower percentage accuracy numbers show that the probability of overrunning the estimate is becoming lower. The diagram in Fig. 5.1 is a graphical representation of the classes of estimates and where they occur in the life cycle of the project. Please remember that the nomenclature, the timing of the estimates, and the duration of the project are highly variable in different project environments. You will have to tailor the diagram to your particular construction project environment.

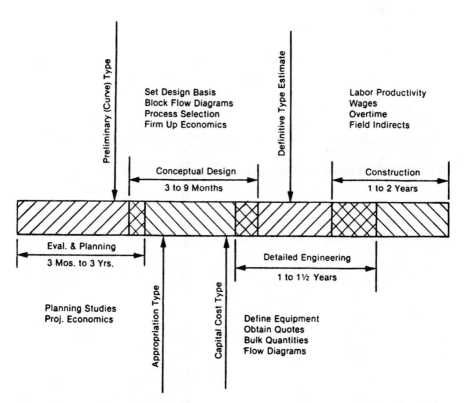

Figure 5.1 Classes of estimates (time line). (*Reproduced by permission from* Chemical Engineering, *July 7, 1975, p. 263.*)

Most larger estimating departments today use computer software programs that offer a more scientific evaluation of a project cost estimate's accuracy. These programs use a blend of Monte Carlo Simulation, Pareto's Law, and proprietary algorithms to quantify the accuracy of estimates. Pareto's Law (also known as the 80/20 Rule) states that a small percentage of the estimating values covers most of the potential cost variables (unknowns) in the estimate. Focusing the computer's attention on the 20 percent of variable activities can lead to a more accurate prediction of the odds on meeting the predicted cost. But remember, these programs do not improve the quality of the estimate, only the predicted accuracy!

The software and running costs for the programs can be substantial compared to the value of the percentage accuracy results. These programs are no substitute for good estimators using proven and current methods, systems, and data to produce the best possible cost prediction.

Project definition

The amount of project definition is the single most important factor controlling the accuracy of any estimate. Table 5.2 lists the various items of design data needed for the four classes. The table is designed to describe industrial type projects. Commercial and A&E projects naturally will not involve the references to process work.

The class 1 feasibility estimate requires only the minimum design data, such as the products (or services), a brief process (or use) description, the unit capacity of the facility, and a general location. Because of the minimal design data available at this stage, the time spent to estimate the cost is low and the percentage of possible error is high.

For a class 2 appropriation estimate, the available design data becomes greater with the addition of block-flow diagrams, general layouts, building sizes, preliminary equipment lists, and a specific site location. The design becomes somewhat more refined with a preliminary plot plan, basic mechanical and electrical designs, and an estimate of the design costs. The additional project definition increases the expected accuracy of the estimate. Some very early construction cost factors emerge at that time, which can affect the construction portion of the project estimate.

The major risk factor in the estimate's accuracy at this point is that the data available are still subject to revision during the schematic design phase. Watch closely estimates made during this early phase for design changes that increase or decrease project costs. Set up a monitoring system to flag significant changes immediately, and determine their effect on the final cost estimate.

TABLE 5.2   Estimate Data Checklist

	Types of estimates			
	1	2	3	4
General requirements				
Process or services description	X	X	X	X
Facility capacity				
Location, general	X			
Location, specific		Xp	X	X
Government regulations	Xp	X	X	X
Preliminary building sizes	Xp	X		
Detailed building layouts			X	X
Process design package (for process plants only)				
Block process flow diagram	X			
Process flow diagram (with equipment size and material)		X		
Process P&IDs		Xp	X	X
Utility P&IDs		Xp	X	X
Equipment list (with motors and prices)		Xp	X	X
Site conditions				
Soil test report		X	X	X
Site clearing		Xp	X	X
Geological and meteorological data		X	X	X
Roads, paving, and landscaping		Xp	X	X
Access to site (road and rail)			X	X
Shipping and receiving conditions			X	X
Major costs are factored	Xf	Xp		
Major process and utility equipment				
Preliminary sizes and materials		Xp	X	
Finalized sizes and materials			X	X
Accessories and drivers			X	X
Equipment foundations, piping and wiring		Xf	X	X
Bulk material quantities				
Preliminary design quantity takeoff		Xp	Xp	
Finalized design quantity takeoff			Xp	X
Engineering data (Process plant)				
Plot plans and elevations		Xp	X	X
Piping and electrical routing diagrams			X	X
Electrical single lines and area classifications		Xp	X	X
Piping isometric drawings and erection diagrams		Xp	Xp	X
General and equipment specifications		Xp	X	X
Home office services estimate		Xf	X	X
Catalysts and chemicals		Xp	X	X
Fire protection systems		Xp	X	X
Underground systems		Xp	X	X

TABLE 5.2   Estimate Data Checklist (Continued)

	Types of estimates			
	1	2	3	4
Design package (Commercial Project)				
Architectural layouts and elevation		Xp	X	X
General specifications and bidding documents		Xp	Xp	X
Structural design and arrangements		Xp	Xp	X
Plumbing and HVAC plans and specifications		Xp	Xp	X
Electrical plans and specifications		Xp	Xp	X
Construction data				
Contracting plan	Xp	X	X	X
Labor wages and fringes		Xp	X	X
Labor productivity and local practices			X	X
Detailed construction execution plan		Xp	X	X
Field indirect costs		Xp	X	X
Subcontract bids		Xp	X	X
Safety programs and cost estimate		Xp	X	X
Constructibility analyses		Xp	X	X
Heavy-lift studies		Xp	X	X
Schedule				
Strategic planning dates	Xp	X	X	X
Project-execution master plan		Xp	X	X
Planned estimate dates		Xp	X	X
CPM schedule			X	X
Miscellaneous				
Freight cost		Xp	X	X
Start-up and training scope and costs		Xp	X	X
Insurance and taxes			X	X
Royalties		X	X	X
Import-export duties			X	X
Financing plan costs		Xp	X	X
Escalation				
Escalation analysis		Xp	X	X
Contingency				
Accuracy analysis		Xp	X	X

NOTES: Xp = Preliminary or partial data for that element
Xf = Factored estimate
1 = Feasibility estimate
2 = Appropriations estimate
3 = Capital cost estimate
4 = Definitive estimate

By the time of the class 3 capital cost estimate, most of the design data are so well defined that the accuracy improves rapidly. Try to freeze the design at this point and permit only minor changes that have a minimal effect on the cost estimate. The desire to eliminate design changes at this point can prove optimistic in the face of a hyperactive owner.

In nonprocess projects, this estimate corresponds to the so-called *engineer's estimate,* which is made by the A&E at the lump-sum construction bidding stage. This is the estimate used to compare with the contractor's lump-sum bids to see that the A&E has designed within the owner's proposed budget.

The class 4 definitive estimate occurs on a design-build job when the design is almost finished. By that time most of the materials have been ordered, and only unknowns in the construction cost area are still outstanding. The major difference between the class 3 and class 4 design data is the improved accuracy of the material takeoff and the pricing information. That permits accurate pricing of the equipment and materials budget. The definitive estimate (Class 4) also compares with a contractor's estimate for a lump-sum A&E type of project. It is a carefully detailed estimate based on a completed design-data package.

Table 5.2 provides a good checklist for the information that your estimating team must have available to produce the desired estimating accuracy. If most of the data are missing, you will not produce the desired accuracy, regardless of how much time you spend on the estimate.

How to Make the Four Classes of Estimates

The approach to the different classes of estimates varies considerably, and it is worthy of some general discussion. Each estimate serves a different purpose and depends on the available project definition. The discussion points out why construction managers should be wary of the accuracy of the early classes of estimates.

Feasibility estimates

During the conceptual phase of most projects, owners must know how much the expected capital investment is going to be if they are to test the commercial viability of the project. These early estimates are also known as *screening estimates,*[1] because owners often use them to compare the cost of alternate capital projects or improvements.

Estimators are usually very short on information about the project at this point. They may have only the proposed capacity of the plant, a block-flow diagram of the processes, and an approximate location.

For an A&E type of project, the estimator may have only some general information such as total square footage of buildings, number of rooms or beds for an office or hospital, KW capacity of a power plant, or cubic yards of fill in a dam project.

One approach to feasibility estimating is to plot the capacity or size versus the actual cost data of previously built projects with similar characteristics. If enough units of different capacities or sizes and their cost figures are plotted, a practical cost-per-unit capacity or size curve results. Plotting the data on a logarithmic scale will yield a straight line that makes it easier to read the data. To improve the accuracy of this method, one must factor out the effects of inflation and project scope differences from the data used.

The total installed cost of the units must reflect the same scope of work, or the curve will be distorted. One cannot compare a grassroots project with an addition to an existing facility with a developed site and infrastructure already in place. The site and infrastructure costs must be estimated and factored in or out, as required, to get comparable costs.

In many industries there are several good sources of unit costs per stream-day of capacity for various types of processes. However, the most reliable ones are those that can be generated within your own company, and that are based on the type of facility built to your firm's standards. These numbers are generated by dividing the plant cost by the tons, barrels, pounds, KW, etc. of daily throughput. The dollar value will increase over the years as inflation affects overall plant construction costs. Be sure your estimator is using current figures.

The high percentage of inaccuracy (plus or minus 25 to 30 percent) expected with screening estimates is a direct reflection of the approximate nature of the pricing methods. It's unfortunate that owners must use the most inaccurate cost estimate for the go/no-go decision at this early stage. The project's commercial viability or its cost/benefit ratio can be based only on this relatively inaccurate estimate of the expected project cost. On the other hand, the flexibility of the cost estimate allows those evaluating the project some leeway in calculating the project's return on investment.

CMs must use good judgment at this stage. An overly pessimistic estimate can lead to the passing up of economically sound projects. Likewise, overly optimistic estimates can lead to wasted investment in uneconomical projects. This early in the game, I lean toward the optimistic school of thought. Several other cost checkpoints along the way will provide more accurate estimates before we reach the point of no return on project funding.

The good news about feasibility estimates is that they are very inexpensive to make while we are still low on the project cashflow curve. One can't spend too much money getting some basic curve or unit-cost figures together to arrive at a feasibility cost estimate.

Appropriation estimates

As we move to the right along the time scale of the project life bar diagram shown in Fig. 5.1, more detailed project definition becomes available. This permits some refinement of our estimating procedures. Because we are going for board approval of a capital project request (CPR) with that number, it had better be reasonably accurate.

Looking under the heading of Class 2 estimates in Table 5.1, we see that some very important decisions further defining project costs have been made. A specific site has been selected, and preliminary plot plans have been developed. Flow diagrams, major equipment specifications, building sizes, and overall layouts are available. Process flow diagrams with preliminary heat and material balances have been developed, and they permit the preliminary sizing and specification of the major process equipment items.

Over the years many cost factors have been developed for various processes, and they make it possible to factor the total plant cost from the process equipment cost. We will discuss that estimating approach in more detail later in the chapter.

The additional design data available can open the door to at least two different estimating approaches at this stage. It may be wise to run an approximation by two different methods and use the resulting figures to check the earlier feasibility estimate. If the numbers jibe pretty well, confidence in the estimate will be increased. If the numbers do not reinforce one another, it's time to find out why.

For A&E type projects that are not equipment-oriented, appropriation estimates can be made from rough quantity takeoffs of bulk materials by applying historical unit-cost figures to them. Other project costs such as site development, HVAC, special finishes, electrical, and plumbing can be factored into the estimate to obtain the total facility cost.

This class of estimate still carries a fairly high percentage of inaccuracy to cover errors of omission and commission due to the preliminary nature of the design documentation. Some of the doubt over the accuracy of the estimate can be removed by carrying a higher percentage of contingency to offset the potential errors. The amount of contingency carried is often a function of the owner's policies on such matters; it could run as high as 20 or as low as 10 percent.

The appropriation estimate is considered a budget type of estimate because it does contain a breakdown of several major project cost elements. For example, the estimate may have been made from separate cost estimates for design, site development, foundations, structural, electrical, and mechanical systems etc. to arrive at the total figure. It is usually the first estimate with enough detail to use the applicable code of accounts to break down and organize the various cost categories. A well-organized estimate can be converted into a preliminary project budget. The design team also uses that budget to control cost factors for subsequent phases of the detailed design.

Capital cost estimates

The next estimate along the time line in Fig. 5.1 is the capital cost estimate, which is sometimes a further refinement of the earlier appropriation estimate. On process projects, we now have fully developed P&IDs, firm prices for major equipment, and preliminary piping models or drawings. The civil, architectural, and structural (CAS) design is well enough along to support some bulk material takeoffs and pricing of the work in that area. Pricing of preliminary piping takeoffs, or factored estimates from comparable projects, give us the first refined look at the piping cost figures. The large amount of piping in a chemical plant represents a large cost item and is often a budgetary problem area.

At that point, we can approach an accuracy of about plus or minus 10 to 15 percent. With further refinement, we are also able to evaluate the accuracy of each major subdivision of the estimate to get a better handle on the overall contingency allowance. That estimate is refined enough to be used as a detailed budget for controlling project costs. We will discuss how that happens later in the chapter, under budgeting.

Definitive estimates

The definitive estimate is the final estimate in the series for most projects. It also corresponds to the contractor's estimate made during the bidding phase on A&E types of projects. All the major equipment has been quoted or ordered, and firm bulk material prices have been established and applied to actual takeoff quantities. The only remaining major cost unknowns are the rate of field labor productivity, field indirect costs, and any possible construction change notices. The contingencies for these outstanding items can be readily evaluated, however, which brings the percent accuracy for the definitive estimate down to plus or minus 5 percent of total project cost. On A&E types of projects, where the construction is being bid lump-sum, the percentage accuracy of the final cost estimate will even drop below the 5 percent figure after the bids have been received and the successful bidder accepted.

The figures developed in the definitive estimate are used to fine-tune the existing cost-control budget for the whole project. Naturally, the contractor's lump-sum bid price will become the project construction budget for controlling the construction costs on those sorts of projects.

Cost trending

Cost trending is a procedure that has been used over the years to predict how the project costs are moving during the time period between estimates. On larger projects, the project team is likely to find itself in "blind spots" when predicting project costs during the design development stage. The project scope may have undergone a series of growth changes during the period between the Appropriations Estimate and the Capital Cost Estimate. That can lead to financial shock when a later estimate turns out to have grown beyond the feasibility cost for the project.

That rude financial shock can be prevented by using a cost-trending procedure. A group of project and cost engineers is set up to monitor the cost of any design changes that may occur during the design development phase. Details of the establishment of a cost-trending group are given by Clark and Lorenzoni[1] starting on page 151.

The trending group issues a biweekly or monthly cost-trend report, which in effect updates the latest estimated project cost to the current design basis. That relatively inexpensive system gives the PM or CM early warning of any adverse cost trends while there is time to act on them. It also permits CMs to be more confident about the project cost forecast when responding to their management's inquiries between estimates.

The series of estimates and the cost-trending procedure give owners a set of control points by which they can recheck the financial feasibility of the project while it is still low on the project resources commitment curve. With design costs running from 6 to 12 percent of total project cost, the owner can cancel or downscale the project scope before passing the financial point of no return. That can be done before placing equipment orders or starting construction. When a project is canceled after the commitment of equipment and construction funds, however, substantial cancellation costs are likely to be incurred.

Phased release of funding based on those key estimating milestones is fairly common in capital project execution. The need for that type of controlled project release is often set out in the contract documents and should be covered in the Field Procedure Manual.

Estimating cost effectiveness ratio

The cost of performing a project estimate is generally directly proportional to the degree of accuracy desired. As pointed out in our previous discussions, a project may require several estimates if the desired control of the project costs is to be maintained. Therefore, making those detailed estimates can be a significant cost factor in the home office construction services budget.

Figure 5.2 is a graph of the cost of estimating versus the accuracy of the result. The time line at the bottom shows the three basic phases of the project, and the vertical axis shows the cost and percent accuracy. In the planning stages, where percent accuracy is high, the estimating cost is quite low. Even if we have to develop some basic cost curves, the amount of estimating labor is not significant. As more detailed design data becomes available, we can spend more labor doing detailed material takeoffs, getting equipment bids, and extending the costs for the various parts of the project. As the estimating cost rises, the percent of estimating accuracy can be expected to improve. Please note the phrase "can be expected to improve." The expected improvement of the accuracy will happen only if we do the estimate right!

In the case of the construction firm that normally bids its work lump-sum on fully designed projects, the estimating department is

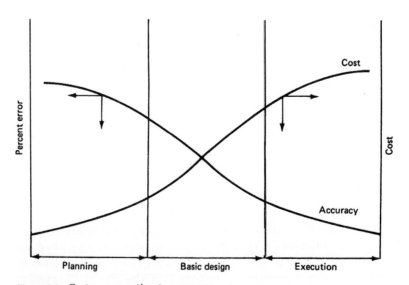

Figure 5.2 Cost versus estimate accuracy.

the lifeblood of the organization. Most of that work is bid by law on a strictly competitive basis, so it's very difficult to sell a project without the low price. Often, the low price results from the estimators leaving out part of the work or underestimating the labor hours. Estimating under those conditions places a great deal of stress on the estimating department.

Companies operating in that area must hire the best estimators available and give them the best tools to produce effective lump-sum estimates. To be effective, they must *win* at least one-third of their bids without erring on the low side. If the firm doesn't get its share of bids or loses money on most jobs, it can't stay in business. The estimators' goal must be to place the construction manager in a position where he or she can perform a profitable job for the contractor on every project!

Estimating Methods

Because the construction estimates set the financial baseline against which the CMs execute their projects, it behooves them to be conversant with the estimating methods used. That interest is in addition to the CM's vital input on the proposed contracting plan and the construction technology used as the basis for the estimate.

The tools and methods available to the estimator should be tailored to the type of projects the company does. Their application is also a function of the estimating team's capabilities and the data available. Most contractors have complete in-house estimating departments, with extensive files of historical cost data relating to the specific types of projects done by that firm. Owners may have only a minimum number of reference books containing average unit costs across the country.

Most large engineering and construction contractors have extensive estimating capabilities that can be drawn on by an owner who has such a firm under contract. Accordingly, most contractors regard their estimating data files as highly proprietary. It's virtually impossible for an owner to have access to those files, except as they pertain to the specific contract being estimated. Developing and maintaining historical cost files is expensive, and it is one of the key advantages that a contractor can offer an owner.

As an alternative to building a costly in-house historical estimating file, owners can use outside estimating services. Some such services can be purchased on an annual subscription basis to supply a battery of cost files at a nominal price.[3,4] An example of such a service is Richardson Engineering Services Inc. of Mesa, AZ. Its present-day cost data are computerized, updated regularly, and tailored to various areas of the country. There are also firms available to assist with estimating costs for foreign projects.

One example of a complete outside estimating service that will do a specific complete estimate for a fee is the Icarus Corporation of Rockville, MD. You give them your design data and estimate basis and they give you the complete estimate, broken down to suit your code of accounts. Icarus also offers a software program called *Questimate* that owners and contractors can use in-house.

On A&E types of projects, the lead architects and engineers are often responsible for preparing the project cost estimates, in the absence of a resident estimating department. In that case, standard cost indexes such as Means[3] and Richardson[4] are used, along with specific local knowledge from similar recent projects that may be available within the firm. The accuracy of the unit prices used in that type of estimate is critical; the estimate will be used to compare the contractors' bids, which are often made in more detail than the engineer's estimate. If the engineer's estimate and the low contractor's bid aren't within plus 10 percent, the owner will probably throw out the bids and rebid the project on a revised design. As we discussed earlier, that has very embarrassing consequences to the project schedule.

Estimating Tools

There are many estimating tools and shortcuts that have been developed over the years and that are described in the literature listed at the end of the chapter. The comprehensive bibliography of cost engineering literature included in Appendix E of Humphreys and Katell,[2] expands the list of available cost engineering literature even further.

Computerized estimating tools

The advent of the computer, and especially personal computers (PCs), has added to the estimator's arsenal of tools. With the use of a mouse and an estimating menu, material takeoff, listing, and pricing have become easier and more accurate. Combination programs can also develop lists for the materials associated with the basic item being taken off. For example, when formulating piping takeoffs, insulation and painting costs can be taken off simultaneously by indicating whether the line is insulated or painted. Trained estimators using these kinds of tools can produce more accurate takeoffs in less time when making a detailed estimate.

Another example of computerized estimating tools is the use of computerized design programs for the preliminary sizing of vessels and heat exchangers before vendors' quotes are available. When the basic design data and code information are input, the program will print out the area of the heat exchangers and the weight of the vessels. On a chemical plant, use of that information with the proper material-cost multipliers gives an excellent handle on the equipment cost. In turn, the equipment cost forms the basis for a more accurate factored type of estimate for associated construction costs early in the project. Using that kind of tool can give a more accurate and cost-effective total-cost estimate earlier in the project.

One might think that the use of the new computerized estimating tools would reduce the cost of making an estimate. That doesn't appear to be true, because the tradeoff has been to generate more takeoff and cost data with improved accuracy for roughly the same cost.

Perhaps the best tool of all for CMs is developing a gut feeling they have developed about their project estimates. Unfortunately, that comes only with time and experience in dealing with the estimating environment. As a CM you will never have time to become an accomplished estimator yourself, so don't try to dig into all the details of any estimate you are reviewing. Look at the various major cost accounts to see how logical the numbers appear to be. Also, compare the interrelations of the various major accounts by using accepted rule-of-thumb values applicable to your area of practice. Any accounts that appear out of line should be investigated in detail to confirm or deny your suspicions.

Estimating Escalation and Contingency

Escalation and contingency are the last two major estimating items that we must deal with before we wrap up our estimating discussion. They are both somewhat esoteric items that seem to defy any fixed rules for accurately estimating them. Because upper-management plays such a large role in setting their values, they are often beyond the control of the CM.

Escalation

Price inflation has been with us for such a long time that we now consider it a normal way of life, but the rate of inflation can swing wildly with economic conditions. To improve our cost projection, we must make some sort of educated guess as to the escalation rate over the expected life of our project.

Some people try to take the easy way out by writing escalation for wage-price inflation out of the estimate. That is done by stating that the estimate covers "the cost in effect at the time of the estimate"—not at the time of project completion. That's fine, but I have never found an owner who would even let their own people off the hook that easily—let alone the contractor. The owner will almost certainly insist on a "best estimate" of the anticipated escalation of project costs, regardless of the prevailing economic conditions. Lump-sum bidding documents are written to include all cost factors, including escalation.

Estimators have a number of ways, including a variety of computer programs, to project the cost escalation. The advantage of a computer program is its ability to evaluate individual accounts and apply a system of factors to get an overall weighted average escalation for the whole project. In any case, any projection of escalation has to be based on past history, coupled with a forecast of economic conditions over the expected life of the project.

Finding past history is easily done by examining indexes of past escalation trends, which are available from a variety of sources. The Bureau of Labor Statistics (BLS) is probably best known for a broadly based national index. Since we are interested in indexes pertaining to capital projects, we will want to look at those that cover the engineering-construction industry. Examples are those published by *Chemical Engineering* and *Engineering News Record.*

Extending the past inflation curve into the future is probably the most simple approach to projecting escalation. The practice can be hazardous, however, if there happens to be a sudden shift in the rate during the course of the project. An outstanding example of that type of situation was the escalation experienced by the construction industry during the oil embargo of 1972–73. In a matter of a few months, the normal escalation of about 5 percent per year vaulted to 20 percent. In addition to the sudden increase in prices because of the OPEC oil embargo, companies were buying up pipe and fittings just to hoard them for future possible needs. That created further shortages of piping materials that drove prices even higher.

I was involved with a lump-sum project to deliver engineering and hardware for a petrochemical plant to an eastern bloc country. The materials had to be procured in the United Kingdom, because of the local financing arrangements. Project costs were suddenly running $4 million over budget, and rising because of the unexpected escalation. Fixed-price bids on the equipment were suddenly out of the question, and on some items we were unable to get bids at all. All prices were quoted with price in effect at time of shipment. Fortunately, there was no field labor involved!

Projects with Eastern Bloc countries had longer-than-average durations due to the red tape involved, so we were caught by the sharp

inflation spike brought on by the oil embargo. Since there was no escalation clause in the contract and the firm did not choose to press a
claim for additional costs brought on by the unusual market conditions,
there was no recourse but to absorb a substantial loss on the project.

I don't recall hearing many people predicting the oil embargo before
it hit, so I doubt that many estimators or CMs could have been
expected to allow for such a wide swing in their estimates. Using
20/20 hindsight, however, the contract should at least have had an
escalation clause to cover the lump-sum cost exposure on a project
that could be expected to experience long administrative delays.

In the final analysis, it is virtually impossible to accurately predict
the rate of inflation over a long period of time. The best way to
approach the matter is to study the recent escalation history and then
weigh that figure against an economic evaluation of the market conditions that can be expected to prevail in the project location. Are many
projects going to be competing for the available material and labor
during the execution of your project? If you are executing a project in
a tight sellers' market condition, you can expect adverse escalation
pressure on your project costs.

On the plus side, remember that the inflation rate does not have to
apply to the total project cost. Monies committed and spent early in
the project do not carry the inflation rate for the duration of the job;
only those monies spent later in the project are subjected to the full
exposure of price increases.

Construction labor can be a problem, depending on whether the
fixed-price work is being done union or merit shop. On union projects,
the owner usually assumes the exposure by accepting the higher
rates if a new union contract calling for a wage increase is negotiated
during the execution of the contract. A cost increase would be allowed
for the increased labor cost on the unexpended labor hours affected by
the wage increase in the union contract.

In the case of merit shop projects, the contract may call for the contractor to include the labor escalation. In that case the extra cost is
estimated and included in the contract price at the contractor's risk.
If the labor cost does not rise, the labor-rate contingency goes into the
contractor's profit. Therefore owners will sometimes assume the risk
of increased labor cost by paying it as an extra as it may occur.

Contingency

The amount of contingency allowance in any estimate is always a
point of discussion. We must remember that contingency is a factor
added to cover the two major unknowns present in all estimates:

- Errors due to inaccurate or incomplete design data
- Errors of omission and commission in estimating

Regardless of how hard we try for perfection, we are likely to experience some estimating errors that are due to materials being taken off inaccurately or omitted, along with mathematical errors. This is not as bad as it sounds, since those types of plus or minus errors tend to offset each other and have only a minor effect on the final number. The percentage allowance for inaccurate design data is a function of how far along with the design we are. At the beginning of the project the number is higher, and it gradually tapers off as the design phase develops.

Contingency factors do differ for various types of capital projects, but they usually run about 15 percent early in design and can drop to as low as 3 or 5 percent when the design is complete. In many cases we don't have a free hand in applying contingency to our estimates. Often the estimated cost is being compared with other costs to determine the selection of a project, a process, or a contractor. In order for the estimate to be competitive, the contingency must be a factual representation of the anticipated accuracy of the estimate.

Estimates can sometimes be subject to a pyramiding of contingencies, which makes the final estimate far too high. A common time for that to happen is when making labor estimates. They are based on using historical *standard labor hours,* which already contain a nominal allowance for contingency. A craft supervisor estimating labor hours will generally add further contingency in the form of extra hours to the estimate. The next higher manager may crank in more safety factor, based on the group's recent poor performance on a project. By the time the final figures go to estimating for pricing, where another contingency factor is added, the total contingency can reach a figure that includes up to 15 to 25 percent fat.

That type of problem can best be detected by applying some industry-standard rule-of-thumb comparisons that reflect the interrelations of labor costs to the overall construction cost. Another check can be made on the amount of labor used in typical field operations such as concrete placement, masonry, steel erection, and the like. When rule-of-thumb comparisons turn up suspicious numbers, check them out in detail for contingency pyramiding or just plain estimating errors.

When escalation and contingency are considered in the aggregate, they can easily amount to 15 to 25 percent of project costs in the early stages of design-build work. That definitely makes them worthy of a PM's or CM's consideration when he or she is striving to successfully meet the project goals. Later, in the construct-only stages, escalation and contingency may be as low as 6 to 12 percent.

Foreign currency fluctuations

The foreign currency contingency arises only on those projects that have an international flavor. We may be doing a project in a foreign country or procuring major equipment and materials from overseas, which brings the factor of foreign currency exchange into the estimate.

In the case of buying equipment and materials overseas, we can protect against currency fluctuations by buying "futures" in the currency of payment for the goods. If the value of the seller's currency goes up and that of ours goes down, we can still pay the original price with the prepurchased funds. If the seller's currency declines, we can sell the currency, pay the original purchase price, and pocket the difference!

If you ever have the opportunity to manage a project with a large budget in foreign exchange, you should get the best advice possible as to setting up the financing and payment terms. In larger firms the whole matter of foreign exchange will normally be handled by the finance department. It's critical for CMs to ensure that the terms of all the agreements for the project are set up to protect your firm's investment and project budget, to cover changing conditions.

Estimating Field Direct Costs

Field direct costs are made up of the material, labor, and subcontracts required to execute the project as delineated in the plans and specifications and as defined in the scope of work. These items form the heart of the detailed cost estimate that is based on the itemized take-off of the material and labor for the job.

Large-project estimates should be broken down to match any definitive site-specific areas such as various buildings, processing areas, systems, and the like. This approach helps to break larger projects down into more manageable units that are easier to visualize and evaluate.

The estimate also should use the contractor's code of accounts at the start of the estimate. That helps to organize the estimate and makes it easier to execute, evaluate, and update during its life. It also makes it easier to convert to the project budget once it has been approved and the contract signed. That approach at least eliminates any transmission errors caused when converting from a random-estimate format to the code of accounts used in the project budget.

Using codes of account

Codes of accounts (COA) have been used in the construction business for years, and are so basic to construction industry practice that I won't show them in detail here. If your firm is not using a COA to control its projects, I suggest you introduce one at the first opportunity and keep it current for continued use on future projects. Just keep in mind that COAs come in many levels of detail, with the detailed ones occurring on large projects. Don't make your code of accounts any more detailed than is needed to control the project. Various models of COAs are given in Appendixes A though D of the *Contractor's Management Handbook* by O'Brien and Zilly.[4] Some adaptation to your construction project environment may be required to arrive at an effective system.

The advantages of a good workable COA are endless, so never underestimate its value to your project. The COA threads itself through the entire project, from the first semidetailed estimate to the payment of the last account invoice. After that, the final costs in the COA are used to establish or update the firm's estimating reference file for use as the basis for pricing the next project.

The Role of the CM in Making a Detailed Estimate

Depending on the size of the company and the project, the role played by the CM can vary from being an estimate coordinator to leading the overall estimating effort. How the assignment is handled depends largely on the stature and leadership exhibited by the firm's estimating department, and the degree of authority with which the CM has been endowed.

We have already cited some of the roles the CM plays in preparing the estimate. The following checklist of estimating activities is a minimum for the CM's role in the estimate:

- Make in-depth review of bidding documents and contract.
- Participate in the preestimate strategy meeting.
- Lead the construction site survey team and report preparation.
- Attend site visitation, management, and client meetings.
- Review the estimating plan with lead estimator.
- Establish the site labor posture with top management.
- Develop the scope of field indirect services and costs.
- Develop the lowest-cost construction technology plan.

- Evaluate heavy-equipment and small-tools requirements.
- Develop a contracting plan with upper management.
- Suggest construction cost reductions where possible.
- Evaluate the proposed construction schedule.
- Evaluate project regulatory requirements and cost impact.
- Develop the field organization and chart.
- Prepare a project initiation and execution plan.
- Assist in supplier/subcontractor sourcing plan.
- Research and review prior similar job histories.
- Respond to estimator information and decision requests.
- Periodically review estimate progress and methods.
- Make in-depth review of final estimate.
- Participate in final pricing meeting.
- Attend the bid opening if required.

This is an imposing list of estimating activities for the CM on *any* size project, either as an estimate leader or only a contributor. On larger projects, the CM will have to delegate some of the duties to others if they are to be covered in sufficient depth. The list focuses on the fact that the CM is management's chief representative in the preparation of a sound, business-oriented construction estimate. If this *management* input is lacking, the estimate will likely miss the intended goal.

Some of the items on the list are self-explanatory, while others warrant a few lines of expansion. A prime example is the CM becoming thoroughly expert in the ramifications of the bidding documents and the pro forma contract, so as to be able to advise upper management in the preestimate meetings. Bidding document packages on large projects can be in the range of a 100 or more drawings and up to 12 volumes of specifications and bidding information. With a decision to submit a price required in a matter of days or a few weeks, fast and effective action by the CM is required if the information is to be digested.

As the technical representative in the preestimate meeting, the CM has to answer any technical questions raised by the management, estimating, and marketing people. These can involve project scope, schedule, technology, company capabilities, contract, suppliers, subcontractors, and the like. This is also a good time to develop a feel for management's commitment to getting this project.

Attendance at site inspections, bidding, and management meetings present a great opportunity for CMs to further their company and personal images. This is done by presenting themselves to their clients and managements as competent, mature, can-do professionals. CMs who present themselves in a professional manner bring to the table countless assets when it comes to preparing the winning estimate and *selling* the job.

Construction site survey and report

The CM, along with the lead estimator, plays a key role in performing the site inspection and report. The report is a fundamental document used in the estimate and in formulating the project initiation plan. This section considers the worst-case scenario of the project site being located in another city. A checklist of the basic points that the site-visitation team must investigate and report on is as follows:

1. Contacts with the potential client team.
2. Site labor survey.
   a. Local labor supply—union or nonunion.
   b. Determine local labor rates and fringes.
   c. Union versus open-shop environment.
   d. Determine local labor productivity rates.
   e. Local labor training facilities.
   f. Determine needs for local site agreement(s).
   g. Labor requirements affecting cost and/or schedule.
3. Evaluate local site conditions affecting project cost/schedule.
4. Evaluate effect of ongoing facility or unit operations.
5. Review local sources of suppliers and subcontractors.
6. Evaluate the affect of the project on the local community.
7. Evaluate local living costs for expatriate field staff.
8. Evaluate local transportation links—air, sea, road, rail, etc.
9. Evaluate local communication system availability.
10. Evaluate local and federal government regulatory climate.
11. Check on local health care and hospital services.
12. Visit local chamber(s) of commerce for area business information.
13. Investigate any other local factors affecting construction costs or schedule (i.e., weather, law enforcement, legal factors, etc.).

Yet again we have an impressive list of important activities to be coordinated by the CM. The costs for this exercise must also be included in the estimate-preparation budget. On larger out-of-town projects, the site-report data gathering can take four to five people the better part of a week. Much of the above information for smaller in-town jobs may be already in hand from ongoing operations.

The site inspection and labor survey report is even more important if the designated project CM comes on board after the estimate is finished. The new CM can study the existing site-survey report along with the design and bid documents to get up to speed in a relatively short time.

The project contracting plan and labor posture

These two interrelated items are usually developed with the contractor's upper management. The labor posture refers to the decision as to whether the project will be done with union or open-shop labor. Usually the contractor works one way or the other, but some firms can go either way. Sometimes no option is available, because the owner has called for one or the other based on in-plant conditions.

The tense construction labor-posture atmosphere has eased somewhat in recent years with the growth and acceptance of open- or merit-shop practices. This decision influences the estimate in the areas of union wage rates and working rules as to labor cost and field productivity. Merit-shop labor estimates have usually been about 5 percent lower than equivalent union-based estimates, but the gap has tended to narrow somewhat due to the easing of union work rules in the face of open-shop competition.

The contracting plan is part of the overall project initiation and execution plans (discussed in detail in Chapter 3) that are developed by the CM. The plans have a strong bearing on how the estimate is prepared, so they must be formulated and approved before estimating starts. They are a key part of the preestimate meeting and play an important role in the bid/no-bid decision. The CM formulates preliminary execution plans right after the bidding documents have been assimilated. The project execution plan may even be finalized and approved as part of the preestimate meeting.

While reviewing the bidding package, the CM is automatically evaluating the best-available construction technologies for building the project at minimum cost. If *selling* the proposed methods to management is necessary, the ideas must be well researched and presented by the CM to get them incorporated into the project-execution plans. Looking at areas of improving labor productivity is fertile ground for that type of idea.

Project regulatory requirements are a vital area of interest for CMs and the estimators. This has grown to be a major area of cost over recent years, as well as contributing to lower overall efficiency of the construction process. Its effects must be investigated early in the estimate, and the cost effect reduced as much as possible. However, those costs should never be intentionally underestimated!

The CM's final review and approval of the estimate before submitting it to management for pricing is important. If management doesn't understand the estimate, or if significant errors show up, it's bound to reflect poorly on the CM as well as the estimators.

The pricing part of the management review is also vital to the CM. If management decides to arbitrarily cut the estimate to ensure that the company gets the job, it may upset the estimators but it affects the CM even more. The CM has to execute the job at the lower figure and still try to make a profit. It's not a problem if management is cutting only fat from the estimate, but if they also take some bone, the CM could be in trouble. If the latter is true and the job is won, the CM needs to start damage-control measures immediately to protect the project's financial goals. An early memo from the CM to management, pointing out the money problems and the planned damage-control efforts, is a good way to alert key people concerning the problem.

All the items on the checklist are important activities for the CM, because most of them are sure to arise on a given construction project sooner or later. Ignoring any of them in the early stages of the project can lead to serious problems later on. There will be enough problems arising by accident later without creating some that could have been avoided with proper foresight.

Estimating Field Indirect Costs

A significant factor in the overall construction estimate are the field indirect costs, which are known as field overhead costs for running the project. The designated project CM is responsible for giving the scope and required details of the field indirect cost estimate to the lead estimator for pricing. Some major items of the field indirect cost account are:

- Temporary site improvements
- Field supervisory staff salaries, burdens, and fringes
- Field office staff salaries, burdens, and fringes
- Field office equipment and supplies
- Site utility systems and bills
- Site safety and security costs
- Materials management and warehousing
- Vehicles and construction equipment
- Cost for small tools, supplies, and consumables
- Staff relocation and living allowances

- Communications lines and services
- Building, utility, roads, etc. maintenance costs
- Legal and insurance fees
- Labor and public relations costs
- Laboratory, field testing, and inspection costs
- Other miscellaneous field costs
- Temporary heat, dewatering, and weather-related costs
- Government, OSHA, EPA, etc. costs

Even this list would not be exhaustive for a very large construction project, so it is easy to see that the field indirect costs are a substantial contributor to the constructed cost. Effectively predicting the field indirect costs has a substantial effect on the estimate and on the total cost for constructing the project. This is especially true in the area of competitive fixed-price bidding.

In a small construction firm, organizing and estimating the cost of these items could fall to the construction manager. In any size firm, the CM must supply the field personnel structures, the extent of the physical facilities, and the scope of the field operations to the estimators if effective pricing is to be achieved.

Given the broad range of construction project size, the scope of field indirects can vary from the CM operating out of a pickup truck to a complex industrial facility and town site for a grassroots megaproject. Construction companies operating in a given market environment often develop a percentage of total project cost for estimating the field indirect costs. If good job-cost records are maintained for a number of similar projects, it's possible to develop a reliable percentage figure. It can at least be used as a check number to evaluate a detailed estimate for field indirect costs.

The contracting basis has a large effect on the field indirect estimate. The greatest difference occurs between a general-contractor execution approach, with a minimum of subcontracting, and the construction-management approach where the contractor acts only as manager. In the latter case, most of the field indirect costs are delegated to the subcontractors.

Actually, the total field indirect costs still exist, but they appear in different contractors' estimates. The managing contractor has a small staff of construction management people overseeing the work of a cadre of subcontractors. The CM must be aware of the contracting environment for the project and be guided accordingly. Table 6.1, typical size of field supervisory staff, shows some field staff personnel loadings for different contracting plans.

Temporary site facilities

The size and complexity of the temporary site facilities vary greatly from site to site. Small modification jobs are usually minimal, because existing facilities and utilities are available to satisfy most of the field operating needs.

For large grassroots projects, temporary site facilities such as offices, warehouses, roads, drainage, fencing, and the like can be quite extensive and costly. The best solution to that cost problem is to schedule the work so that the applicable new facilities can be used as temporary facilities for construction. This is especially true for site improvements, utilities, offices, and warehouses.

On large projects involving extensive temporary site facilities, a sound management approach from the construction management staff is required. The layout of the temporary facilities in relation to the work must be studied from an industrial engineering perspective to maximize productivity and to reduce cost at the work site. Wasted time in repetitive operations can wreak havoc with the best planned construction budget.

Field staff salaries and expenses

This cost category is usually the largest item in the field indirect budget. The category includes a number of highly paid supervisory staff and runs for almost the duration of the field work. The project leaders come on board early and are among the last to leave at completion.

That's one reason for the CM to do a thorough job of evaluating the number of people required for the field office and their qualifications. We will discuss these factors later, under Staff Mobilization in chapter 7, Project Organization.

Vehicles and construction equipment

The cost for this category of field expense is closely tied to the construction technology used in project execution. The construction technology must be studied, evaluated, and decided on at the beginning of the estimating process. The original technology plan is sometimes revised during the course of the estimate as improvements are suggested and evaluated.

At a minimum, the CM should study, as a means of facilitating construction, the field transportation needs, cranage, heavy lifts, field shops, welding, concrete batching, forming, and placing requirements. Use of high-tech construction equipment can be costly, so each application must be evaluated and used only if it reduces the overall constructed cost.

Small tools are considered almost an expendable item on construction projects, because not many of them are around at the end of the work. They are really a cost factor on process-type projects where much of the work is done by the prime contractor. Subcontractors typically furnish their own tools, so their tools are not included in the general contractor's small-tools allowance. Generally this item, when it does come into play, is handled as a company standard percentage of the project cost.

Field indirect general costs

Each of the other field indirect general costs should also be evaluated as it relates to the specific project. In the aggregate these could represent a considerable cost, and a unique project condition could vault any of them into the position of significant cost item.

The field indirects are a fertile area for application of the project cost-control system. The CM should audit them at least monthly when evaluating the field cost-control report.

Requirements for a Good Estimate

Each reference book listed at the end of this chapter includes specific details for making the wide variety of estimates we have discussed here. You may want to refer to them to broaden your specific background in cost estimating and to supplement the management approach to estimating given here.

We can summarize the discussion of project cost estimating with a checklist of the basic requirements for a good estimate:[1]

- A sound, fixed design basis
- A realistic project execution plan
- Good estimating methods and accurate cost data
- Neatly documented detail
- A reasonable estimating budget
- A knowledgeable estimator

Figure 5.3 shows a diagram of data flow for the preparation of a detailed design-build capital cost estimate for a typical industrial project. The diagram also serves as a checklist for the types of input and their sources, which you will need to consider for any capital cost estimate. If you can satisfy the listed requirements for a good estimate of your project, you should be able to develop a sound estimate to serve as a basis for a reliable project budget. Success in those two

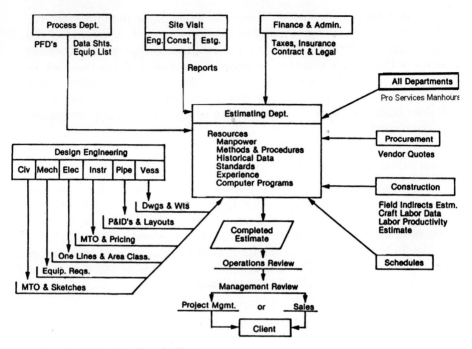

Figure 5.3 Estimating flow diagram.

areas will provide the basis for an effective financial performance on your next project.

Project Budgeting

The purpose of the project budget is to set a cost, or money target for each material, labor, and subcontract cost on the total project. Since all the financial aspects of the project revolve around it, the project budget must be realistic when compared to the actual expected cost of the project.

We have had a good deal of discussion about how to make an accurate estimate of the anticipated project costs to arrive at a realistic budget. A poor estimate will lead to an unrealistic budget. That in turn will lead to loss of control, cost overruns, and financial problems for all concerned with the project. Loss of cost control will surely prevent our reaching the project goal of *finishing under budget.*

The project budget is the baseline for the cost-control program. It is of the utmost importance that we set our sights on beating, or at worst meeting, each item in the budget. Note that I said "beating" first because we want to be under budget if possible. We know that the odds are against our batting 1000 on any given budget, but we will need to have some of our winners offset the losers. The only way to meet or exceed our project financial goals is to win more items than we lose, especially the big ones. With a conservative budgetary stance and prudent management of our escalation and contingency accounts, we should have every chance of meeting our expectations to finish the project *within* budget.

Budget breakdown

To organize the budget for better control, it is broken down into a code of accounts as we discussed earlier. A typical summary of major budgetary accounts for a self-perform industrial construction project might be as follows:

1.0 Home office services (3 to 5 percent)
    1.1 Engineering services
    1.2 Procurement services
    1.3 Out-of-pocket expenses
    1.4 Support services salaries and expenses

2.0 Major equipment accounts (20 to 25 percent)
    2.1 Vessels and exchangers
    2.2 Rotating equipment
    2.3 Packaged process and utility units
    2.4 Materials handling systems

3.0 Bulk material accounts (20 to 25 percent)
    3.1 Piping
    3.2 Instrumentation
    3.3 Electrical equipment and materials
    3.4 Structural and reinforcing steel

4.0 Site development and buildings
    4.1 Grading, roads, and utilities
    4.2 Buildings
    4.3 Equipment foundations

5.0 Field construction costs (45 to 50 percent)
    5.1 Field labor
    5.2 Field purchased materials
    5.3 Subcontracts

6.0 Field indirect costs (8 to 12 percent)

7.0 Contingency

8.0 Escalation

A typical summary of detailed account codes for a typical A&E non-process construction project budget might be as follows:

1.0 Home office services (3 to 5 percent)

    1.1 Procurement services

    1.2 Support services salaries and expenses

    1.3 Project accounting

2.0 General construction trades (65 to 70 percent)

    2.1 Site development

    2.2 Foundations

    2.3 Structural framing

    2.4 Architectural trades

    2.5 Roofing

3.0 Mechanical trades (10 to 15 percent)

    3.1 Plumbing

    3.2 Heating, ventilating, and air conditioning

    3.3 Fire protection

4.0 Electrical trades (10 to 15 percent)

    4.1 Power supply

    4.2 Power distribution

    4.3 Lighting

    4.4 Communications, alarms, etc.

5.0 Field indirect costs (5 to 7 percent)

6.0 Contingency

7.0 Escalation

The percentages shown in the parentheses are order-of-magnitude numbers given only to indicate the relative portion of the whole project cost that each section might represent. The percents will vary somewhat from those listed, depending on the type of facility being built. None of the above figures includes the project design costs, as might be encountered in a design-build situation. The percents for escalation and contingency are left out, since they are so highly variable and are calculated as a percentage of the actual accounts.

The construction manager has responsibility for all the accounts, except perhaps the home office services, which are relatively minor. The breakdowns are given to highlight the major cost areas on a typical construction project. The larger accounts require most of the CM's attention.

Code of accounts

In a detailed budget, each of the major divisions given in the above lists is further broken down into much greater detail according to a standard code of accounts. Good accounting practice demands that a standard code of accounts be used for effective cost control. If the code-of-account format was not introduced during the detailed estimating stage, it *must* be used at the budgeting stage.

Codes of account are accounting systems in which some CMs tend to take little interest. Nevertheless they are important in cost control, as we will learn later, so it behooves the CM to become familiar with their use. Chapter 15 in *Contractor's Management Handbook*[5] includes a detailed discussion of the reasons for having a workable code of accounts for effective project cost control.

A word of caution on account codes is necessary, because some accounting people tend to go into too much detail. The purpose of the account code is to sort and record costs in separate accounts, for use in controlling the budget and establishing a cost history for the project. The history function of the account code provides the best estimating data source for subsequent project estimates. Any cost-account breakdown beyond that needed to satisfy the above needs is useless, and wasteful of project resources. If your code of accounts is overly detailed, you might want to start a campaign to simplify it for better workability and cost savings.

Budgeting Escalation and Contingency

The handling of the contingency and escalation accounts in the budget is critical to effective project cost control. Remember that we said that these two accounts must be prudently managed if we are going to control project costs. The most prudent way to manage them is to distribute them no lower than the major account codes. If the escalation and contingency are distributed to each individual account item at the outset, we will be committing funds against inflated budget numbers. That means we are setting the target cost higher than we really hoped to pay, which leads us into a feeling of complacency about meeting targeted costs as we travel through the project. The CM evaluates the escalation and contingency allowances monthly, and allocates funds to individual accounts when warranted by actual conditions.

Our first effort in committing funds for project resources should be against the uninflated number. That applies the necessary pressure on the project staff to buy the best value, and also to ensure that the project staff is operating within the minimum specifications suitable for meeting the project's quality requirements.

Sometimes company policy demands that the contingency and escalation be prorated among the accounts at the start of a project. In my opinion, having people committing funds against a fat budget is not the best way to *beat a budget*. All efforts should be made to change such a flawed policy, if you hope to have financially successful projects.

Holding the two blocks of funds in separate accounts allows the CM to use them to offset inflation and contingency costs as they arise during the project execution. For example, field office salaries usually escalate only once or at most twice a year, at scheduled times. Thus, any actual escalation of unexpended base salaries and fringes can be factored into the estimate and budget as required.

Escalation on hardware purchases depends somewhat on the speed of placing the purchase orders and getting fixed-price bids. If the market or project conditions are such that firm prices cannot be obtained, the escalation account must be handled very carefully to avoid financial surprises at the end of the project. When we buy at fixed prices we are prepaying the escalation, but at least the vendor is assuming some of the escalation risk—especially if the delivery is late.

If needed, contingency is fed into the budget at certain milestone points in the project. The timing for that activity depends on how well the project was estimated in the first place. A less accurately estimated project would have a larger contingency account, which might have to be fed into the budget earlier as required. If early use of the contingency fund is not required, it can be adjusted downward as the project progresses.

Conversely, a well-defined estimate and budget might not require the release of any contingency funds until the 50 percent commitment point has been reached and a significant overrun condition perceived. Each area where contingency is to be applied must be analyzed separately, and the proper amount of funds must be added to each account where the need is justified.

The release of the contingency account is usually considered a very important action, and it is taken only with the concurrence of the upper operating managements of the contractor and the owner on reimbursable work. On incentive-type contracts, those two accounts may be part of the *upset price* in which case the owner and contractor share the savings, so careful administration of the funds is critical to calculating the final fee payment.

The escalation and contingency accounts should be evaluated each month against the remaining work to be done on the project. They can be adjusted downward as the project nears completion and the window of exposure to escalation and contingency narrows. Any funds remaining in the accounts at project closeout are added to any budget underruns to arrive at the total budget underrun for the project.

Construction managers of lump-sum contracts have a special need to properly manage the escalation and contingency accounts, because any amounts left over accrue to the project profit column. Since the owner agreed to pay a lump-sum price, including any escalation and contingency allowances, the owner can make no claim on any unused balances in those accounts. If the balances come out on the negative side, the contractor can make no claim against the owner for additional costs for escalation and contingency, unless there is an escalation clause in the contract or unless some unusual conditions arose during the project.

The key point for CMs to remember when carrying out any type of contract is not to use the budgeted reserve funds for legitimate project scope changes and field change notices. Construction managers must initiate change orders to cover the new costs and not tap their contingency funds, which may be needed later for their intended use.

Computerized versus Manual Budgets

The decision of whether to use computerized or manual methods for setting up the budget was once based on the size of the project. In those days we did small projects manually to avoid the high cost of mainframe computer charges. That decision is no longer necessary, because even the smaller offices and construction sites have access to personal computers. With the use of one of today's standard spreadsheet programs, it is relatively simple to set up a computerized budget for a small to medium-size project. Today only megaprojects are forced to use mainframe computers, because of databank size and processing speed.

The main problem with using the computerized format is keeping the budget posted as commitments are made and funds are expended. Updating PC data is really no different from making manual budget entries, but we tend to think that somehow the computer is able to do the ongoing entries by itself. Be sure that you make one member of the project team responsible for making the entries on a routine basis as the paperwork on the project is generated. That ensures that other members of the team will get accurate, current information when they call up the budget on the project PC.

Budget format

Budget formats can vary widely depending on company standards. A few things, however, are required to make any budget work. Figure 8.5 illustrates a budget format suitable for control of capital project costs. The header contains the necessary identification data and calls out the title of the report, project name, number, location, report run, and period-covered dates, along with a description of the costs covered in that section of the report. Once the original budget has been approved, it generally takes on the title "Project Cost Report," since it is a *living* document subject to monthly updates.

The account code and the description of each cost item are listed in the left-hand columns. The items are usually grouped by cost codes in sequential order of the account numbers. Using the same format as in Fig. 8.5, each major equipment account is further itemized by equipment tag number. In that way each item purchased can be compared against its original budgeted cost.

On most projects, the individual cost code groups are "rolled up" or summarized into total costs, which are then shown in executive summary report as seen in Fig. 8.6. For example, each piece of process equipment will be itemized with the budgeted cost in the equipment cost account. Then the total cost of all equipment will be shown on a roll-up summary sheet as the equipment account, along with the cost of all the other materials such as concrete, steel, and piping to give a total equipment and material cost for the project. Using that method makes it possible to show the total cost for the project on a single executive summary sheet, for a quick overview by the CM and upper management.

The next column, Original Budget, lists the approved estimated cost budget obtained from the approved cost estimate. The numbers in the original budget should never be changed, because they form the agreed baseline for the financial plan against which the final cost report will be compared. Any authorized changes in the original budget can be handled in the next column, Trans (for transfers). That column is used if it becomes necessary to transfer any funds from one account to another for any reason.

The next column to the right is called Approved Extras (or Change Orders); it covers any approved changes in scope or other approved extra costs. The sum of all the columns to the left is then equal to the column entitled Current Budget, which represents the latest budgeted cost for the project. Those columns to the right of the Current Budget column represent the recording of project commitments, funds paid out, and the estimated cost to complete. We will discuss the detailed workings of those columns later when we cover the subject of cost control and reporting.

I strongly recommend that all CMs pay close attention to the conversion of the cost estimate to the original project budget. Your acceptance of the original budget means that you agree with the financial baseline for the project costs. Those are the numbers against which your performance on meeting the project's financial goals will be judged. If you feel that the numbers are seriously flawed, now is the time to speak up about it. Protesting later in the project that you had a bad budget will not improve your image!

The Project Cashflow Plan

The final leg of the financial plan is the project cashflow plan. It is the projected rate at which the project's cash resources will be spent. Note that the cashflow plan is based on monies actually *spent,* not just *committed.* The project financial people must know when the actual funds are needed to support the project. If the funds for the project are set aside at the start, any funds not immediately required for the project must earn the highest rate of return prior to their being used. The people who administer the construction loan also need a schedule for the expected flow of the funds.

The cashflow plan is developed from the cost estimate and the project schedule. The overall shape of the cashflow curve will follow that of the project life-cycle bell curve. Payments start slowly at the front, rise to a peak near the midpoint, and taper off at the end. Plotting money spent versus time will result in a classic S curve. The projected monthly expenditures also can be shown as a series of vertical bars, which will result in a typical bell curve when the tops are connected. See Fig. 8.11.

The first expenditures pay the costs of the field indirect charges. As we noted earlier, these are a relatively small part of the overall project costs. The cashflow curve falls behind the normal percent progress curve because the purchased equipment, materials, and subcontractor invoices come in after the hardware has been delivered or the services rendered. That makes the initial slope of the cashflow curve fall behind physical progress on the project.

An exception to the general trend of the cashflow lagging the progress curve would occur if the equipment vendors required advance and/or progress payments for their work. That does happen on some foreign projects, where it is customary to have the owner finance the fabrication of the equipment with down payments and progress payments.

The cashflow curve is usually evaluated each month, along with the project budget and schedule. Any significant changes to the schedule and budget are incorporated into the cashflow curve to update the curve each month. Reviewing the cashflow is a fairly routine activity, handled by the cost people on the project. The CM needs to get into the act only if serious cashflow changes, cost overruns, or cash shortages suddenly appear.

Summary

The CM plays a key role in generating the project financial plan, which establishes the baseline for the control of all project costs. It is the part of the overall project plan that enables us to meet our goal of finishing the project under budget. The CM must become well informed on the art of cost engineering to ensure that a sound estimate of the project costs is made. The project cost estimate must be converted into an effective project-control budget, which serves as a basis for the cost-control system. A project cashflow plan is a valuable tool for projecting the cash needs for the project.

References

1. Forrest D. Clark and A. B. Lorenzoni, *Applied Cost Engineering,* 2d ed., Marcel Dekker, Inc., New York, 1985.
2. Kenneth H. Humphreys and Sidney Katell, *Basic Cost Engineering,* Marcel Dekker, Inc., New York, 1981.
3. Robert Snow Means, *Building Construction Cost Data,* Robert Snow Means Company, Inc., Duxbury, MA. (Latest edition issued annually.)
4. *Richardson Rapid System,* Richardson Engineering Services, Inc., Mesa, AZ. (Updated per annual schedule.)
5. James J. O'Brien and Robert G. Zilly, *Contractor's Management Handbook,* 2d ed., McGraw-Hill, New York, 1991.
6. Sidney M. Levy, *Project Management in Construction,* McGraw-Hill, New York, 1987.

Case Study Instructions

1. Make a feasibility (Class 1) type of estimate for your selected project. State the basis of the estimate, (i.e., curve, unit tonnage, unit area, etc.), any assumptions made, and the expected accuracy of the resulting estimated cost.

2. Make a dummy mock-up of a lump-sum estimate for the construction costs on your selected project based on the value obtained in instruction 1 above. Use a suitable code of accounts to organize the estimate, including escalation and contingency. Include a statement of the basis of the estimate.

3. Prepare a project budget format based on the above project estimate. Make an executive summary sheet showing the major cost accounts. Present a proposal to your management (or client) showing how you plan to administer the escalation and contingency accounts in the project budget.

4. Make a cashflow curve for your project based on the estimated cost and the project schedule developed earlier.

6

Project Resources Planning

Up to this point we have prepared a project-execution plan, a time plan, and a money plan for the project. Now we are ready to turn to planning the human and physical resources needed to implement the construction project execution.

There are seven key project resource areas that require early evaluation and forward planning on the part of the construction manager. Simpler projects may require only some of these areas, whereas complex projects may involve them all. The seven key areas are:

- Human resources
- Engineered equipment and materials
- On-site facilities
- Construction equipment
- Project services and systems
- Transportation arrangements
- Project financing

The first two areas are the most important from a financial and planning standpoint, and they have the greatest influence on performance early in the project. Also, they are used on virtually every capital project.

Human Resources Planning

At this stage we are discussing only the types and numbers of people required, not the way in which we will organize them into a team. Project organization will be handled as a separate subject in Chapter 7.

Human resources for construction project planning breaks down into three major categories as follows:

- Home office personnel
- Construction personnel (field supervision and labor)
- Construction subcontractors

The personnel segment of the construction industry can best be described as bordering on chaos. We are talking about a major industry segment that hires large numbers of talented and highly skilled people on a largely temporary basis to perform dangerous work to a strict schedule anywhere in the world. That employment environment is the antithesis of virtually every other industry in the world. If it were not for the chutzpah, dedication, and high morale of the average construction worker, the industry could not survive in such a rough-and-tumble labor environment.

Effectively manning the world's construction projects makes the construction contractor's personnel department a key to successful contracting. Anything that CMs can do to facilitate staffing their projects like personnel planning will make their personnel departments' work easier and more effective.

Theoretical personnel-loading curves

The simplest form of a personnel-loading curve is a trapezoid, as shown in Fig. 6.1. The curve plots personnel required versus the scheduled time to accomplish the work. Actually, the most efficient way to staff a project would be to immediately staff the average num-

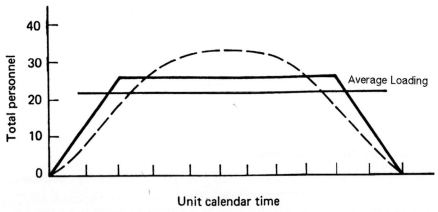

Figure 6.1 Theoretical personnel loading curves.

ber of people required on day one, continue to the end of the work, then drop to zero. The average number of people is arrived at by dividing the total hours by the total calendar time, and it is shown graphically by the solid horizontal line in Fig. 6.1.

We know that such ideal personnel loading is not feasible for many reasons, such as not having the site ready, not having all the materials and equipment available, and not having enough places for the people to work. Because we have to start from zero and assign people gradually, the next most efficient theoretical personnel loading curve is the trapezoid. It has a uniform buildup, a level peak, and a uniform builddown.

In actual practice the personnel loading curve takes the shape of a bell curve as shown by the dashed line in Fig. 6.1. Because the bell curve tends to fall inside the trapezoid at the start and the finish, its peak must extend above the peak of the theoretical curve to account for the lost hours. Remember, the area under the curve is a value that is set by the total labor hours estimated to perform the work.

A good rule of thumb to remember is that the bell curve peak usually exceeds the average personnel loading curve (line) by about 20 to 30 percent. That ratio allows you to make an approximation of peak craft personnel requirements as soon as the personnel estimate has been completed.

The elapsed time for construction execution runs longer than shown on the theoretical curves, so it's possible to get multiple peaks in certain crafts, as shown in Fig. 6.4. Typically, construction schedules run two to three times longer than corresponding design schedules. As we said earlier, that's why we try to overlap construction with design and procurement as much as possible.

Practical personnel loading curves

So far we have been discussing theoretical curves, but now let's look at some more likely loading situations that are useful in the practical planning of our human resources. In Fig. 6.2, I have shown the theoretically ideal bell curve as a dashed line, and the forward- and backward-loaded curves as solid lines. The latter two result when the personnel loading occurs earlier or later than planned on a project.

The significance of these conditions becomes apparent when we look at the set of S curves resulting from plotting percent of hours expended versus scheduled time as shown in Fig. 6.3.

The S curve for the ideally-loaded project has a gradual start and finish, which indicates smooth starting and finishing conditions. The forward-loaded curve shows a rapid project start-up and an even more gradual than normal phaseout at the end. The backward-loaded project indicates a more relaxed start and a very steep finish slope on

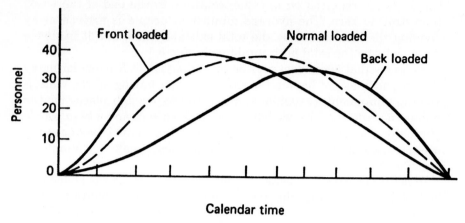

Figure 6.2   Practical personnel loading curves.

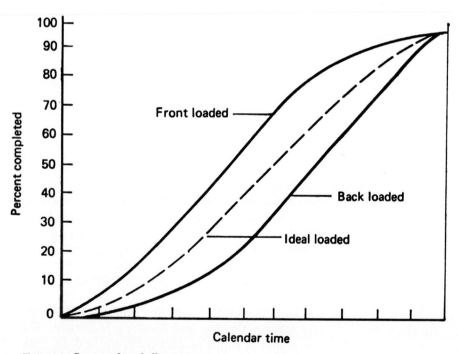

Figure 6.3   S curves from bell curves.

the S curve. The steep finish leads to such problems as inefficient use of personnel and overrunning the budget. The inefficiency results from having too many people working on only a few remaining tasks.

Normally, projects cannot phase personnel off the job so quickly, which means the project will overrun the schedule, finish late, and likely run over the budget. The simple lesson to be learned here is that front-loaded projects may slip to a normally loaded mode and still finish on time. There is little or no hope that backward-loaded projects will finish on time. Any slippage during execution further exacerbates the phase-out problems and makes the project finish still later and more over budget than planned.

CMs must remember, however, that front-loaded projects don't happen just by drawing the personnel loading curves that way. All necessary start-up requirements of design documents, facilities, personnel, and/or materials and equipment must be available to support an early labor pool.

The classic form of bell and S curves introduced here is used throughout the project, as we will see in succeeding chapters. For a more detailed discussion of bell and S curves and their varied applications to project planning and control, see the chapter devoted to that subject in Kerridge and Vervalen's *Engineering and Construction Project Management*.[1]

Planning the Construction Project Personnel

Personnel planning for the field work involves detailed craft loading curves and the field supervisory team. Although the home office support personnel play an important role in the project effort, their numbers are usually relatively small, so loading curves are not practical. Those personnel may be assigned for a fixed duration of time or even on a part-time basis. Also, the home office people are not under the direct administration of the CM. It is important to keep track of their project activities, however, if they are charging time to the CM's project budget.

To arrive at the total project personnel distribution for the craft labor, it's a good idea to plot a curve for each major craft to be used in the construction. A typical example of such a plot is shown in Fig. 6.4, which is actually a composite of all crafts required to build a process-type facility.

The basis for making the curves is the number of labor-hours for each craft originally estimated in the construction cost estimate. Each craft supervisor (or the field superintendent) projects the number of people required to carry out the craft's work scheduled for that week. The number of people is obtained by dividing the estimated hours to

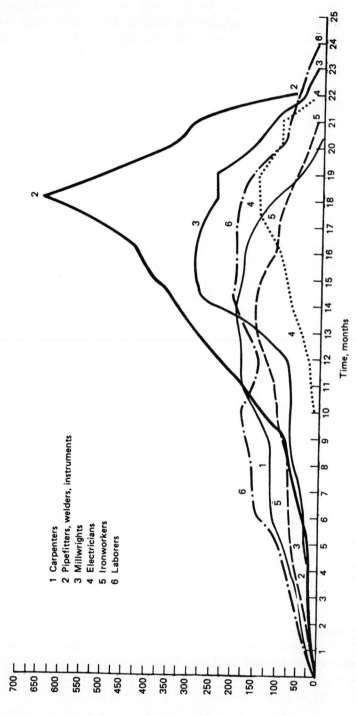

Figure 6.4 Construction labor-planning curves.

1 Carpenters
2 Pipefitters, welders, instruments
3 Millwrights
4 Electricians
5 Ironworkers
6 Laborers

Time, months

be expended that week by the hours in the work week. The number of people for each week is plotted on the graph, using a different line code for each craft.

The craft personnel projection can also be done on computer to generate the graph. If the project is to be run on a CPM program, the human resources requirements can be entered electronically. The CPM software usually offers the option of human resources leveling by taking advantage of the schedule float in certain noncritical activities.

The weekly value for each craft curve is added to get the composite total craft personnel count. You will note that each craft tends to peak at different times in the schedule, but that the total personnel peaks at about 1430 in about the eighteenth month of the schedule. Peaking in the last third of the project makes this a back-end-loaded project, which is not unusual for a process-oriented project. The main contributors to the late loading are the so-called *mechanical trades* of pipefitter/instrumentation, millwrights, and electrical, which peak late in the project. That back-end loading on process-type projects is what makes them very difficult to finish on schedule. The nature of the work gives those key crafts a steep build-down curve at the end of the project.

The early availability of the composite craft numbers gives the CM an indication of the type and size of facilities needed for the project as it develops. The S curve, which is developed from the total personnel plot, is used by the control people to track the overall construction team's progress during the control phase of the project.

The start and finish dates for drawing the curves are derived from the project schedule. Thus we find the so-called building trades of carpenters, laborers, and ironworkers starting early, and building to a relatively flat peak during the site development, foundation, and steel erection work.

Piping shows an early start with a slow buildup during the underground piping phase. Later, piping peaks during the installation of the process and utility instrumentation and piping. Instrumentation and electrical crafts are generally late starters, because they must await the installation of the process and utility equipment, buildings, etc. before they can start their lighting and interconnection work. Their final activity of final system checkout and calibration can be extensive and time consuming in the latter stages of the fieldwork.

The most striking thing about the composite curve is its sharp peak. The sharp peak is a definite cause for concern in a construction labor curve, because it could lead to an overcrowded condition on a limited-access job site just when we are looking for maximum productivity. As we said earlier, the actual curve will tend to roll to the right of the planned curve due to scheduling problems. Unfortunately, that makes a late project finish even more likely. That's a very good reason to consider craft personnel peak-shaving, reducing major peaks by using early start and late finish dates on noncritical activities.

On very large construction projects, the total elapsed time is great enough to have multiple personnel peaks. The longer elapsed time in the construction schedule also gives the possibility of the "roll-to-the-right syndrome," making an earlier peak roll over onto a later peak and thereby creating a superpeak. That is another reason why using personnel peak-shaving is very critical in planning construction personnel loading curves.

Remember, Fig. 6.4 is for a *process* project. A nonprocess project will have another set of discipline curves, depending on the type of facility being designed. The curves should reflect your labor-hour estimate (or budget) and project schedule.

Another factor to consider in the field labor curves is the use of construction subcontractors for a substantial portion of the work. For example, steel for the project may be purchased on a fabricate-and-erect basis, which causes the steel-erection labor to be reflected as a subcontract. If the job involves a large amount of steel work, it may be desirable to include the iron workers in the overall field labor planning chart in Fig. 6.4. On the other hand, if the work of a ceramic tile subcontractor is minor, it may not be worth putting in the diagram. The important thing to remember is to account for a high percentage of the total craft labor on the site to give the CM an idea of just how crowded the work site will become.

Planning field supervisory and staff personnel

The field personnel break down into two groups of people: field supervision and craft labor. Percentagewise, the field supervision labor-hours are very low compared to the craft labor hours, so personnel loading curves are not usually required for the supervisory staff. A simple list of the staff and their proposed duration of assignment is usually sufficient. The number and quality of the field supervisory staff are the most vital activities contributing to the success of any construction project. The CM must expend plenty of time and effort to make a good selection happen. We will be discussing this in Chapter 7, Project Organization.

The number and type of field supervisory staff varies greatly depending on the size, contracting plan, and type of project involved. Table 6.1 shows a range of field supervisory staff numbers and types one might expect to see on various types of projects.

TABLE 6.1   Typical Sizes of Field Supervisory Staff

Contracting basis	Number of people		Types of people
	Process	Nonprocess	
Self perform (Direct-hire craft labor)	30–50	15–25	Managers, craft supervisors, foremen, administration
Construction management (All subcontracted)	10–20	5–10	Manager, supervisors, control people, etc.
Third-party constructor	25–40	12–30	Managers, craft supervisors, foremen, administration

The self-perform format means that the contractor is hiring most of the field craft labor directly with a minimum of subcontracting. This results in the largest field staff, because the prime construction contractor furnishes most of the supervisory staff.

A construction management approach requires less supervision, because the subcontractors furnish the craft labor and field supervision as parts of their contracts. The construction management firm need only supply the management and administrative personnel to administer and control the field work. The third-party constructor mode requires only slightly fewer numbers than the self-perform contractor mode.

Subcontracting Design and Construction

A convenient way to expand the human resources on a project is through the use of subcontractors or consultants. This means is often used in the areas of design and construction to expand a firm's capability in a particular area of expertise, as well as in the area of people. The CM plays a major role in implementing the project subcontracting plan and in overseeing the subcontractor's performance. This is another area of construction management that sometimes does not get the attention it deserves.

Subcontracting design

The matter of subcontracting design work occasionally arises in design-build contracts, as well as construction contracts involving on-site design services. The latter case occurs on major rehabilitation projects involving major demolition and rework of existing systems, and on structures where it is more practical to do the redesign in the field.

Naturally, the necessary design people will be furnished by the design portion of the firm if the contractor offers both design and construction services. Another variation of that scheme is to send out key design group leaders and hire the main portion of the designers locally. That system can be used if the project is out of the company's home city or country and local designers are available. Using local design people can save on travel and living expenses if the design team is moderately large.

Designers for almost any discipline can be obtained from agency firms specializing in that sort of business. Fortunately there are a lot of qualified people who prefer job-shopping to regular full-time employment. The services of these people can be added and deleted at very short notice without obligation to pay any benefits or termination pay. The markup on these services is a nominal 25 to 30 percent of base salary to cover all taxes, benefits, and profit. The base pay for shoppers does tend to be slightly higher than that for regular employees, and all rates are usually quoted with premium rates for overtime. On average, however, the overall hourly cost for contract versus captive personnel is about a wash. One thing that does tend to push overall contract personnel rates higher is the payment of per diem allowances for people brought in from out of town.

Because the CM may not have design supervisory experience, it is vital to have a capable design supervisor to run the on-site design effort. The on-site design work will require daily contact with the owner's on-site design representative, and the CM can't get bogged down in that sort of detail and run the construction at the same time. Also remember, if the design work is in your contractual scope, you are responsible for its quality, cost, and schedule.

Another case of on-site design subcontracting occurs on design-build projects bids by contractors without design capability. That problem can be solved by subcontracting the design or forming a joint-venture partnership with a design firm. That means the construction-only firm must get some design subcontractor bids or a joint-venture design partner during the estimating phase. I would say that this sort of situation could occur if the design effort is fairly complicated and/or is expected to be over 20,000 design hours. A design effort less than that could be handled as described above.

Subcontracting significant design work can be a complex process that may not be of interest to the average CM reader. Those readers interested in learning more on the subject are referred to Chapter 2 of my book *Total Engineering Project Management*.

Construction subcontracting

Such trades as insulation, painting, electrical, sheet metal, roofing, etc. normally are subcontracted on most construction jobs.[2,3] Since the subcontractor has to come to the job site to perform the work, the problems of communications, cost control, and quality are relatively easy to handle. The prime contractor's field supervisors are available on the job site full-time to oversee the work and to administer the subcontract.

The prime contractor's project-control people also supervise the activities of the subcontractors for conformance to schedule, progress payments, and handling change orders. Subcontracting in construction reaches its peak when a construction management format is used. In that case, a construction management team is assigned to the project to totally subcontract the work for the owner. That arrangement was discussed in earlier chapters.

The subcontracting plan is originally formulated during the estimate, so that valid subcontract prices can be obtained. Prices should be solicited only from those subcontractors with whom the firm is willing to contract. There is no point in getting prices from undesirable subcontractors, for their low prices will only complicate the estimating process. If the estimator uses the low price for the bid and you get the job, you may not want to use the low-price subcontractor because it will cost more in the end.

As with any subcontract, it is critical to have a good set of subcontract documents if the work is to be controlled. CMs should review the subcontract documents in detail to ensure that they are applicable to their specific project. Especially look for areas that might fall between the cracks of the prime and the subcontract. Anything left out of the subcontract that is included in the prime contract falls to the prime contractor's account. Make sure that the ancillary contractual requirements of schedule, guarantees, warranties, quality, payment schedules, change orders, and the like are extended to the subcontract. If material is being furnished as part of the subcontract, ensure that the quality standards are equal to the prime contract requirements.

Most subcontractors are major players on the construction team, so be sure to foster mutual respect and project goal participation among the field staff and the subcontractors. If any subcontractors fail to perform properly, do not hesitate to discuss the problem with their management and to make personnel changes if necessary.

Construction Material Resources Planning

The basis for construction material resources planning is in the project materials plan.[2,3] That document is created by the CM and the project purchasing agent (or manager). The report resulting from making the materials resource plan should include the division of procurement responsibility, current delivery data for engineered equipment, bulk materials, and subcontracts. That's in addition to a survey of current market conditions, pricing trends, bulk materials availability, and vendor lists. Those data are invaluable in formulating the project material resources plan and the project schedule.

The division of procurement responsibility is defined in the scope of work and scope of services sections of the contract. In most industrial work, the engineered equipment is purchased by the design firm, because of the lengthy delivery times involved. In that case the constructor is involved only in buying the bulk materials such as concrete, masonry, bulk-piping and electrical materials, and architectural items. Much of the material may also be furnished as part of the work subcontracted to other contractors. It's important that all the required material and equipment needs for the contract be covered in the construction materials plan. It is vital to project success that the proper equipment and materials be available in time to support the field construction schedule.

Today's materials plan is usually a detailed document listing by account code the quantities of the required materials and equipment, a description, their field-required date, responsible supplier, and the like. Columns are also shown for vendors, purchase orders, promised delivery, partial shipments, shipping mode, etc., as that information becomes known. A computerized spreadsheet or database is ideal for handling the large quantity of information required to control the material on a medium- to large-sized project. The daily update of the materials-tracking document makes computerizing a must.

I recommend that CMs be assigned during the estimating stage, so that they become familiar with the construction material requirements early in the project. If they did not participate in the cost estimate, formulation of the material plan should occur right after project assignment and learning the contractual terms.

The CM's role in procurement

The location of the actual procurement functions plays a major role in the project procurement function. The overriding goal of any procurement plan is to buy the necessary equipment, materials, and subcontracts within the proposed project budget and deliver them on time. Experience has shown that effective buying can involve many facets, such as getting the best price and delivery, combined order volume, knowledge of the local market, and delivery. Some items are best bought in the home office, others are best bought in the field, and some are best furnished by a subcontractor. As one facet of organizing the project, CMs must make the field versus home office decision early, to ensure that there is a prompt, well-organized start to the key procurement function.

Because we are looking only at the relationship of the CM to procurement, a detailed description of the construction procurement function is beyond the scope of this book. The books listed in the reference section at the end of this chapter describe a variety of approaches to sound construction procurement systems.

Because CMs rarely rise through the ranks of procurement, they sometimes feel that buying out the job is basically a staff function bordering on clerical work. Nothing could be further from the truth. In the engineering-construction business, procurement is a line function involving the expenditure of 25 to 40 percent of the project budget. Adding timely ordering and delivery of the construction materials and services makes the procurement activities even more critical to project success. CMs must *never* neglect the procurement portion of any construction project.

Buying out the materials and subcontracts is especially critical to success on lump-sum bids on completed designs. The bids received during the estimating stage are already in hand and are subject to a limited validity period by the vendor. That means the procurement people must swing into high gear to buy within the bid-validity period, assuming that the vendor won't extend it. On the other hand, the bids given for the estimate may actually be a little high, based on the vendors' knowledge that the price is being used for an estimate. My experience has been that vendors are likely to drop their prices from 5 to 10 percent after the prime contract has been awarded. Although haste is required if the materials and subcontracts are to be ordered, CMs must not overlook the potential savings to the budget.

Any good procurement effort must be supported by sufficient technical effort to ensure that the equipment and materials are being ordered to specifications whether ordered in the field or the home office. The technical support is responsible for preparing the requisitions, technical bid evaluation, and review of vendors' drawings. Nothing is more devastating to a construction project than tearing out improperly specified materials. CMs are responsible for ensuring that qualified technical people are available to complement the procurement effort, both in the home office and the field.

Long-delivery materials

Ensuring that the material resources for the project arrive on time involves these important planning areas:

- Long-delivery equipment
- Special materials and alloys
- Common materials in short supply

- Heavy construction equipment and tools
- Services and system requirements
- Transportation systems
- Financial resources

It's vital that the CM make an early review of the project's physical resources, to give those items on the critical path special attention. They must be recognized early on to preserve any available schedule float or to keep them from slipping into negative float.

We generally consider any equipment with a delivery of 10 or more months as potentially *long-delivery equipment*. Delivery times in that area or longer usually place the equipment on the critical path. Examples of long-delivery equipment for industrial projects include such items as high-horsepower centrifugal compressors, turbogenerator sets, heavy-walled vessels, field-erected boilers, paper-making machines, rolling mills, or other complex engineered systems. You may not find any of those or similar items on your equipment list, and therefore assume that you don't have long-delivery equipment to worry about. That is a false assumption; every project has long-delivery equipment. On any project, the items with the longest delivery dates are the long-delivery items. If schedule improvement is required, it can be done only through improving the delivery of equipment falling on the critical path.

Long-delivery items must be given top priority in the schedule, starting with the first operations in the project design. If they are engineered equipment, the data sheets must be generated first and pushed into the procurement phase for taking bids. Each step should be expedited to get the order placed. Continued expediting through the vendor data approval stage also is advisable. The vendor will not start working on the equipment until approved shop drawings have been received.

Long-delivery equipment is sometimes preordered by the owner on preliminary sizing criteria that are refined by change orders as the design progresses. If that method is used, unit prices for changes should be obtained from the vendor with the bid price, so as to gain better control over the vendor's change orders. A typical example of the owner prepurchase approach is the preordering of paper machines in the pulp and paper industry.

If the long-delivery equipment or materials purchasing is left to the construction contractor, it is really late in the scheme of things, with any time savings during the design and construction bidding stage lost to the schedule. This places the added pressure of late delivery of critical equipment on the already critical construction schedule.

Owner CMs must keep that factor in mind when making their contracting plans.

Special materials and alloys

It is a good idea to review all materials specified for the project that might not normally be stocked because they are so special. Lately, many manufacturers and vendors have taken special pains to reduce inventories, which makes availability of unusual materials even more scarce.

In the process industries, special alloys fall into that category. Some of the rarer ones include the refractory metals, titanium, Hastelloy G, carbate, and to a lesser degree monel, inconel, and some special stainless steels. Some of these special materials may also require special treatment such as casting, tube-bending, welding, stress-relieving, and testing, which tends to delay their delivery even further. Since most CMs are not strong in that highly specialized area, it is best to have your technical staff or consultants thoroughly investigate any potential delivery problem areas. Those long-delivery item reviews should be made early enough that there is still time to deal with the problem. After the bids have come in, it might be too late to improve the delivery date to meet the schedule.

Problems of that nature can also be found in nonprocess projects. The design might call for a special aggregate in precast wall panels, special window systems, people-movers, a rare quarried and polished stone, or any one of many others that are likely to fall on the critical path. Early planning for the delivery can get such an item off the critical path, and that can pay a dividend of earlier job completion.

Common materials in short supply

Common materials in short supply often include such ordinarily mundane items as structural steel, concrete, and reinforcing bars. In a large high-rise building, the structural steel is heavy and high in tonnage. Early design and takeoff for placing mill orders for the heavy steel is critical to maintaining the schedule. Each section of the structure must be closely scheduled, if the steel is to be delivered at a time that suits the erection sequence. A large dam project uses huge quantities of reinforcing steel, forms, earth fill, and concrete over long periods of time. These relatively common materials must be planned for and delivered on time if the schedule is to be met. These commonly used materials sometimes tend to be overlooked on larger projects, so don't pass over them lightly. That advice is especially valid if the project is being built in a remote location.

Special construction equipment

Unlike the other materials, special construction equipment does not enter into the final product, but I like to consider it a physical resource worthy of early planning. The operation of the heavy-equipment pool on large projects is a separate subject that will be covered later, so the discussion here deals only with the planning effort.

In design-build work, the heavy-equipment requirement is something that should be investigated by the construction people during the design phase. I have seen many cases where it was more economical to redesign the project than to use more costly erection equipment than might otherwise be needed. That is difficult to do in fully designed lump-sum-bid project environment. The design manager might suggest bringing in an outside construction consultant as constructibility problems arise during design, but that is not done very often.

The above problem areas come into focus and can be more readily solved when preconstruction planning sessions are used on the project. At the very least it is a good idea to have the owner's construction people make a constructibility analysis near the 50 percent design stage, during their review of complex projects.

Services and Systems

Strictly speaking, project services and systems are not physical resources either, but they must be planned for at about the same time as the physical ones. Planning for them is particularly critical on large projects.

The project scheduling, accounting, cost-control, and administrative systems must be decided on at this time, so that the necessary manual or computerized project-control systems can be effectively implemented. If computers are to be used, the required hardware, software, and communication resources must be planned and implemented.

Even such ordinary resources as the office service and the site facilities for construction have to be planned early. What form of security system will govern the project operations? How much space do we need, and when? What design documents or models are available, and when? Is construction to be in a remote location? What special climatic conditions are going to prevail?

These are just a few typical questions that must be investigated and answered by CMs. Some items will fall on the critical path and some will not, but the critical items must be discovered early enough to allow for their resolution. Formulation of the CM's priority list is vital during this early phase of the project.

Transportation systems

Usually this category is involved only on large construction projects that may have an international flavor, either in site location or in purchasing sources. Another special case are construction projects that embody prefabricated modules built offsite for erection at remote locations. Any of that type of project will involve special loading, transportation, and receiving facilities to handle unusually large pieces of equipment.

These facilities and services are uniquely specialized, and must be planned and organized well before they are needed. Poor timing in this area can incur high costs and lost time if any details are overlooked. The CM must see that the project traffic people are doing a thorough job in this vital area.

In the case of using premanufactured modules, the CM has two sites to oversee and coordinate. The manufacturing site will probably be run by a subcontractor, but the overall CM must monitor the output quality and schedule in order to support the erection site. The overall CM should insist on having a qualified full-time representative at the manufacturing site. Frequent CM visits to the prefabrication site are also very much in order. The nuances of this type of construction are so different from stick-built construction that the CM should have prior lower-level experience or in-depth training before attempting it.

If the project involves a significant amount of foreign or imported equipment and materials, the overall marshaling of shipments, import licenses, custom regulations, dockside security, and invoicing should be developed by the project purchasing manager. The traffic portion of the materials management plan must include all the legal and quasi-legal factors involved in getting the goods to the construction site on time. The CM must review the complete plan to assure that the field construction needs are being fully supported. If the overall system doesn't function smoothly 100 percent of the time, the field construction will be delayed.

Financial Resources

It has always been the owner's responsibility to arrange for the project financing. To a degree, that excludes it from the CM's responsibilities, but the CM cannot be blind to the fact that the project must be backed by sufficient financial resources. Usually, CMs have the authority to approve all reasonable expenditures required to execute the construction. They also have the responsibility to see that vendors and subcontractors are being paid for goods and services rendered. Underfunded projects can create many problems for the CM.

When the project is not adequately financed, I would be very careful about being the final link on the chain of those committing the project funds. I advise you to check your legal position if such a situation should arise.

I was involved with two major international petrochemical projects that were underfinanced, and to my knowledge neither has ever been completed. One was canceled before start of construction, the other stopped with construction about 35 percent complete. Fortunately, I was the contractor's design-build project manager in both cases, so we had no financial responsibility in letting the purchase orders.

In the absence of solid financing for the second project, it became increasingly obvious that the work could not go on. I was really just as happy not to be around when the project finally collapsed, after the owner could not raise the funds for materials and construction.

My experiences over the years on a large number of projects have taught me at least two things:

1. Projects can eventually be finished *even without* such things as planning, organization, control, computers, leadership, sophisticated project systems, and even talented people.

2. Projects *cannot* be completed without money!

Summary

This chapter wraps up the planning portion of the CM's activities. Sound planning forms the foundation for everything that comes later. Effective CMs must train themselves to plan all facets of the project, as well as their day-to-day work activities.

We must remember, however, that a plan is only a proposed baseline for the execution of a project. Any plan is subject to possible changes along the way. Although it's often necessary to change our plans, I don't recommend making any *radical* changes to your original plan. If your plans were well thought out, you should resist pressures to change them.

References

1. Arthur E. Kerridge and Charles H. Vervalen, *Engineering and Construction Management,* Gulf Publishing Company, Houston, TX, 1986.
2. James J. O'Brien and Robert G. Zilly, *Contractor's Management Handbook,* McGraw-Hill, New York, 1991.
3. Sidney M. Levy, *Project Management in Construction,* McGraw-Hill, New York, 1987.

Case Study Instructions

1. Prepare a project personnel resource curve for the construction trades for your selected project. Use the breakdown of labor hours from your construction cost estimate and/or budget. Discuss any unusual peaks in the curves and say what you would do to eliminate them. Include any major subcontractors.

2. Based on the contracting plan, describe the size and makeup of the field supervisory staff you would anticipate having on your selected project. Make an estimate of the total field indirect costs, and check the percentage cost against the total construction cost.

3. Assume that your craft labor will be short 50 percent in the area of several skilled trades. How do you plan to make up the shortfall? Give your reasoning in support of your plans.

4. Review your selected project for expected long-delivery items of any kind. Discuss your recommendations for minimizing the adverse effects of these on your project's strategic end date and schedule.

5. Make an outline of your materials management plan to ensure that the necessary equipment and materials arrive at the field site to support the construction activities. Make a sample page format to show the data the material plan will cover.

7

Project Organization

With planning out of the way, we are now ready to get into the organizational part of our project-execution philosophy of *planning, organizing and controlling.* I am sure you understand that we don't perform those three activities in a compartmentalized fashion as I present them in this book. All the activities in fact overlap and proceed concurrently. We are continually reviewing all three management functions simultaneously and modifying them as necessary to meet day-to-day operating conditions.

Organizational Overview

Organizing can be defined as *the function of creating in advance of execution the basic conditions that are required for successful achievement of objectives.* The operative phrase here is *achievement of objectives*—the objectives being the *project goals.* Never allow the organizational structure to get in the way of meeting the project goals!

My first law of organizing is: To meet the objectives, design the organization around the work to be done, not the people available. The law is based on the assumption that you have access to a bank of skilled personnel to perform the necessary activities. However, we know that it rarely happens, and that some compromises have to be made in selecting a project team.

The general goal of any organizational structure is to establish the proper relationship among:

- The work to be done
- The people doing the work
- The workplace(s)

Later we will work our way into building some typical project-organization charts based on the above principles.

Organizational design

By way of building some background in organizational design, it will be valuable to review a bit of the basic theory and practice used to design organizations. Most project organizations tend toward the *functional* or military style, using a combination of vertical and horizontal structures. As with the military, we also use line and staff positions.

In recent years other forms of organization charts have been put forward, but they don't seem to have caught on in the capital projects business.[1] Circular arrangements, such as the one shown in Fig. 1.2, are an example of a different type of chart. An advantage of the circular layout is its ability to show the relationship of the construction manager to all the other operating departments in the company and the project. Despite that advantage, it doesn't seem to be used much.

A vertical structure, as the name implies, places one position (block) over the other as shown in Fig. 7.1. The vertical dimension establishes the number of layers in the organization. Keep in mind that every layer introduces a communications filter into your organization, and each filter is a potential choke point for the necessary flow of vital project information.

The horizontal dimension is shown in Fig. 7.2, with a number of horizontal positions reporting to a single block above. The horizontal dimension introduces the term *span of control*. In a vertical organization (Fig. 7.1), the contacts are one-on-one; in a horizontal one, they are n on 1. If n becomes too great, supervisor A will not have enough time to devote to each of the supervised blocks. There is no fixed maximum for n, since the ability to supervise depends on the nature and complexity of the contact. The number 6 is often mentioned in management circles as a normal maximum span of control, and it's one I

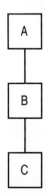

Figure 7.1 Vertical organization.

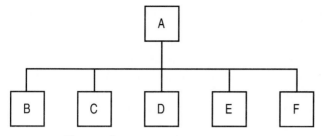

Figure 7.2 Horizontal organization.

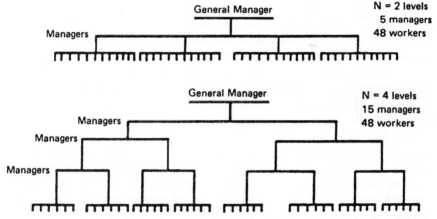

Figure 7.3 Vertical and horizontal organizations.

wouldn't exceed without giving the matter a lot of thought. Remember too that communication filters are added in the horizontal direction as well as the vertical.

Figure 7.3 compares layers of tall and flat organizational structures. The tall (lower) organization shows the rapid growth in the ratio of intermediate supervisors to workers. The horizontal structure has fewer managers, but each manager has a much broader span of control.

As we can see, the lower format shows a ratio of 15 managers (supervisors) to 48 workers, whereas the upper chart shows only 5 supervisors for 48 workers. Keep in mind that the labor-hours are the main part of a project budget and that the supervisory hours are relatively nonproductive. If supervision runs over 10 percent, your budget is likely to be in trouble.

Obviously, we will want our organization to have an optimum balance of vertical and horizontal dimensions, to give us the most effective control of the project work. I prefer to have my project supervisors slightly overloaded as this encourages them to extend their capabilities and to have an opportunity to grow with the job. Remember that some people exceed their span of control because they are not organizing their time effectively. I don't consider that a sound reason for revising the organizational structure to eliminate the problem. Some on-the-job training or a shift in personnel is definitely a better solution to such an apparent span-of-control problem.

Line and staff functions

The line and staff functions in your organization are best defined by looking at the project goals that we discussed earlier. Our main goal is a quality facility built on time and within budget. To build a quality facility, we must produce a sound design and deliver the project resources in a timely manner. That means that all design, procurement, and construction teams must perform in a focused capacity if they are to meet the main project goals. The key line functions in each of these organizations must be effectively supported by efficient staff functions.

In a business or manufacturing setting, we normally look on procurement as a staff function. How about our project procurement function—is it line or staff? In the project mode, procurement delivers the project's vital physical resources. That definitely makes it a line function because they also exercise a good deal of line-type decision-making power in the procurement cycle.

How would you classify the functions of scheduling, cost engineering, and project accounting? Although they play a major role in meeting the project's time and money goals, they more closely fit the definition of staff functions. They monitor and report on the performance of, but have no direct authority over, the various line functions. The authority of the staff groups must be exercised through the project or construction managers.

The analysis of line and staff function can be extended to the structures of owner and contracting companies. In owner companies, the prime function of the line organization is the production of the product (or service), be it paper, chemicals, electricity, or health services. Therefore the group responsible for creating the owner's facilities is functioning in a service or staff mode. That is why the owner's operating division (operating as a profit center) has the final say on facility design and construction.

In the case of a contracting organization, the engineering, procurement, and construction groups make up the line organization. They are the profit centers for the contractor's operation. Their prime duty is to provide services to their clients at a profit. This explains why contracting companies have been forced into using a stronger project/construction management approach to improve their company profitability. Project and construction managers in a contractor setting usually have direct access, and often report directly, to their top managements. That's an important aspect to remember when owners and contractors are working together on a project.

The Corporate Organizational Structure

Although project organizations are set up in such a way as to serve for the duration of each project, the project organization also must blend in with the corporate organization and policies. Corporate organizational structures and policies also vary greatly, depending on the size and the type of market the firm serves. Differences due to size tend to reflect themselves in the number of duties assigned to individuals. Small firms use multifunctional people, whereas larger firms have enough volume to afford specialists or departments to handle the work.

Both line and staff home office construction people usually move into the organization from service in the field. That should make it easier for the CM to relate to the corporate structure. Having smooth relations with the home office structure always is important, but especially so when the field team must depend heavily on home office staff functions. Most CMs prefer to have a fully staffed field team for greater independence and control of their project destinies. The shorter lines of communication, especially from distant construction sites, also constitute a big plus.

Figure 7.4 shows a typical corporate organizational structure for a design-build firm. This structure may vary, depending on whether the firm is design-driven or construction-driven (Fig. 7.5). That in turn depends on the firm's beginnings and markets, which in turn control the flow of personnel into top management positions.

Organizational structure often is affected by where the firm stands in the business cycle. Home office organizations tend to grow in good times and shrink in bad. The size of the home office organization has a direct bearing on the general and administrative (G&A) expense added to contract bids, which can directly affect the number of projects won. These economic factors, combined with the whims of top management, contribute to an almost constant state of flux in the home office organizational structure.[1] It is important for CMs to keep the current structure in mind as they go about executing their projects!

A good understanding of the services available from and the structure of the home office is important to the practicing CM. Detailed descriptions of typical contractor's home office operations are given in the referenced handbooks[2,3] devoted to that subject.

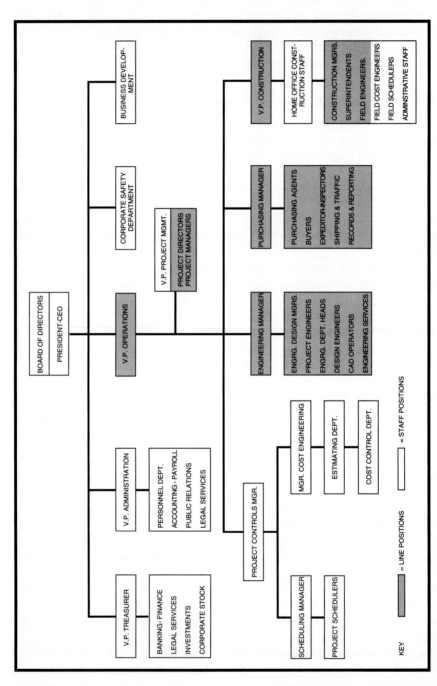

Figure 7.4 Design-build firm corporate structure.

KEY ▮ = LINE POSITIONS ▢ = STAFF POSITIONS

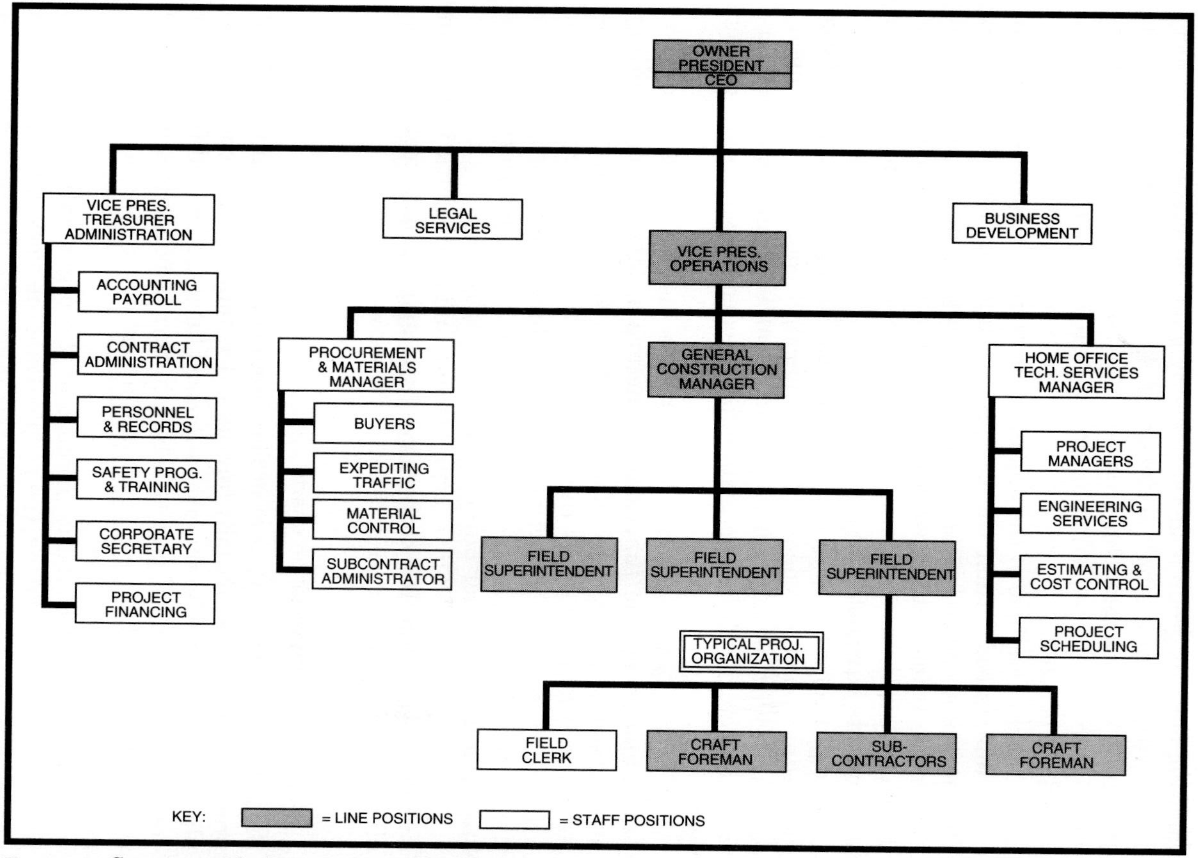
OWNER
PRESIDENT
CEO
VICE PRES.
TREASURER
ADMINISTRATION
LEGAL
SERVICES
BUSINESS
DEVELOPMENT
VICE PRES.
OPERATIONS
ACCOUNTING
PAYROLL
CONTRACT
ADMINISTRATION
PERSONNEL
& RECORDS
SAFETY PROG.
& TRAINING
CORPORATE
SECRETARY
PROJECT
FINANCING
PROCUREMENT
& MATERIALS
MANAGER
BUYERS
EXPEDITING
TRAFFIC
MATERIAL
CONTROL
SUBCONTRACT
ADMINISTRATOR
GENERAL
CONSTRUCTION
MANAGER
HOME OFFICE
TECH. SERVICES
MANAGER
PROJECT
MANAGERS
ENGINEERING
SERVICES
ESTIMATING &
COST CONTROL
PROJECT
SCHEDULING
FIELD
SUPERINTENDENT
FIELD
SUPERINTENDENT
FIELD
SUPERINTENDENT
TYPICAL PROJ.
ORGANIZATION
FIELD
CLERK
CRAFT
FOREMAN
SUB-
CONTRACTORS
CRAFT
FOREMAN
KEY:
= LINE POSITIONS
= STAFF POSITIONS

Figure 7.5 Construct-only firm corporate structure.

Constructing the Project Organization Chart

Any operating project is worthy of an organization chart showing positions, lines of authority, titles, group relationships, and even the names of current group leaders. Depending on the size of the project, it may vary from a single-letter-sized sheet to one or more E-size drawings. When approaching the structure of your project organization, I hope that you will remember that *simplicity is beautiful,* regardless of size!

The main advantage of the organization chart is that it shows in an easily understood format the key project functions and the players. This makes it easier for new people reporting to the project to get a feel for the team and the work assignments. It's a good idea to have the chart available early in the project, even if all the personnel assignments have not been filled.

My main concern as to the effectiveness of organization charts is the feeling of compartmentalization and *empire-building* they engender.[1] Any chart tends to set up psychological and communication barriers that detract from the team effort needed for successful project operations.

The best way to overcome those obvious disadvantages is to have superimposed on the functional structure a network for handling interpersonal contacts and information flow. A good analogy is to consider the functional chart the skeleton of the project organization, with the *unseen network* acting as a circulatory system through which the lifeblood of the project flows. A good CM must work hard at breaking down barrier-building tendencies by fostering the *unseen network,* which is done by exerting strong and charismatic leadership on the project team. If a cooperative team spirit is not created across all blocks on the chart, project performance will suffer.

Typical project organization charts

Since the number of types of project organizations used in the capital projects business is virtually limitless, it is impossible to show a sample chart for each. The samples discussed here can be modified to suit your particular type of project, which in turn depends largely on your contracting plan. Figure 2.2 shows the four major contracting strategies possible.

Earlier we said that the purpose of the organization chart is to define the work to be done, the people doing the work, and the location of the work. The work to be done is a function of the project scope. A turnkey-type project involves design, procurement, and construction activities in one organization. We will use that scope, as well as a construction-only project, to build two sample organization charts.

Organizations involved in a given project include owners (clients), central engineering departments, design consultants, construction contractors, and subcontractors, in a wide variety of possible combinations. This is in addition to the various specialty construction groups such as engineering, project controls, procurement, personnel, and the administrative functions. The chart should clearly define the working relationships and the duties of the personnel within the participating groups.

Remembering that our *overall project goal* is to produce a quality facility that meets the owner's needs, it might be a good idea to start at the bottom of the chart with the human resources required to reach that goal effectively. The project-execution plan also goes a long way toward shaping the overall organization chart, because it defines how the job will be done.

Where the work will be performed also provides our chart with some *givens*. The field work usually is remote from the design office, and that calls for a separation between design and construction. Procurement can be done in the design office or in the field. Design work can be split between offices and even among specialty design consultants, owners, licensors, subcontractors, and R&D. The various combinations of participants affect how we arrange the blocks on the chart.

Keeping the three factors controlling organizational structure in mind should allow you to construct a viable organization chart for any type of project. After all concerned parties have approved the basic structure, you can proceed with reviewing and selecting the key personnel to fill the slots. We will address that important function in more detail later in the chapter.

A typical process project organization chart
for design-build mode

The relationships required on a process project organization chart can be fairly complex, especially where an owner, a design contractor, and a third-party constructor are involved. Each of the major players will have its own multifaceted organization. Figure 7.6 shows one approach to how the three organizations could interact.

The top group of blocks show the main parts of the owner's team assigned to bring a process facility into being. An owner's central engineering department has been assigned the job of designing and building a facility for one of its operating divisions (the true client). Because the owner's central engineering group does not have sufficient design, procurement, and construction resources, the services are farmed out to contractors.

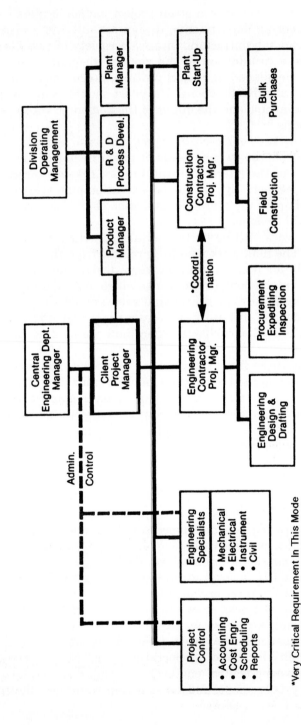

*Very Critical Requirement In This Mode

Figure 7.6 Typical design-build project organization chart.

The owner's central engineering group appoints a project manager (shown as Client PM) to be responsible for supervising and coordinating all aspects of creating the new facility. The Client PM has six major areas of control and/or contact. Fortunately, all six activities will not peak at the same time. For example, the work with R&D will peak early in the project, whereas plant start-up will not occur until late. Also, the design and construction activities will peak at different times, so the project manager should be able to effectively keep the six balls in the air without undue strain.

The division of work calls for the design firm to supply a complete design package, procurement for the tagged equipment, and start-up assistance. The constructor supplies construction services from pre-construction through plant start-up and turnover, including procurement of the bulk materials and subcontracts.

The chart clearly shows that the Client PM is the focal point for all information and communication between the owner and the various groups responsible for executing the work. That means the Client PM has to have a well-organized project team to ensure timely flow of all project data, decisions, and approvals through the project team. If that doesn't happen, a king-sized bottleneck that chokes off progress throughout the total organization will occur.

Obviously, the three project management slots head the main line functions in the overall project organization. The project control and engineering specialists serving the client's project manager, along with the product manager and R&D people, are supplying key staff-support functions to the project. The design and construction functions shown in Fig. 7.6 will be further detailed in separate charts showing some typical organizations required to support those major activities.

You will note that the only horizontal lines crossing between the vertical sections of the chart are at the upper levels. That brings home my earlier statement about the adverse effects of compartmentalization as engendered by most organization charts. The lines of authority show most of the flow up the chart to some point, then back down to the point of execution. We also know that about 80 to 90 percent of the information must flow directly across the chart to make a project really move well. That's what I mean by the project and construction managers' facilitating the flow of project information across the organization, but in a controlled manner.

Sometimes the cross-flow is inhibited by contractual buffers between the design and construction contractors, such as the one shown in Fig. 7.6. Neither of the two prime contractors (design and construction) has any contractual obligation to the other, except through the owner's representative. This means that a specific coordination procedure must be set up to expedite and control the flow of project information across that key interface as shown by the double-headed arrow. If that key horizontal communication channel doesn't exist, or if it breaks down, the project goals will suffer.

A clear statement defining the duties, division of work, and responsibilities of the three key parties is clearly needed. That will ensure early preconstruction input to design, effective flow of design information and materials to the constructor, and a good owner-design-construction interface throughout the project. A good joint effort on the coordination document will keep any potential adversarial relationships from developing along the way to meeting the project goals.

Typical design organization chart

The chart shown in Fig. 7.7 is a typical engineering-construction contractor organization for a design-build or a separate design-and-build project. I include it here as background material for those CMs involved in design-build project situations.

This chart serves as the detailed blowup for the design-and-procurement portion of Fig. 7.6. In that case, the client's organization shown is the same as the one in Fig. 7.6. Figure 7.7 could also represent a central engineering department's project organization for an in-house project. In that event, the client's organization would be the operating division's project group responsible for developing the facility.

In any case, a project manager (director) supervises the functional groups responsible for the line activities such as design, procurement, and construction for the client. That's in line with the strong project-management environment that exists in a typical contracting organization: One person is totally responsible to the client for project execution and to company management for project profitability.

The block to the left of the project manager shows a "management sponsor." This is a top-management executive appointed by the CEO to bring a definitive top-management focus to the project. The sponsor ensures that the project and construction managers are performing up to standard and are getting sufficient top-management review and support in meeting the project goals.

The sponsor also maintains top-level contact with the client project manager and his or her management to take a proactive stance in client-contractor relations. A good sponsor keeps a finger on the pulse of the project throughout its life cycle and judiciously steers the project and construction managers away from the rocks of project-relations adversity.

Although usually it is not a problem, the project and construction managers must not let the sponsor take over active control of the project. Smart PMs and CMs make good use of their sponsors in han-

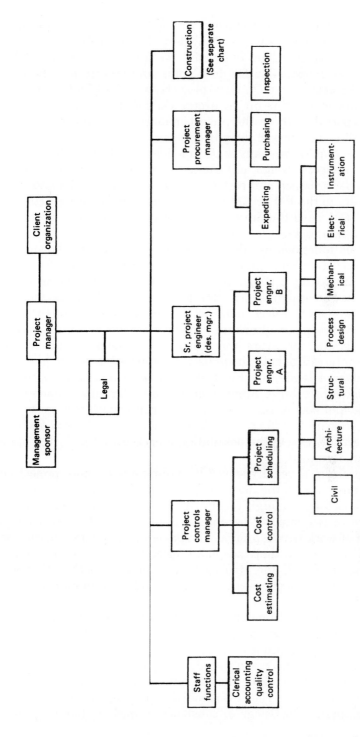

Figure 7.7 Typical design-build contractor's project organization chart.

dling high-level political matters and getting management support *without* surrendering control of any part of their projects.

Moving down the chart in Fig. 7.7, we find the project-control people who are responsible for developing and monitoring the project time and money plans. On large projects, we might find a project controls manager heading up the group. On small projects, those people would report directly to the project manager (or engineer) and could be part-time members of the project team.

The group performs only staff functions and has no line authority over the project's design, procurement, or construction operations. It gathers performance and cost data from the line groups and reports the results to the PM or CM for any necessary corrective action. The PM and CM have the duty and authority to bring off-target performance back into line.

The line functions usually are grouped together as shown on the middle line of the chart in Fig. 7.7. On larger projects the design function is directed by a senior project architect or engineer or a design manager . The design leader often has the additional responsibility of acting as deputy project manager during the absence of the project manager. Acting as alternate project manager offers an excellent training ground for future project managers. The arrangement serves the best interests of both the project and the design leader.

The design leader is the first block we come to that requires supervision of a large number of discipline groups. Depending on the type and size of the project, it can run from 5 to 8 groups containing from 40 to 150 or more design people. Numbers of such magnitude quickly create the possibility of exceeding the design manager's span of control.

The problem is readily solved by adding one or more project engineers to assist with the design supervision. The project illustrated in Fig. 7.7 is large enough to afford two project engineers who are assigned to two major geographical areas, A and B. Sometimes the division of work between project engineers is made on a discipline or *horizontal* basis. In that case one project engineer would handle the civil, architectural, and structural disciplines and the other would handle process, mechanical, electrical, and instrumentation for all areas of the project.

Each design discipline should be headed by a lead person who is responsible for the group's contribution to meeting the project goals. Discipline leaders must be people who have enough management smarts to control the group's productivity and technical output. They are responsible for making the estimates and schedules for producing the finished product. They also monitor the budgets and schedules for the execution of the design or construction work. Comparing them to their military organization counterparts, they are the sergeants who carry out the battle plan in the project trenches.

I have a favorite saying for managers: "Get out into the trenches to see what is actually happening on your project." It pays to get to know your sergeants and to check their performance on a firsthand basis. Discipline lead people are in the natural line of progression for promotion into the ranks of project engineering or department leadership. Improvement in the handling of their project administrative duties will determine just how rapidly promotion will occur.

Construction Organization Charts

The CM interfaces existing in the various construction project settings vary widely and are often confusing. Fifty years ago, CMs used to ride off into the sunrise with a roll of drawings under their arms. Some months or years later they rode out of the sunset saying the project was completed. Increased project size and sophistication along with modern management systems have seen to it that it doesn't happen that way any longer. Although CMs are still left to their own devices executing the field work, now they have several higher authorities such as owners, project directors, and group managements monitoring the overall project goals.

As we have shown in our previous examples, the construction organization is part of the project team reporting to the project manager or director. Therefore the project manager shows up at the top of the typical construction organization chart shown in Fig. 7.8. I want to stress that the owner's or contractor's project manager has *functional control* only of the field work and *not* of the day-to-day construction operations. The project manager is responsible for meeting the overall project goals of *finishing a quality facility on time and within budget*. The PM's responsibility for construction is to ensure that none of the major project goals is placed in jeopardy during the execution of the field work. Since the field work for an integrated project usually involves the expenditure of 40 to 60 percent of the overall project budget, the construction organization plays a major role in overall project performance.

The leader of the construction team can be known by a variety of titles. Years ago he was called a "construction [or field] superintendent." As the management of construction projects grew in scope, technology, and sophistication, the title "construction manager" became more appropriate. Today, the title "field superintendent" usually is used for the person in charge of the actual construction operations under the construction manager. Some heavily construction-oriented companies even use the title "project manager" for the person in charge of the construction operations. In this case, the title

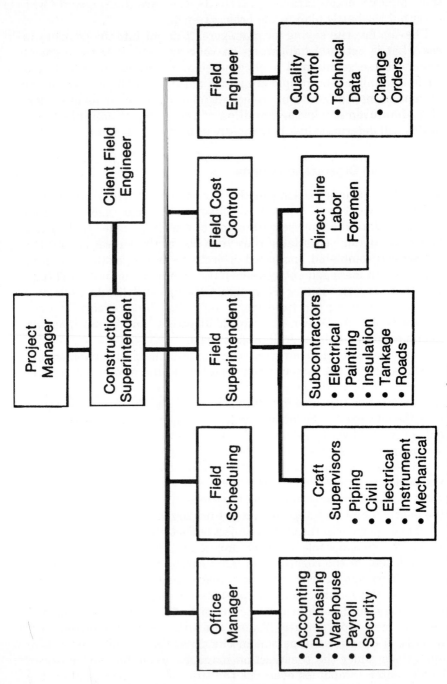

Figure 7.8 Typical contractor's field project organization chart.

"project director" is used for the person in overall charge of design-build projects.

The CM also has an important contact with the client's organization through the owner's field representative. The client's field engineer usually is assigned to the construction site to protect the client's interests in the field. He or she generally reports to the owner's project manager. On larger projects (e.g., $50 million and up), the client's field organization can become quite extensive, depending on the project complexity and level of quality, cost, and schedule control that the owner wishes to maintain.

Another common arrangement that allows the owner to control quality and schedule is to employ the A&E firm that designed the project to also inspect and follow the construction work. Here the A&E firm has no contractual relationship with the contractor and acts only as the owner's representative. Any communications with the contractor affecting contractual matters must be made through the owner's project manager.

Underlying all of the above supervisory inputs, CMs also are scrutinized by their home office construction department management. These people administer the CM's recruitment, salary, training, annual performance review, and promotion. In addition to those personal issues, they also are the source of vital construction department personnel and physical resources needed by the CM. Generally that important link doesn't show up on most project-oriented organization charts, because there is no convenient way to show it.

As shown in the chart in Fig. 7.8, the CM controls the field operations through a group of five or six key line and staff managers. Although the structure shown has a heavy horizontal component, the CM's span of control should not be exceeded if the key managers are well qualified. The chart in Fig. 7.8 is predicated on the construction work's being performed by the general contractor along with a normal amount of subcontracting. A chart showing a construction management organization for a totally subcontracted job would be entirely different.

The field superintendent

The most important line function in the construction organization takes place under the direction of the field superintendent, who manages the largest commitment of human and physical resources on the project. With that sort of major responsibility, the field superintendent must be supported by a staff of area engineers, craft supervisors, and craft foremen. Those people are responsible for planning the work and organizing the personnel to meet the detailed work schedules. They also coordinate activities among the different crafts and are responsible for the quality of the work as well as safety in the workplace. Going back to our military comparison, they are the sergeants directing operations in the trenches where the project battle is being won or lost!

All the other project operations, including the design and procurement efforts, must be directed toward supporting the field operations where the end product is being created. The home office and field operations probably exhibit the worst *compartmentalization syndrome* within the organization. It is a prime concern of the project and construction managers to break down the natural age-old barrier between design and construction so that the overall project goals can be met. Without a doubt that barrier presents the PM and CM with their greatest leadership challenge.

The supervision of the construction activities can be subdivided on either a geographical (vertical) or craft (horizontal) basis. The partition might even follow the arrangement used earlier in the design area. On larger projects the division tends to be vertical by areas, whereas on smaller projects the split tends to division by craft, as shown in Fig. 7.8.

The comment made earlier about developing future project managers applies to the field organization as well. Future CMs evolve from the ranks of construction superintendents, area engineers, and craft supervisors. Again a geographical split of duties results in more rapid personnel development due to the broader nature of the responsibilities involved.

The staff functions of project control that started in the home office (on integrated projects) carry over into the field operations. In fact, as the field operations gain momentum and design activity declines, the center for project-control functions gravitates to the field. The home office control function then becomes one of overall monitoring and reporting, usually through the project manager's office.

Field staff functions

The field scheduling group reports to the construction manager and is responsible for maintaining the construction scheduling activities within the overall project schedule. The key role of the field scheduler is to lay out the work to be accomplished in the field each week. That is usually done in a weekly scheduling meeting attended by all the key construction and subcontractor supervisors. The field scheduler converts the CPM construction schedule to a bar-chart format for the detailed control of the execution of each construction activity in the field.

The field scheduler is in charge of organizing and chairing the weekly scheduling meeting. The CM attends the meeting to monitor progress and arbitrate disputes as they may arise. The field scheduling group also reviews the weekly progress with the field superintendent, area and field engineers, and subcontractor supervisors to monitor and record job progress. The weekly schedule reports are rolled up into a monthly report for inclusion in the CM's field progress report. Problem areas are reviewed at least weekly with the CM to develop effective strategies for solving them.

The field cost-control group reports directly to the construction manager, and is responsible for the field cost-engineering activities. Chief among those duties are monitoring the field cost-control system, field change orders, and preparing the field cost report. In conjunction with the field superintendent, the group also calculates the percent completion of the field work. By balancing the percent complete against the cost, it calculates the productivity of the field forces. If field productivity lags, it is extremely important to get it back on track immediately. Otherwise, the construction budget will be overrun. Monitoring field-labor productivity is probably the most important cost-control activity under the CM's control. We will discuss this in more detail in the next chapter, Project Control.

Although it doesn't show up on the organization chart, both the field scheduling and cost-control groups must have a tieback into their respective home office supervisors. This is necessary to prevent a strong-willed construction manager from dominating the field cost and schedule groups, and forcing them to issue overly optimistic cost and schedule reports. Such an idea may seem farfetched, but exactly that situation occurs quite often. Home office PMs and construction supervisors should monitor that area of organizational conflict of interest very closely.

Field engineering group

Another important staff function is performed by the field engineering group, which directs the technical activities in the field. It receives the technical documentation from the design group and distributes it to the field team. All field revisions and as-built drawing changes are handled by this group, along with any technical interpretation of the drawings for the construction people. The field engineer is the liaison point between the field and the design office for design clarifications and changes.

The group also handles the field survey work needed at the site and sets the lines and grades. It also serves as the quality-control inspector of the ongoing construction work, and provides liaison with the owner's inspectors.[5]

Doing the construction quality control with the field engineering group is another one of those potential conflict-of-interest areas. In the absence of an owner's or outside inspection team, it places the field organization in the position of inspecting its own work. Pressure from field organization peers and the field schedule can sometimes be placed on the field engineer to accept less than the specified quality of work. Those situations place a strain on the ethical practices of the CM and the field engineer. Owners should keep that situation in mind when arriving at the degree of quality assurance they need on their construction projects.

The scope of field inspection services ranges from planning the inspection program to final facility checkout and acceptance. In between, it includes checking incoming materials and equipment, erected equipment and systems, field welding, steel alignment, civil works, and laboratory inspection services. Proper planning of the inspection work and its associated budget is critical to meeting the allocated field indirect costs.[5] The inspection program must be designed to catch any errors early before costly corrections are needed!

Field change orders are an important part of the work handled by the field engineering and the cost-control groups. Engineering makes the technical input, and cost control furnishes the cost-engineering portion for the field change orders. The CM must alert all key staff members to call the field engineer's attention to legitimate field change orders.

The position of field engineer is another possible springboard into the construction manager's position. With some additional training in the construction area, the field engineer gets a good all-around exposure to construction technology and management. The CM's duties in key staff development should not be neglected in that vital area.

The office manager is the final staff person to round out the construction manager's staff. He or she reports directly to the construction manager and handles the administrative duties in the field office. These include field personnel administration, accounting, materials management, subcontract administration, site safety and security, and the field warehouse—seven key areas of administration, which will push the span of control of most office managers to the limit. On a medium-size to large project, the office manager will need a first-class team of section managers to keep the administrative staff on top of its work.

Personnel administration on a construction project is an especially crucial area for project success. The project will be no better than the quality of the people supervising and executing the work. Having sufficient people on the job also has a direct relation to maintaining schedule. In short, having enough of the right kind of personnel at the right time is essential to safely meeting the project goals of specified quality, on-time completion, and meeting the budget!

The field labor posture of open versus union shop has a direct bearing on meeting the craft-labor requirements. Either method usually results in a local labor agreement that outlines the project working rules. The necessary site-labor documents are developed as a joint venture by top management, the CM, the personnel manager, and major subcontractors. On union-labor projects, the documents are also signed by the relevant local and national labor officials.

The factors affecting craft labor supply are infinite, but the two most common are job location and level of economic activity. If the job is located in a labor-rich area in poor economic times, the labor supply will be excellent. Conversely, the worst case occurs in a labor-poor area in booming economic times, making good labor scarce.

Local contractors working in a restricted locale have no problem staying informed on the local labor supply. Larger contractors working a specific country or the whole world have a different problem. In developed countries, state and federal governments usually keep construction labor employment figures, but generally they are not very current or reliable.[4] The reports are often 6 to 12 months after the fact, so they don't represent the current situation, which can fluctuate widely in that period of time.

The best thing a CM can do is to participate with the personnel people in the site-labor survey during the bidding stage, and formulate the project's labor needs. The CM and the personnel department can then evaluate and plan for the potential craft-labor sources. The important thing to remember to do after contract award is to ensure that the craft-labor supply plan is credible. Don't accept a pie-in-the-sky plan that has only a limited chance of working. The plan can vary from sorting out the best people from the stream of applicants at the gate to setting up craft-training programs on the job site. The first is the cheapest and the latter the most expensive. Remember, a poor craft-labor supply can sink the project budget in more ways than one!

The integrated project organization chart

If we assemble all the examples that we have discussed in this section, we can readily see that even a small project can involve a large number of people. On a relatively small project of $10 million, we could have a design team of 25 people. Combine that with a construction team of 75 for a total of 100 people contributing to the project. On a project of $50 million, the number becomes around 500 people. Each person's project activities must be channeled into effective work if project expectations are to be met. This points up the importance of designing an effective project organization chart as the first step in building an outstanding project team. Designing the chart is, however, only the first step in building the team.

Selecting and Motivating the Project Team

We said earlier that we first designed the organization to suit the work to be done; then filled the organization from the pool of people available. In the ideal situation, we would like to have two or three candidates presented to us for each key position. Unfortunately that doesn't happen very often, except in periods of low work load. Even in those periods it doesn't happen often, because the staff has already been reduced to meet the existing low-work-load situation.

Selecting the project team

If you are an owner who has gone into a contractor's shop to get a job done, you should have picked two or three of the top people during the contractor selection process. You then worked closely with those people to fill the rest of the key slots with capable people. If you are a contractor's project or construction manager, you will have to select the people offered by the department heads or manager of the construction department. If you are fortunate enough to have worked with the key people before, you can easily make a decision as to the candidates offered. If you don't know the candidates, you should read their resumés, interview them, and check out any available references.

After you have established a candidate's abilities to do the job, you must look at how the person will fit into the team. Does the candidate subscribe to your management m.o.? Will he or she fit in with the other players and the client? Don't select anyone who you feel has a stronger loyalty to a department head than to you or the project team. Don't, under any circumstances, fill a key slot with anyone who isn't a team player! You will only have to change the person later, at some cost to the organization.

Having said all of the above about selecting only quality people, we also know that pulling it off is very difficult and even a matter of luck. You will have to make some compromises and accept some lesser-quality performers. If possible, it is best to blend them into less critical levels of the organization where they are least likely to hurt project performance.

If you are forced into taking some marginally qualified people at key levels, back them up with strong people who will require less supervision. Also, it may be possible to place them in positions reporting directly to you, which gives you an opportunity to develop them on the job. If certain people show good potential, don't be afraid to take a chance on them.

As the selections are made, add the names to the organization chart. It may not be possible, or even desirable, to fill all the slots on day 1 because of a slow project buildup. Issue the incomplete chart anyway, because it will help to get the organization working together as well as help to orient new arrivals on the project.

Motivating the project team

The first motivational tool is to establish the project goals and instill them into the minds of the key players. In addition to the general goals of quality, budget, and schedule, some project-specific goals need to be formulated and written down. They may involve key milestone dates or budget targets that must be met to earn a bonus. It may be a target to design or construct a higher level of project quality to meet a fixed project financial package. The client may have set some unusually tough aesthetic design and construction standards, or there may be some tough environmental goals that have to be met within a tight schedule and budget.

After the specific project goals have been selected, they should be discussed by the team members to generate ideas on how they can be met and on the role to be played by each member in meeting them. Through those discussions, we can get each team member to buy into the goals and become committed to meeting them.

The early project-goal-setting activity is probably the most overlooked and underrated activity in project initiation and team building. Without the setting of project goals and continuing reinforcement, the likelihood of meeting the goals is indeed slim. I urge you to remember that this goal-setting philosophy applies equally to every facet of the project as well as to the major areas of design, procurement, and construction!

Having a system of project bonuses for key team leaders is also an effective tool for motivation. A program of company-paid social functions can be used as incentives when craft groups have met their goals. The basis for the incentive bonuses should be simple and clearly stated, and the bonuses should be paid at the point in the project when the goals have been met. I am sure we all have seen project incentive bonuses evaporate while we are winding down the project!

Writing job descriptions

The project team job descriptions should not be the standard claptrap found in the departmental files! The format I propose that you use is one that is project-specific to each player's accepted goals on the project team. They will not be usable from project to project, so you won't find them on file anywhere.

An interesting approach to preparing the project job descriptions is to have the team members start by writing their own. When they present the drafts for review, you are sure to be surprised by just what *they* think their jobs are. The chances of developing good job descriptions are improved when you work out the final drafts together. The final drafts must be typed and made a part of both the project organization and the team-member files.

A job description can be designed to be a basis for a management by objectives (MBO) program. MBO is a tool that we can borrow from general management to use in managing our projects. If your firm has a companywide MBO program, so much the better. You can slip your project MBO program right into its niche in the overall company system, and your people will already be knowledgeable as to its use.

Making the MBO system work

Most of the groundwork for a project MBO program already has been done. The objectives for the program already have been set by selecting and stating the specific project objectives (project goals) and getting the team members to buy into them. Writing the objectives into the job descriptions gives everyone involved a clear and permanent statement of what is expected. All that remains is to set up the review procedure and a schedule of dates to monitor the subject's performance against the objectives. It's up to the CM, as project leader, to see that the MBO program is carried out to project completion.

For integrated project results, the CM and the key project leaders should carry the MBO program down through the organization chart to cover all project objectives. This means getting down to the cutting edge of the work, be it in design, procurement, project control, administration, construction, or any other key area of the project. Although project and construction managers ultimately are held accountable for project performance, the MBO system forces them to delegate the responsibility for meeting specific project goals and to force the decision-making process as far down into the organization as possible.

The motivation engendered by delegating the responsibility for meeting the goals throughout the organization goes a long way toward building a spirited project team. That sort of project climate makes the practice of leadership much easier, while keeping a strong hand on management of the project. Naturally that motivational spirit must be revitalized from time to time during the life of the project. A good time for motivation-boosting is during the one-on-one MBO objectives review meetings. Group morale revitalization also takes place through well-managed routine project meetings and goal-meeting reward functions.

Project Mobilization

Project mobilization is a critical time in the birth of any project. Everyone is gung-ho to get the project started. The client, the company management, and just about everyone wants to see some dust flying just to feel like something is going on in the field. Don't let that sort of activity panic you into bringing supervisory or craft people into the project too early. Construction projects are almost universally task-force-oriented, which means that the people who are assigned to your project will be charging full-time to it whether they are working productively or not. A few home office services people may also be assigned to your project on a part-time basis, so it's necessary to monitor their activities as well.

If you happen to be working for a large contracting firm, you are likely to have a large central personnel department headed by an Industrial Relations Manager. That group usually has personnel, safety, security, labor relations, and field construction methods sections. The personnel section of that group maintains a computerized database of past employees and potential hires for all classifications of workers. A good description of the operation of a personnel (industrial relations) department in a union shop setting is given in Chapter 23 of Frein's *Handbook of Construction Management and Organization.*[2]

If the project starts off in a personnel-glut situation, people may be forced on you before you are ready to absorb them. The other side of the coin is seen in periods of personnel shortage, when you are scouring the organization for good people to add to your project. Naturally, each of the conditions requires a different approach.

In a case of oversupply, don't take people early just to satisfy some department head's desire to cut departmental overhead. Bring your key people on board first to assist in the project-planning phase, as we discussed earlier. It's important to have a good and continuing flow of work available before bringing the main body of supervisors and labor. That applies to any phase of the project, be it design, procurement, or construction. If there is a personnel shortage, you may want to consider taking on a few good, key people sooner if they are likely to be snapped up by another project. You will have to weigh cost versus potential benefits. Sometimes the investment will pay off.

The advice given for project mobilization works in reverse when the project is destaffed as it nears completion. People should be released to their respective departments or other projects just as soon as they have completed their assignments. Demobilization is no problem in periods of personnel shortage, but it can be a problem in periods of surplus. In the latter case, the people must be released to protect the project budget. Don't park unneeded personnel on your project just to make life easier for other managers.

The most serious project-staffing problems usually arise during periods of personnel shortage. If the work is available, the schedule clock is running, and the project personnel are not forthcoming, you must take effective action to get your necessary staff. Make your staffing needs known to the personnel providers as soon as possible to give the suppliers ample lead time to locate the human resources. Also, make your request in writing and provide copies as high into the organization as necessary to get the desired results.

Once the field supervisory organization has been staffed, it is difficult to make major shifts in loading during the execution phase. People who are released temporarily in the middle of a project often will not be available when they are needed again. Where the people are working on out-of-town sites, it's not feasible to relocate them to another site and return them later. That's a nasty problem that occurs during unforeseen project suspensions or slowdowns, and it can have disastrous side effects on project budgets, productivity, and morale. There is virtually no way to manage slowdowns out of existence. At best, one can minimize the damage with an intelligent approach to reducing the staff and restaffing when the need arises.

When a major destaffing problem occurs in the middle of a project, all major players—the owner, contractors, and central engineering groups—must participate in finding the best solution to the problem. The project and construction managers must inform interested parties of the short- and long-term effects of early destaffing on the execution of the project's master plan. An agreement whereby the project can be reorganized with minimum damage to all participating project partners must be reached.

If the need for reorganization was not caused by the project itself, the project and construction managers must take immediate action to protect the original project budget with a claim for relief under the contract. As defenders of the project budget, it is no time for the contractor's project and construction managers to be fainthearted!

Updating the organization chart

The organization chart should be kept up to date as project personnel changes occur in the course of the job. As we will discuss later, the organization chart and staffing become part of the Field Procedure Manual (FPM), which also is subject to continuing updating.

The project organization is a living organism, so it can be expected to change as the project proceeds. I don't recommend that wholesale organizational changes be attempted in the middle of a project if they can possibly be avoided. Major organizational surgery should be done only when it is absolutely necessary to correct catastrophic organizational problems. That doesn't mean that personnel changes shouldn't be made if necessary.

Construction organizations

It is necessary to discuss the various methods of project execution available to us because it's impossible to tell how the organization operates just by looking at the organization chart. The presence of the design-build concept forces us to define the terms *task force* and *matrix* organizations.

Construction organizations are automatically forced into a task-force mode by the nature of the work. Because most construction sites are remote from the home office, a project construction group has to be organized to build each specific facility. Some support services for a local construction project, such as personnel, procurement, and accounting, may be done from the home office. The CM coordinates those activities over the phone or through visits to the local headquarters office.

On larger out-of-town jobs, the CM should strive to have all the project functions in the field office. This shortens the lines of communication and gives the CM a totally dedicated task force that can concentrate all its efforts on that specific job. In either case, the CM has complete responsibility for meeting the project construction goals.

When that project has been completed, the construction team is disbanded and the personnel are sent to other construction sites or laid off. A completely new task force is designed and organized for the new project.

Design-build organizations

This contracting mode introduces a significant home office input in the form of the design team required for the project. Design teams have the option of being operated in task-force or matrix mode. Because the supervision of the design group may or may not be the CM's responsibility, I am not including the discussion of that section here. If you become responsible for organizing a significant design effort, you can refer to Chapter 7 of my book *Total Engineering Project Management* for a discussion on organizing a design team. As with construction, I usually recommend that design teams be operated on a task-force basis for the best results.

Organizational Procedures

A key CM responsibility in organizing any construction project is developing and issuing the Field Procedures Manual (FPM). That document lays the ground rules under which the field organization will function in executing the work. Each company will have a different standard to which the field procedures are prepared. The procedures should spell out the minimum regulations under which your company management wants its CMs to operate. They can vary from being too simple to offer good project control to being so ponderous that efficient execution is next to impossible. A good FPM does give the CM a sole source for the procedures in effect on the project and how the various field functions relate to each other. Please remember the KISS principle for this task. The FPM should be as simple as possible to suit the work being controlled.

The Field Procedure Manual and Contents

The heart of the operating procedures for any construction project is the FPM. The CM has the prime duty of seeing that the FPM is produced on time and that it works effectively for the life of the project.

Almost every company has a standard table of contents for the FPM that is geared to its type of work. In the event your firm does not have such a standard, I have included a typical one for a design-and-construct project in Fig. 7.9. The sample can be expanded or contracted to suit the size and complexity of your particular project. The manual for a small, uncomplicated project can be just a few pages, whereas one for a large project usually runs into one or more looseleaf notebooks.

The introduction

The introduction should contain a statement of purpose for the project. What is the owner hoping to accomplish with the project? What needs is the project going to fill in the community, industry, or market? The people involved with the work need to know what the owner's goals are. Also, it is always well to include a statement to the effect that the FPM does not replace the contract and that any conflict between it and the contract will be resolved by the contract.

The project description

A project description gives the location of the project, a site description, an overview of any processes involved, and any other outstanding features of the project. An outline of the scope of work and the

Table of Contents

1.0 Introduction
- 1.1 Statement of purpose
- 1.2 Contract controlling statement

2.0 Project description
- 2.1 Brief description of project
- 2.2 Location and site description
- 2.3 Scope of company services
- 2.4 Work by others
- 2.5 Owner's responsibilities
- 2.6 Owner's project objectives
- 2.7 Contractor's project objectives

3.0 Contractual matters
- 3.1 Type of contract
- 3.2 Secrecy requirements
- 3.3 Checklist of reimbursable and nonreimbursable charges
- 3.4 Subcontracting procedures
- 3.5 Special contractual requirements
- 3.6 Statement of guarantees and/or warranties
- 3.7 Applicable codes and government regulations

4.0 Project field organization
- 4.1 Project organization chart
- 4.2 Brief project job descriptions
- 4.3 Client project organization chart
- 4.4 Brief client job descriptions
- 4.5 Organization chart of companies or divisions
- 4.6 Client's resident field team (if applicable)

5.0 Project personnel policies
- 5.1 Statement of craft labor posture
- 5.2 Site labor agreement
- 5.3 Project hiring practices
- 5.4 Site personnel records
- 5.5 Travel and expense account policies
- 5.6 Labor relations procedures

6.0 Project coordination
- 6.1 Communications procedures
- 6.2 Communications systems
- 6.3 Key personnel names and addresses
- 6.4 Correspondence logging procedures
- 6.5 Document distribution schedule and transmittals
- 6.6 Document approval procedure
- 6.7 Meetings and preparation of meeting notes
- 6.8 Telephone and verbal information confirmations
- 6.9 Project language (international projects)
- 6.10 Project filing system
- 6.11 Reproduction procedures

7.0 Planning and scheduling
- 7.1 Contracting strategy
- 7.2 Project execution plan

Figure 7.9 Sample table of contents for field procedures manual.

Table of Contents, Cont'd.

Figure 7.9 (*Continued*)

Table of Contents

13.3 Assignment of safety program responsibilities
13.4 Local ambulance, fire, health, and police facilities
13.5 Safety training and meetings
13.6 Applicable government regulations
13.7 Project security plan

14.0 Field engineering procedures
14.1 Field surveying procedures
14.2 Field, owner, and design firm interface procedures
14.3 Shop drawing submittals and approvals
14.4 Construction and quality-control procedures
14.5 Inspection and laboratory services
14.6 As-built drawing procedures and records
14.7 Field change-order preparation and control
14.8 Design documents control and records file
14.9 Photographic or videotape records

15.0 Change-order procedures
15.1 Contractual requirements
15.2 Change-order format
15.3 Change-order log
15.4 Approval procedure

16.0 Computer services
16.1 Scope of hardware systems
16.2 List of project software programs
16.3 Computer billing rates

17.0 Design procedures (for design-build projects)
17.1 Drawing and document formats
17.2 Document and equipment numbering systems
17.3 Units of measurement
17.4 Applicable design standards and codes
17.5 Design document schedule and control
17.6 Design quality-control procedures
17.7 Use of modeling or computerized drafting techniques
17.8 Material requisitions
17.9 Technical bid evaluation procedure
17.10 Vendor data procedures
17.11 Design subcontract procedures
17.12 Preparation of mechanical catalogs

Figure 7.9 *(Continued)*

services offered is important to the general knowledge of the team members. It can then be neatly tied into the project objectives, which form the basis for the project MBO program. All the goal-oriented groups involved on the project should be covered in this section, including any project-team performance incentives. Any work by others involved on the project, including major subcontractors or licensors, along with their contributions to the project, should be mentioned here.

Contractual matters

Since the contract is a quasi-confidential document, the key areas affecting project performance should be included in this section. The people who are working on the project but who will not have access to the contract need to know how the contract can affect their work. For example, it makes a difference to the project team's performance whether the contract is on a lump-sum or a reimbursable basis.

Any requirements for project secrecy or confidentiality must be addressed in the FPM. All members of the team need to conform to the regulations for security and secrecy agreements, including the handling of confidential documents and equipment.

Project organization

This is where we cover the project organization charts, work descriptions, and any information pertaining to organizations involved with the project. If there are any special organizational interfaces, they should be described in this section. For example, the design-construction interface should be covered here. You may also want to include the key project personnel job descriptions here.

Project personnel policies

This section covers the project labor policy and the handling of the related personnel policies for the project's craft and supervisory people. The hiring practices can go into such highly sensitive subjects as prehiring and on-the-job drug testing and substance abuse, all of which are key to having a safe project.

Project coordination

The main part of the project coordination section covers the communication procedures for the project. The key names and addresses and the correspondence logs are set up to expedite the handling of project communications. Logging the huge volume of letters, memos, transmittals, and minutes of meetings generated during the project expedites the location of vital correspondence when needed later.

Minutes of meetings and confirmations of project information that has been transmitted verbally are critical to maintaining control over the project scope and design. Often such oral communications result in project scope changes that can have significant effects on the budget.

The document distribution schedule, which sets up who gets copies of correspondence, drawings, specifications, and so on, plays a key role in controlling the project. It establishes the budget for the project reproduction costs, which can be substantial on most jobs. Constant vigilance on the part of the CM is necessary to keep that perennially self-expanding cost item under control.

Document approval procedures are the key to controlling project progress. They should be set up with reasonable but fixed time limits for the approval process. If approval has not been forthcoming by the time the limits have expired, the work should be allowed to proceed without it. Because clients or their agents do most of the approving, they are the ones who need to agree to such an arrangement.

Having a standard project filing system is a big help in organizing the work in the field office. Having a standard project filing system throughout the company makes for easier access to project information by key team members as they move from project to project. It's a relatively simple item that pays big dividends when it comes to meeting the firm's financial goals.

Planning and scheduling

Planning and scheduling comprises a key area that has to be decided on early in the construction project. Quite a bit of generalizing has probably gone into it up to this point. Now is the time to crystallize all the prior thinking about scheduling and to set down the detailed procedures to be followed for scheduling the construction of this project. Agreement with the client also is critical in this area.

Pay particular attention to the item of establishing an *earned-value system* for reporting project completion in the status reports. A simple cost-effective approach is essential to success in this area.

Project procurement procedures

The procurement procedures section lays out the work plan for the procurement and delivery of the physical resources for the project. We are speaking of a procedure to control about 30 to 40 percent of the total project budget, so this area deserves a good deal of management attention. The starting point is an approved vendors list, an often overlooked item. If inquiries are sent to ill-chosen vendors, the whole procurement chain will suffer.

As I will state several times in this book, do not for any reason slight the procurement effort on your project, because it plays *such* an important part in attaining your project goals! I have seen too many CMs mistakenly consider procurement a quasi-clerical function unworthy of their valuable time.

Field warehousing procedures

This section covers that portion of the materials management master plan that begins when the materials start to arrive in the field. It needs to cover the physical storage facilities and the procedures for controlling the materials passing though them. The cost of the materials passing though the field warehouse is a large part of the project budget, so it can really affect the project costs.

Heavy construction equipment and small tools

This is a key section on those projects with a major input of heavy construction equipment and machinery. Heavy-equipment costs are continuous as long as the equipment is on the job, so its management is crucial to good cost control. The timely availability of the machinery is also critical to meeting the project schedule.

Project estimating

Estimating is the foundation of the project financial plan, so it must be well conceived if the money on the project is to be controlled. Many owners do not like to spend money on cost estimating because it adds nothing visible to the finished product. That makes the selection of sound estimating procedures even more critical when it comes to meeting a tight budget. You will have to be creative in developing this section of the procedures to get the best handle on the project cost within the limited estimating funds available.

Project control and reporting

Project control and reporting is generally the largest section in the FPM, because there is a lot of ground to cover. It is also a pivotal section, because failure here can cause loss of project control, which is sure to result in unmet project expectations. We will be covering most of those subjects in more detail in the next chapter, so I will not dwell on them here. The CM plays a key role in all the activities listed, but he or she can also delegate a great deal of the work to project team specialists. In that case, however, the CM becomes the editor of the material generated by the specialists. It is important to read and check the procedures for content, writing style, conflicts, and project goal criteria before releasing them for publication. It will be your first chance to evaluate the ability of your key project staff leaders to communicate! Particularly important items in the section are the cost-control procedures, the project budget, project reporting, and project accounting. At a minimum, they will appear in the FPM for most projects.

Site safety and security

This section spells out the safety program and procedures to be followed for the specific project. It starts with a statement of the owner's policies and then lays out a program to meet those needs. The viability of the safety and security programs starts with good input to this section of the FPM.

Field engineering procedures

These procedures define the operations to be followed by the field engineering group, which plays such a key role in quality control. It also plays a key role in managing the design-construction interface and change-order procedures, which are very project-cost-related factors.

Change-order procedure

The change-order procedure could be included in the FPM controls section, but some people consider it important enough to give it a section by itself. Change orders are the bane of a project's existence. No one connected with the project likes to talk about them, and some project participants even refuse to believe they exist. Like most other problems, however, change orders cannot be swept under the rug, and they do have to be disposed of before the project closeout. Even in a relatively painless cost-reimbursable contracting environment, change orders must be recorded to keep the project budget current and to calculate the fee.

The legal language for changes in scope used in the contract is usually clear enough and does admit the existence of change orders. That language is a good starting point for writing a detailed procedure for handling changes. Perhaps the key clause to use here is the one that states: "No work will be started on the change until the parties agree on the scope, schedule, and cost of the additional work." Since changes happen after the job starts, they usually have a considerable impact on schedules. The above clause gives the CM some leverage in forcing a decision on acceptance or cancellation of the proposed change because it is holding up the project schedule.

In actual practice, however, the revised work often does proceed in order to avoid delaying the schedule. When contractors proceed with unapproved change orders, they are placing themselves at financial risk. I will expand on the processing of change orders in later chapters.

Computer services

The growth of computers in construction work has led us to make computer services a separate section. For small projects, it could fit in the project-controls section. For larger projects, it usually rates a section of its own. In any event, money is involved, so give computer services plenty of thought before deciding on the scope that the project will bear.

Design procedures

The design procedures section plays an important part in any FPM involving design work done in the field. The first part of this section covers such mundane matters as drawing formats and numbering systems. However, when they are not properly thought out to suit the particular type of work being done, the problems will nag you throughout the entire project. That is enough reason for the CM to give them proper attention.

Selecting the applicable project design standards and codes involves legal matters and money, so it must concern the CM. The design firm and the owner are legally bound to meet the minimum code requirements of the area in which the work will be done. If codes and standards are improperly selected or applied, expensive rework can result—with disastrous consequences to project performance. As we discussed in Chapter 1, the CM stands at the center of the construction team target!

Quality-control procedures are established in the FPM for all to read and subscribe to for the duration of the project. The company's reputation is riding on this one, so the procedures must be both results-oriented and cost-effective. This section of the FPM, like most sections, must constantly be monitored for performance.

The ground rules for the critical interface between design and procurement must be covered in this section. Work in that area involves technical bid evaluations and approval of vendor drawings, both of which are critical to equipment and material delivery and therefore to project schedule!

Project relations among design, procurement, and construction personnel sometimes get edgy because of turf disagreements over supplier contacts. Many members of the project team need to have contact with subcontractors and vendors, so some diplomacy on the part of the project and construction managers is required if the groups are to keep working effectively toward project goals. Design and field engineers should stick to the technical aspects of the buying activities and leave the commercial aspects to procurement.

As was mentioned several times in Chapter 4, Construction Scheduling, the timely availability of vendor's data is the single most critical need for meeting design and equipment delivery schedules. Here is where the procedure to make that happen is developed. Make

sure the procedure is simple and workable, and follow up on it periodically to make sure that it keeps on working.

If there are going to be any engineering subcontracts, the coordination between the prime and subcontract groups must be worked out
here. Don't assume that it will take care of itself, because it won't!
Here it is best to assume the worst that can happen and to try to
develop procedures for minimizing any potential hassles before they
happen. The method for setting up and monitoring the design document schedule and the control system falls into this area. If the
method is standard within your organization, just make sure that it
will work on your particular project. Sometimes it will require modification if it is to meet your project's special needs.

Issuing the FPM

A key factor to remember in issuing your field procedures is to get
them published as early in the project as possible. An issue date more
than four or six weeks after project kickoff is too late. Issuing the
FPM with "holds" to be cleared up in later issues as the information is
finalized is quite normal, so don't be made late by trying to perfect
the first issue.

Early issue of the FPM is an excellent project personnel indoctri-
national tool: it gets the new team members up to speed in a hurry.
It's essential that they learn the "who, what, when, where, how, and
why" of the project without any false starts. This is especially true
when a particular group of people may not have worked together as a
team before.

Interoffice Coordination Procedures

I am including a section for interoffice office procedures, because at
some time you may be involved with a construction project that is
being run on a split-office or site basis. We have already noted the
potential for an increase in problems when a split basis is being used.
An example of that situation occurs when modular units are being
used to build the job. The only way to minimize those problems is to
put in place a good coordination procedure to organize and harmonize
the work at both sites.

A good starting point for the interoffice procedure is to use the FPM
and tailor it to suit the interoffice operations. The satellite office has
to perform the same functions as the prime office, so the systems
should be made compatible at the outset. In areas in which the same
systems will not fit for some special reason, a workable adaptation
must be made. The differences must be minimized to the highest

degree possible, in order to maximize the opportunities for meeting the project goals.

A typical case for an interoffice procedure occurs when a third-party constructor has a contract directly with the owner. When the design firm has a commitment through the construction stage, a detailed coordination procedure is needed. The division of work for handling material deliveries, design modifications, drawing interpretation, responsibility for start-up, and so on, must be resolved early in the project.

Summary

This chapter has given CMs an insight into the organization of the human resources, materials, and procedures crucial to meeting the project goals through successful execution of a capital project. The section dealing with building the project organization involves many human factors that sometimes do not come easily to CMs. Leadership in those areas cannot be delegated to subordinates, so CMs must train themselves to overcome any weakness in that area.

The area of project procedures is an excellent one for delegation of the many detailed procedures that need to be developed. A strong input from the CM is needed to guarantee a uniform, effective set of working rules designed to make the project run smoothly.

References

1. Robert D. Gilbreath, *Winning at Project Management,* John Wiley and Sons, New York, 1986.
2. Joseph P. Frein, ed., *Handbook of Construction Management and Operation,* Van Nostrand Reinhold, New York, 1980.
3. James J. O'Brien and Robert G. Zilly, *Contractor's Management Handbook,* 2d ed., McGraw-Hill, New York, 1991.
4. The Business Roundtable, CICE Report D-5, 1982, *Labor Supply Information,* 200 Park Ave., New York, NY 10166.
5. Dan S. Brock and Lystrel L. Sutcliffe, Jr., *Field Inspection Handbook,* McGraw-Hill, New York, 1986.

Case Study Instructions

1. Draw up an overall organization chart for your selected project. Include the key interfaces such as: owner, designer, procurement, and construction pertinent to your project.

2. Draw a second organization chart showing the details of the field execution team. Write brief job descriptions for the key members of the organization including line and staff groups.

3. Apply the job descriptions of (2) to a proposed MBO program for your project. Explain how you would set each individual's goals and the schedule for reviewing them.

4. Explain how you would go about filling the key field positions from the human resources available to you from your organization. Explain how you would handle requisitioning the necessary craft personnel in accordance with your craft-labor posture.

5. Prepare the table of contents for the Field Procedure Manual for your selected project. Discuss why you included or deleted specific items from your table of contents.

6. Assume that you were forced to split the construction work on your project between two sites. How would you handle the procedures necessary for such an operation?

8

Project Control

Project control is the pivotal activity that ties all the previously discussed project management techniques together. Planning and organizing are certainly important in leading us toward meeting our project goals, but effective project control is absolutely essential. We might be a little off-target on planning and organizing and get away with it, but we can't fail even a little bit in control and hope to come through in one piece.

A definition of control that I think is particularly appropriate for project work is *the work of constraining, coordinating, and regulating action in accordance with plans to meet specific objectives.* We have set our objectives of safely building a *quality project, on time, and within budget.* We have made our time and financial plans, and created an organization to execute them. Now all we have to do is proceed with the constraining, coordinating, and regulating activities that will deliver the desired results.

The Control Process

The basic mechanism of the control function is shown in Fig. 8.1. The diagram is a good analogy for technical people, because we readily recognize it to be like the simple functioning of a wall thermostat. The project-control function is the mechanism that keeps the work of the project on target to meet the goals.

Basically, we start the cycle in the upper-right corner measuring actual performance, which is then compared against planned performance. If there is any deviation (or variance), we analyze the causes. We formulate corrective actions and implement them to correct the variance, then repeat the cycle by measuring the revised performance and comparing it to standard. The process is repeated until the variance has been "tuned out."

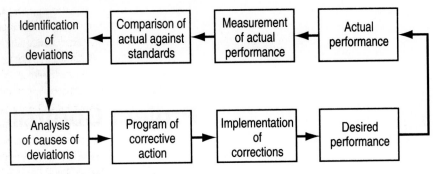

Figure 8.1 The control process.

Areas of control

Which key project activities must we *constrain, coordinate, and regulate* to reach our project goals? The primary control areas are the ones developed in the project planning phase, namely:

- The money plan (the project budget)
- The time plan (the project schedule)
- Quality standards
- Material resources and delivery
- Labor supply and productivity
- Cashflow projections

By concentrating on control of those six key areas, we should be able to meet our project goals successfully. Of course there are a myriad of lesser areas of control, but most of them, with the exception of the *human factors,* are related to the above six areas.

Controlling the Money Plan

The cost-control system lies at the heart of controlling the money plan. Many well-designed cost-control systems have been developed over the years to suit the broad spectrum of the capital projects industry. Regardless of the system used, it will take your single-minded devotion throughout the project to make any system work!

Cost-control definitions

Cost control, despite its simple name, means a lot of different things to different people. Some often-heard synonyms are *cost engineering, cost reporting, value engineering,* and *cost reduction.* None of them alone is equivalent to *cost control.* Let's define a few of the synonyms in order to understand the differences in meanings.

Cost engineering is a generic term that covers the total field of cost estimating, budgeting, and cost control. It is too general a term to use for real cost containment.

Cost reporting consists of gathering the cost data and reporting the actual versus planned results without mentioning the operative word *control.*

Value engineering gets closer to cost control because it looks at ways to reduce costs on specific items or activities. However, it does not look at the total project picture or check the daily performance; it focuses only on specific items in the design, procurement, or construction area.

Cost reduction also gets closer to cost control, and would be fine if it included cost reporting. The result would then be evaluation and containment of costs on a complete project.

As it turns out, true cost control for capital projects involves all of the above activities at various times. To me, cost control means *the purposeful control of all project costs in every way possible.* That means that every member of the project team has a part to play in reducing and controlling costs. The project and construction managers are the leaders of the cost-containment program, and they must constantly reinforce that philosophy throughout the life of the project![1]

Cost-control philosophy

A comprehensive philosophy for cost control that I have developed over the years is based on three building blocks:

- The encouragement and promotion of cost-consciousness in the performance of all phases of the work
- The provision of accurate and timely data on cost status and outlook, and the highlighting of any unfavorable cost conditions or trends
- The taking of prompt and effective action to correct problems and to provide positive feedback for continuous evaluation of those problem areas.

A major problem area in most cost-control systems arises under the second point: "the provision of accurate and timely data on cost status and outlook." Most good project and construction managers are willing to sacrifice minor differences in the accuracy of the cost data if they can get their cost data in a "timely" manner. Unhappily, most cost-reporting activity takes on an *accounting* level of accuracy, which often delays the publishing of the data.

Effective managers are looking for trends in the control of the financial plans, so accuracy within a range of plus or minus 2 to 3 percent is close enough. (Accounting practice could never countenance such inaccuracy in the formal cost report.) Although the age of computers has speeded up the number-crunching a lot, cost reports still take time to check, print, and deliver. Issue dates of 7 to 10 days after the report cutoff date are not uncommon.

This built-in delay forces the CM to develop an "unofficial" system to keep a running monthly approximation of the project cost status while he or she awaits the official cost report. That's not an altogether bad situation, because the construction manager should learn to develop a "continuing feel" for all facets of job progress even before the formal status reports have been issued. If CMs are continually finding *surprises* in their project reports, they are not staying on top of their jobs.

In that regard the CM comes very close to being an entrepreneur running a small business. Successful operators of small businesses are those who develop a sense that *internalizes* their bookkeeping without waiting for monthly statements from their accountants. It is a matter of learning the ratios and rules-of-thumb that apply to your particular area of capital projects.

If the project is overstaffed and people are underutilized, the next cost report will surely show that you are overrunning the personnel budget. That will also apply if the work quality is poor and unbudgeted hours must be spent doing rework. If you wait four to six weeks for cost information highlighting the problem, you will have lost valuable time for taking any corrective action. That *sixth business sense* is one of the chief attributes that separates superior CMs from the herd.

Despite the secondary advantages of untimely cost reports, it's important for the CM to keep pressure on the field control staff to issue reports promptly. Examine the report cycles closely to eliminate any delaying factors inherent in the production cycle. The CM's unofficial gut feeling must still be verified regularly to eliminate possible accumulation of any error in the CM's internalized system.

Cost-control system requirements

If the cost-control philosophy set forth above is to be successfully implemented the control system must include these basic features:

- A simple but comprehensive code of accounts

- Assignment of specific responsibilities for controlling costs within the field organization
- Use of standard forms and formats based on a standard code of accounts throughout the estimating, procurement, design, construction, and cost-control groups
- A sound budget (based on a sound estimate)
- A mechanized system for handling the data on medium-size and large projects

In Chapter 5 we discussed the need for a standard code of accounts, which deals with the formulation of the project estimate and financial plan. If you feel rusty on that subject, it would be a good idea to go back and review the advantages of a standard code of accounts.

The assignment of specific responsibilities for cost control should be delineated in the project cost-control section of the Field Procedure Manual. The description should include the forms to be used, the formats for any reports, assignment of duties, liaison with home office, and any other matters needed to create an effective field cost-control system.

The CM also should stress the cost-control theme when assigning responsibilities in the pertinent job descriptions and the goals in the project MBO program. These are specific assignments to ensure that the philosophy of cost consciousness *is interjected into all phases of the construction work.*

Part II of *Applied Cost Engineering* by Clark and Lorenzoni[1] is an excellent in-depth treatise on cost control from the cost engineer's standpoint. It also shows the working relations between the PM/CM and the cost-control group on the project. I have included a number of references to that book for more detailed descriptions of various points that I want to stress. Since space limitations preclude reproducing those descriptions here, I recommend that you add the book to your construction management library.

The use of the same standard forms and formats throughout all of the project and field groups protects against the introduction of transposition errors when dissimilar forms are used. The extra cost incurred by the errors must also be added to the actual transposing cost. The additional work inflates the home office and field indirect budgets unnecessarily.

If a cost-control system is to be effective, it *must* be based on a sound cost estimate and project budget. If the cost figures being controlled are bad to start with, no amount of cost control can make them right. If that situation occurs, the heart will go out of the cost-control program, and the morale of the field supervisors and the cost-control team will decline sharply.

If the bad budget is discovered early enough in the job, the project and construction managers should press for a new cost estimate and a revised construction budget. It will not be obtained without a great deal of turmoil, but it is the only way to properly control the costs for the remainder of the project. If the bad estimate is discovered on a lump-sum project after the contract is signed, there is not much left to do but to try to stop the bleeding.

With today's easy access to personal computers and standard software, there is no reason not to have a mechanized system for maintaining the cost data on any small or medium-size project. Using your cost code system along with any standard spreadsheet programs will allow you to set up an inexpensive mechanical system for developing and monitoring the construction budget. As I will repeat many times in connection with the use of PCs on the job, you must make sure that a particular person has been designated to keep the input data current. Bad computer output is even worse than bad written information!

If you don't have a mechanized system for a small to medium-size project, do not despair, because a manual system can work just as well. The computer doesn't add anything to the validity of the numbers, it only massages the numbers faster (or so we hope!). Somehow, we all feel that numbers generated on a computer are more believable and have an aura of reliability. Nothing could be further from the truth!

A Typical Cost-Control System

Figure 8.2 shows how the work flows through the major elements of a typical cost-control system. The idea here is to have all the necessary cost information flowing to the cost-control group on a routine basis so that it can be assimilated and organized into accurate and timely cost reports. The diagram includes the design function as it would occur on a design-build project.

The process starts with the project estimate passing through client and management reviews before it is converted into the project budget. As we said earlier, the budget is the basic *money plan* for the project and it becomes the baseline for the cost-control system. The approved budget (the original cost baseline) should never be revised; changes to the *baseline* budget are shown in a modified or *current* budget.

On the left side of the diagram, we show the normal staff departments that feed routine construction costs and expenditures into the cost-control center on a regular basis. On the right side we show the design, procurement, construction, and estimating functions feeding commitments, project change orders, and estimating data into the cost-control center. At the lower right we show that the schedule is

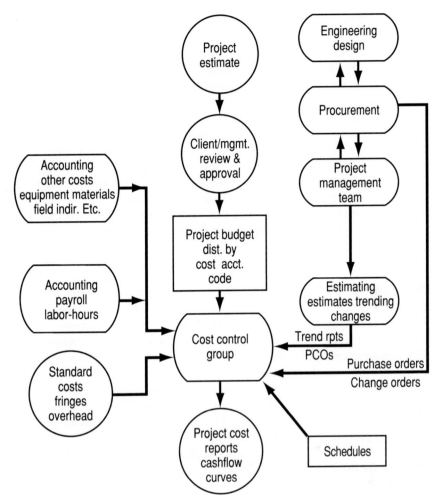

Figure 8.2 Cost-control flow diagram.

regularly issued to the cost-control group so that it can evaluate the effects of schedule changes on project costs.

At least monthly, the cost-control center issues the project cost report and cashflow curves. The idea behind the flow diagram is to have the large volume of routine data flowing in standardized normal channels to ensure that nothing affecting construction cost is overlooked. Missing data can result in inaccurate reports, which lead to loss of control of the project money plan. This in turn will result in unmet project expectations for finishing the project under budget.

Staff functions

The staff functions on the left side of Fig. 8.2 are recording the monies paid out for the human and physical resources that flow into the project. These are mainly accounting functions that are paying for the commitments made by the operational groups on the right side of the diagram.

In cost control, it is important to differentiate between *commitments* and *expenditures*. Commitments are made when we order materials and equipment and have personnel who charge their time to a project. Expenditures occur when the bills for goods and services are paid and payrolls are met. Payrolls convert from commitments to expenditures fairly rapidly, usually in a matter of days or weeks. The time lag between commitment and expenditure for material and equipment, however, can stretch out for months.

Line functions

The activities of the line functions on the right side of the diagram are more variable and much less routine. The progress of the job can affect schedules, which in turn affect costs. Project change orders, with their adverse effect on budgets and schedules, will always arise. The procurement program is the chief source of the longer-term project commitments that show up in the cost report.

The project and construction teams are the center for cost-control activities on the project. As we said earlier, the PM and CM, as the project team leaders, must generate the project cost-control philosophy. The project and field groups provide the coordination needed to make the other right-side groups perform effectively. All the corrective cost-control actions and feedback are processed through the project and construction teams.

In addition to coordinating the design and procurement efforts, the project supervisory teams initiate and process all the design and construction change orders. They are responsible for getting the change orders estimated, approved, and fed into the cost-control center for use in revising the *current* budget.

The project and construction teams also are continuously monitoring the cost-trending reports that are maintained between the various project estimates as the design develops and the construction progresses. Close coordination between the execution teams and the respective cost-engineering groups is vital if the project cost-control system is to be effective.

The visible products of the system shown in the flow diagram in Fig. 8.2 are the monthly cost reports and the cashflow curves. These are usually very detailed reports that account for all commitments and expenditures on the project. Because these cost reports are issued to the client and analyzed by the client's cost-control people on CPFF projects, they must be accurate. The accuracy requirement is what makes them less timely for the day-to-day needs of the project team's cost-control program.

Any off-target trends on cost and cashflow must immediately be made known to the PM/CM on an informal basis, either through a memo or personal contact. After the problem has been investigated and some possible solutions have been developed, the client and the company management should also be informed prior to the issuing of the formal cost report. A quick informal notice saves valuable time in solving the problem instead of waiting for the detailed cost report.

It's always better for the PM/CM to get a problem out into the open early, along with some suggested corrective action, rather than to wait for the client's cost people to find it in the cost report. The first approach will surely make you look like a more effective PM/CM than the latter.

Another key activity that doesn't show up on the flow diagram is the execution supervisor's *estimate to complete* either the labor or materials for the project. This key value, added to the commitments and expenditures, allows the *estimated completed cost* to be compared with the baseline budgets. Because these values are still estimates, their generation must be pragmatic rather than overly optimistic!

How a Cost-Control System Really Works

Let's look into some of the details on just how a typical cost-control system can work on a design-build project. We have already gone through the estimating and budgeting phase; now the project is in the early stages of execution.

The budget is the baseline of the cost-control system, so we must refresh our memories as to what it contains. We can expect the major cost accounts for a turnkey industrial project to be as follows:

Account	Percentage
1. Design services	6–10
2. Major equipment accounts	25–30
3. Bulk material accounts	15–20
4. Construction and field costs	45–50
5. Contingency and escalation	Variable

I have not included a percentage for the contingency and escalation items, because they vary widely from project to project and over time. They should be determined as part of the budgeting process and added in at the end. Adding together items 3 and 4, which are field costs, shows us that most of the project budget is expended in the field. Naturally, the largest cost accounts offer the CM the greatest cost-control challenge on the project.

Controlling labor costs

The first and fourth items just listed consist almost entirely of human resource costs. Variances in labor budgets can stem from three possible sources:

- An original hourly takeoff error
- A variation in the assumed labor rates
- A variation from standard in actual labor productivity

Design and construction labor budgets usually are controlled in the same way on a monthly or weekly basis. Any possible error in the original estimate is evaluated by projecting the number of hours needed to complete the work during each reporting period. If overruns start to appear early in the reports, it's a strong indication that the hours in that area may have been underestimated. The CM's early investigation into such a symptom is vital. Later in the project, labor-hour estimating errors are harder to detect and prove.

All discipline or group leaders are required to estimate the number of hours needed to complete the unfinished work in their areas. It is vital that the estimates be based on real facts about how much work actually has been completed and how much is left to do. The estimates to complete must be based on earned value and *physical percent complete,* not *labor-hours expended.* We will discuss the measurement of physical percent complete and earned value later in the chapter.

Figure 8.3 is a sample computer printout used for controlling field indirect costs in the construction of a typical project. In this case the contract is lump-sum, so cost control is of particular concern to the contractor's CM. The labor estimate was based on the contractor's prior experience on several earlier projects of the same type, and the lump-sum price was based on the contractor's estimate.

The original budget values as shown in the first column in Fig. 8.3 were obtained from the approved field labor estimate. The differences between columns 1 and 3 reflect approved change orders, which add or subtract hours from the original budget.

DIAMOND SHARP CONSTRUCTION COMPANY
FIELD INDIRECTS COST REPORT
LABOR HOURS & FIELD COSTS

CONTRACT NO: S-3030
CLIENT: ABC CHEMICALS, INC.
REPORT NO: 7
TYPE: LUMP SUM CONSTRUCTION

CUTOFF DATE: OCT. 24, 1993
RUN DATE: OCT. 25, 1993
PHYSICAL % COMPL.: 55%

PERSONNEL HOURS

COST CODE	DESCRIPTION	ORIGINAL BUDGET	CHANGES	CURRENT BUDGET	HOURS EXPENDED TO DATE LAST PER.	THIS PERIOD	TO DATE THIS PER.	ESTIMATE TO COMPLETE	PROJECT FINAL COST	CURRENT PROJECT VARIANCE	PERCENT OF TOTAL	PERCENT EXPEN-DED
1000	CONSTRUCTION MANAGEMENT	2500	80	2580	1259	172	1431	1275	2706	-126	4.82%	55.47%
1100	FIELD SUPERVISION	8575	350	8925	3499	860	4359	4200	8559	366	16.68%	48.84%
1200	PROCUREMENT - EXPEDITING	7450	175	7625	3684	675	4359	3140	7499	126	14.25%	57.17%
1300	PERSONNEL ADMINISTRATION	4000	-135	3865	2010	234	2244	1745	3989	-124	7.22%	58.06%
1400	ACCOUNTING - PAYROLL	3690	0	3690	1945	165	2110	1700	3810	-120	6.90%	57.18%
1500	FIELD SCHEDULING	3200	80	3280	1456	185	1641	1620	3261	19	6.13%	50.03%
1600	COST CONTROL - ESTIMATING	3500	150	3650	1687	180	1867	1750	3617	33	6.82%	51.15%
1700	FIELD ENGINEERING	5400	-235	5165	2679	243	2922	2130	5052	113	9.65%	56.57%
1800	STENO - CLERICAL	4500	80	4580	2359	344	2703	1876	4579	1	8.56%	59.02%
1900	FIELD MAINTENANCE LABOR	3500	-25	3475	2243	195	2438	1123	3561	-86	6.49%	70.16%
2000	SAFETY PROGRAM	6550	120	6670	3546	350	3896	2675	6571	99	12.47%	58.41%
2900	CONTINGENCY	2500	0	2500	0	0	0	0	0	0	0.00%	0.00%
	TOTAL FIELD HOURS	2854710	560	53505	26367	3603	29970	23234	53204	301	100.00%	51.84%

FIELD EXPENSES - DOLLARS

COST CODE	DESCRIPTION	ORIGINAL BUDGET	CHANGES	CURRENT BUDGET	DOLLARS EXPENDED TO DATE LAST PER.	THIS PERIOD	TO DATE THIS PER.	ESTIMATE TO COMPLETE	PROJECT FINAL	CURRENT PROJECT VARIANCE	PERCENT OF TOTAL	PERCENT EXPEN-DED
3000	TEMPORARY FIELD OFFICES	20000	9500	29500	27850	276	28126	1000	29126	374	8.41%	96.34%
3100	TEMP. FIELD WAREHOUSE	23450	4500	27950	25675	775	26450	2100	28550	-600	7.97%	94.63%
3200	FIELD UTILITY INSTALLATION	17890	2345	20235	18775	321	19096	785	19881	354	5.77%	94.37%
3300	TEMP. SITE IMPROVEMENTS	26750	-2350	24400	23070	1254	24324	1688	26012	-1612	6.96%	99.69%
4000	UTILITY COSTS	12500	1235	13735	6555	897	7452	6500	13952	-217	3.92%	54.26%
4100	OFFICE MACHINE RENTAL	12450	0	12450	6233	875	7108	5788	12896	-446	3.55%	57.09%
4200	OFFICE SUPPLIES	7500	576	8076	3465	234	3699	3976	7675	401	2.30%	45.80%
4300	LAB & TESTING SERVICES	22540	0	22540	16567	1085	17652	3800	21452	1088	6.43%	78.31%
4400	COMPUTER & SOFTWARE	40000	-1500	38500	32988	1254	34242	4500	38742	-242	10.98%	88.94%
5100	HEAVY EQUIPMENT RENTAL	88900	1200	90100	45688	7659	53347	37500	90847	-747	25.70%	59.21%
5200	SMALL TOOLS	15760	575	16335	14555	875	15430	980	16410	-75	4.66%	94.46%
5300	CONSUMABLES & SERVICES	18000	-1230	16770	8975	1474	10449	6500	16949	-179	4.78%	62.31%
5400	SAFETY EQUIPMENT	9500	0	9500	7566	743	8309	1950	10259	-759	2.71%	87.46%
9000	CONTINGENCY	25000	-4500	20500	13545	750	14295	5560	19855	645	5.85%	69.73%
	TOTAL INDIRECT EXPENSES	340240	10351	350591	251507	18472	269979	82627	352606	-2015	100.00%	77.26%

Figure 8.3 Field indirect cost report on electronic spreadsheet.

DIAMOND SHARP CONSTRUCTION COMPANY
FIELD INDIRECTS COST REPORT
LABOR HOURS & FIELD COSTS

CONTRACT NO: S-3030
CLIENT: ABC CHEMICALS, INC.
REPORT NO: 7
TYPE: LUMP SUM CONSTRUCTION

CUTOFF DATE: OCT. 24, 1993
RUN DATE: OCT. 25, 1993
PHYSICAL % COMPL.: 55%

PERSONNEL HOURS

COST CODE	DESCRIPTION	ORIGINAL BUDGET	CHANGES	CURRENT BUDGET	HOURS EXPENDED TO DATE LAST PER	THIS PERIOD	TO DATE THIS PER	ESTIMATE TO COMPLETE	PROJECT FINAL COST	CURRENT PROJECT VARIANCE	PERCENT OF TOTAL	PERCENT EXPENDED
1000	CONSTRUCTION MANAGEMENT	2500	80	2580	1259	172	1431	1275	2706	-126	4.82%	55.47
1100	FIELD SUPERVISION	8575	350	8925	3499	860	4359	4200	8559	366	16.68%	48.84
1200	PROCUREMENT - EXPEDITING	7450	175	7625	3684	675	4359	3140	7499	126	14.25%	57.17
1300	PERSONNEL ADMINISTRATION	4000	-135	3865	2010	234	2244	1745	3989	-124	7.22%	58.06
1400	ACCOUNTING - PAYROLL	3690	0	3690	1945	165	2110	1700	3810	-120	6.90%	57.18
1500	FIELD SCHEDULING	3200	80	3280	1456	185	1641	1620	3261	19	6.13%	50.03
1600	COST CONTROL - ESTIMATING	3500	150	3650	1687	180	1867	1750	3617	33	6.82%	51.15
1700	FIELD ENGINEERING	5400	-235	5165	2679	243	2922	2130	5052	113	9.65%	56.57
1800	STENO - CLERICAL	4500	80	4580	2359	344	2703	1876	4579	1	8.56%	59.02
1900	FIELD MAINTENANCE LABOR	3500	-25	3475	2243	195	2438	1123	3561	-86	6.49%	70.16
2000	SAFETY PROGRAM	6550	120	6670	3546	350	3896	2675	6571	99	12.47%	58.41
2900	CONTINGENCY	2500	0	2500	0	0	0	0	0	0	0.00%	0.00
	TOTAL - FIELD HOURS	2854710	560	53505	26367	3603	29970	23234	53204	301	100.00%	51.84

FIELD EXPENSES - DOLLARS

COST CODE	DESCRIPTION	ORIGINAL BUDGET	CHANGES	CURRENT BUDGET	DOLLARS EXPENDED TO DATE LAST PER	THIS PERIOD	TO DATE THIS PER	ESTIMATE TO COMPLETE	PROJECT FINAL	CURRENT PROJECT VARIANCE	PERCENT OF TOTAL	PERCENT EXPENDED
3000	TEMPORARY FIELD OFFICES	20000	9500	29500	27850	276	28126	1000	29126	374	8.41%	95.34
3100	TEMP. FIELD WAREHOUSE	23450	4500	27950	25675	775	26450	2100	28550	-600	7.97%	94.63
3200	FIELD UTILITY INSTALLATION	17890	2345	20235	18775	321	19096	785	19881	354	5.77%	94.37
3300	TEMP. SITE IMPROVEMENTS	26750	-2350	24400	23070	1254	24324	1688	26012	-1612	6.96%	99.69
4000	UTILITY COSTS	12500	1235	13735	6555	897	7452	6500	13952	-217	3.92%	54.26
4100	OFFICE MACHINE RENTAL	12450	0	12450	6233	875	7108	5788	12896	-446	3.55%	57.09
4200	OFFICE SUPPLIES	7500	576	8076	3465	234	3699	3976	7675	401	2.30%	45.80
4300	LAB & TESTING SERVICES	22540	0	22540	16567	1085	17652	3800	21452	1088	6.43%	78.31
4400	COMPUTER & SOFTWARE	40000	-1500	38500	32988	1254	34242	4500	38742	-242	10.98%	88.94
5100	HEAVY EQUIPMENT RENTAL	88900	1200	90100	45688	7659	53347	37500	90847	-747	25.70%	59.21
5200	SMALL TOOLS	15760	575	16335	14555	875	15430	980	16410	-75	4.66%	94.46
5300	CONSUMABLES & SERVICES	18000	-1230	16770	8975	1474	10449	6500	16949	-179	4.78%	62.31
5400	SAFETY EQUIPMENT	9500	0	9500	7566	743	8309	1950	10259	-759	2.71%	87.46

Figure 8.3 (Continued)

Columns 5 and 6 show the hours expended during the period (one month) and the total to date, respectively. The hours expended during the period should compare favorably with the CM's *unofficial* prediction based on the number of people working on the project for the reporting period.

The time required to complete the remaining work is evaluated by each group leader and entered in column 7 as the Estimate to Complete (ETC). The computer then adds columns 6 and 7 to get column 8, the Projected Final hours to complete the activity. Any change-order hours approved during the reporting period would be shown in column 3 and added to or subtracted from the Current Budget values.

The computer then compares the Projected Final Hours to complete to the Current Budget to find the variance. In that case, projected overruns show up as positive numbers and underruns as negative numbers. A quick review of the Current Projected Variance column gives the cost engineer and the CM an insight as to where the cost-control problems are in the labor budget. In this case there are overruns in the field supervision, procurement, field engineering, and safety program accounts that indicate potential problems. Although the overruns are not major budgetary problems yet, they cannot be ignored. Also, they probably did not show up in one reporting period, so the CM, cost engineer, and the discipline leader already should have taken some corrective action to reverse the trend.

Although the underruns in other areas of the variance column help to offset the overrun of the total labor budget, the hours cannot be switched now; each discipline's labor budget must stand alone. The overruns may have been caused by work done on unapproved change orders, poor productivity, or project delays. The construction team has to work with the discipline leaders to determine the cause and to take corrective action, or the labor budget may seriously overrun. If so, the result will be a losing project and unmet project goals.

The last two columns of the report in Fig. 8.3 give percents of the discipline hours to the total budgeted hours and the percent of the budget expended. Since the project is about 55 percent expended, there is still time to do something about any overruns. If overruns occur after budgets are 75 percent expended, there is little that can be done to stop the bleeding!

An important rule-of-thumb number, *average hourly labor rate,* can be obtained by dividing the total direct payroll cost by the total hours expended. The number is used to check on how your project is loaded from the standpoint of the personnel mix of craft and supervisory people. If your average project hourly rate is not within the normal range for your firm, your project could be suffering from a poor selection of personnel mix required for the work.

Obviously, the average rate is dependent on the type of project being performed. A high-tech project, for example, usually requires a preponderance of higher-paid crafts to do that type of work. An average construction project in a given construction business sector, however, usually will not stray very far from the norm for the type of project. That sort of ratio must be averaged over a large number of projects and be adjusted, as required, for inflation.

The out-of-pocket costs included in Fig. 8.3 are monitored and controlled similarly to the labor portion. Monthly expenditures are posted, calculated, and monitored in the same way.

Controlling construction labor budgets

The system for controlling construction labor-hours is very similar to that discussed for the field indirect hours. The only difference is that the work is broken down into more detail by craft for better control. The estimate to complete and the physical-progress and earned-value systems are used to calculate the physical percent complete. The craft labor from the foreman level downward is budgeted to the actual construction operations.

Estimating the hours to complete the various labor tasks is again the key number to controlling craft-labor costs and calculating craft productivity. The estimates must not be overly optimistic or pessimistic to effectively evaluate the current situation as to meeting the labor budget. The CM should ensure that the supervisors doing the estimate to complete understand that productivity inevitably declines near the end of the work. Also, many field supervisors believe they can somehow make a miraculous recovery to turn around a losing situation.

Changes in labor rates

Changes in labor rates usually are not a problem in capital project work, because they are fairly predictable. Home office and field supervisory rates change no more than once or twice per year, when salary reviews are scheduled. Since cost-of-living raises are subject to the inflation rate, the amount of increase is predictable. The relatively small percent of merit increases usually averages out against the normal personnel turnover rate within the organization. The potential increases in those rates are factored into the estimate, depending on how long the project schedule has to run. Missing an estimate in this area is likely to occur only in periods of unexpected high inflation or job stretch-out.

The craft-labor rates are even more predictable, because they are often controlled either directly or indirectly by union negotiations with the contractors. Sometimes these labor agreements are negotiated for several years in advance and provide for fixed increases from year to year. Again, the rate increases are fairly predictable, and are factored into labor cost estimates and budgets.

Variations in productivity

Worker productivity is the most likely area in which labor-hour budgets will go astray. Most standard construction unit labor costs are based on average hours used on a number of previous projects. They do allow for an inherent variation in the productivity of various work forces. The labor productivity prevalent in the area of the job site is also evaluated during the estimating stage to arrive at the standard labor hours for a given job. The competitive environment in which most capital projects are estimated and performed, however, works against having a very high safety factor in the area of field productivity.

Most managements will not allow any padding of the standard labor hours, unless an unusually good seller's market exists. Therefore most *standard hours* are assumed to be at 100 percent efficiency. If the construction team's average productivity (efficiency) drops much below 100 percent, the labor budget is going to be overrun!

When we think about loss of productivity, our first thought is that the problem lies in poor quality and performance of our field or craft personnel. That is logical enough, since even CMs are human. However, it is also unlikely that *all* (or even a high percentage) of the people are of poor caliber and not performing up to standard.

We like to overlook the possibility that *management* factors might be contributing significantly to poor labor productivity. It's only human nature to want to find fault with someone else. In the construction area we must look at improper supervision, personnel shortages, missing materials, poor operating systems, incomplete design documents, poor communications, and so on as management's possible contribution to low productivity. The foregoing list includes only a few of the hundreds of possible management failures that can adversely affect productivity. To the list we can add some items peculiar to the construction environment such as unusually bad weather, strikes, unexpected obstructions, labor shortages, and remote site access.

On international projects we can list even more typical productivity killers, such as cultural differences, political unrest, onerous government regulations, and lack of tools and equipment—to name but a few. A case in point in this area is the fall of the Shah of Iran, which caught many foreign companies operating there by surprise. I read some articles that blamed company losses on the failure of the resident CM to foretell the collapse of the regime! Given that sort of thinking, project and construction managers should not be surprised at being blamed for anything at all that goes wrong on or off their projects.

In any event, the above examples point out the importance of the CM getting out into the project trenches to see how the actual work is going. Experience should sharpen a CM's entrepreneurial sixth sense for evaluating how the work is progressing and for judging overall labor productivity *before the labor productivity report has been issued.*

Large construction projects usually can afford a methods engineer to study the overall production methods and special areas needing attention. Smaller contractors usually aren't able to justify the additional expense of the IE approach, so they must depend on their field supervisors for productivity enhancement. However, taking even a limited IE approach to construction productivity can pay for itself by making craft and field management productivity as good as or better than planned.[2,6]

Underrunning the labor-hours

Some people don't consider it a problem when the fortuitous circumstance of underrunning the labor-hours arises, but some comment on how to handle it does seem to be called for. Remember, one of Murphy's laws is that "the amount of money (or time) expended will always rise to the amount allotted."

Some supervisors (like many politicians) think that the hours budgeted to them at the start of the project are theirs to spend no matter what. They can waste the hours in a variety of ways such as getting plain sloppy with executing the work, keeping unneeded people on the job, and doing work outside the scope. CMs must be ever-alert to those situations and preserve any underruns to improve overall project cost performance or to offset overruns elsewhere in the project. Since CMs are judged only on overall project performance and not on individual discipline performance, they have to impress on all field supervisory people that vital motto: *Preserve underruns wherever they occur!*

CMs also are responsible for the efficient operation of their subcontractors. Lump-sum subcontractors are responsible for their own profit goals, but the CM is responsible for any schedule delays subcontractors cause through inefficient performance. This factor is especially critical in the construction management mode using 100 percent subcontracting.

Summary of labor cost control

The total cost of human resources for an integrated project may run as high as 60 percent, including design and field labor. On a design-build project, the design labor cost will be at least 60 percent of the design fee. With field labor running five to seven times design hours, controlling field labor is even more important to overall cost control.

Periodic reviews, estimates to complete, and timely cost reports are critical to getting good results from any cost-control system. The most likely factor causing labor budgets to overrun is poor productivity, and poor productivity more often results from ineffective supervisory management than from unqualified workers. PMs and CMs must find and eradicate the causes of poor management practices that contribute to low productivity. Methods improvement must be used in executing field operations to fine-tune field productivity. Labor underruns must be guarded and nurtured through the closeout of the project.

Controlling Material Resource Costs

The physical resources on a given project can run from 40 to 60 percent of total installed project cost, which makes it a budgetary force to be reckoned with. The human element is present to a much lesser degree than in the labor portion of the budget. We don't have to deal directly with personnel productivity factors when we are dealing with equipment and materials. Prices are controlled by market forces, but people do participate in the timely delivery of the physical resources, so controlling material cost is not entirely devoid of human influences.

The overall philosophy for controlling the physical resources budget is much the same as for controlling human resources. We start with an estimate of the physical resources cost, and convert it into a budget that becomes our baseline for buying them. As the project progresses, we check the actual delivered cost against the estimated cost of each item. The estimate of the cost to complete the delivery of each physical resource occurs in the same way as the control of our human resources. Any differences between "predicted" and "planned" are reflected in a variance column for the project team's use to control equipment and material costs.

Reviewing a cost report

The best way to get an overview of cost control of the materials budget is to review a typical capital project budget as shown in Figs. 8.4 and 8.5. The sample budget that I have selected is from a large petrochemical project that was handled in a computerized format. Manual budgets work the same way, except that they are posted and calculated by people, which makes the manual work a bit onerous on larger projects.

Actually, the total budget for the project consisted of about 100 pages of printouts, but I have selected the two most critical ones that

CONTRACT:
CLIENT:
PROJECT:
LOCATION:

*** COST PROGRESS REPORT ***
MATERIAL STATUS REPORT
SUMMARY ALL AREAS

DESCRIPTION	ORIGINAL BUDGET	TRANS	APPR EXTRA	CURRENT BUDGET	COMMIT PERIOD	COMMIT TO DATE	COST TO DATE	EST TO COMPLETE	PROJ COST	PROJ VAR	X COMMIT
MAJOR EQUIPMENT	2582	0	0	2582	0	0	0	2998	2996	414	0.
0111 TANKS	138	0	0	138	101	172	0	198	370	232	46.5
0112 VESSELS AND DRUMS	1363	0	0	1363	857	1447	0	773	2220	857	65.2
0113 TOWERS AND REACTORS	6646	0	0	6646	0	4030	0	3641	7671	1025	52.5
0114 INTERNALS	1435	0	0	1435	0	0	0	1519	1519	84	0.
0115 HEAT EXCHANGES	4460	0	0	4460	-25	1072	0	4018	5090	630	21.1
0117 COMPRESSORS/DRIVERS	3490	0	0	3490	0	2235	0	356	2591	-899	86.3
0118 PUMPS AND DRIVERS	1674	0	0	1674	4	4	0	1745	1749	75	.2
0119 SPECIAL EQUIPMENT	4958	0	0	4958	634	636	0	3871	4507	-451	14.1
MAJOR EQUIP ESCAL	1246	0	0	1246	0	0	0	998	998	-248	0.
*TOTAL EQUIPMENT	27992	0	0	27992	1571	9596	0	20115	29711	1719	32.3
0121 A/G PIPE	2285	0	0	2285	0	0	0	4685	4685	2400	0.
0122 A/G FLANGES	1281	0	0	1281	0	0	0	1226	1226	-55	0.
0123 A/G FITTINGS	1715	0	0	1715	0	0	0	1695	1695	-20	0.
0124 U/G PIPING MATERIAL	957	0	0	.957	0	0	0	957	957	0	0.
0125 SHOP FABRICATION	2379	0	0	2379	0	0	0	2379	2379	0	0.
0126 A/G VALVES	2661	0	0	2661	0	0	0	2137	2137	-524	0.
0127 PIPING SPECIALTY	1326	0	0	1326	0	0	0	1340	1340	14	0.
0131 FDNS/STRUCT/PVG MAT	1016	0	0	1016	0	0	0	1095	1095	79	0.
0138 STRUCTURAL STEEL	2095	0	0	2095	0	0	0	2131	2131	36	0.
0140 MAJ ELECT EQUIP	3286	0	0	3286	0	314	0	3033	3347	61	9.4
0141 ELECT MATL/DEVICES	1583	0	0	1583	0	0	0	1822	1822	239	0.
0150 INSTR/CONTROL DEV	4020	0	0	4020	0	0	0	4249	4249	229	0.
0151 INSTR VALVES	1242	0	0	1242	0	0	0	1242	1242	0	0.
0152 INSTR/BULK MATL	160	0	0	160	0	0	0	160	160	0	0.
0165 BULK FRIEGHT	390	0	0	390	0	0	0	390	390	0	0.
0166 VENDOR REP	192	0	0	192	0	0	0	192	192	0	0.
0199 STEAM PIPELINE	5040	0	0	5040	0	0	0	0	0	-5040	0.
BULK MATERIAL	6957	0	0	6957	0	0	0	7360	7360	403	0.
BULK MATL ESCAL	3659	0	0	3659	0	0	0	3765	3765	106	0.
*TOTAL BULKS	42244	0	0	42244	0	314	0	39858	40172	-2072	.8
**TOTAL DIRECT MATERIAL	70236	0	0	70236	1571	9910	0	59973	69883	-353	14.2

Figure 8.4 Cost progress report, equipment and materials.

CONTRACT
CLIENT
PROJECT
LOCATION

** COST PROGRESS REPORT **
MONTHLY JOB PROGRESS REPORT
SUMMARY ALL AREAS

DESCRIPTION	ORIGINAL BUDGET	TRANS	APPR EXTRA	CURRENT BUDGET	COMMIT PERIOD	COMMIT TO DATE	COST TO DATE	EST TO COMPLETE	PROJ COST	PROJ VAR	% COMMIT
**TOTAL DIRECT MATERIAL	70236	0	0	70236	1571	9910	0	59973	69883	-353	14.2
**TOTAL SUBCONTRACTS	37633	0	0	37633	0	0	0	26448	26448	-11185	0.
**TOTAL DIRECT LABOR	22203	0	0	22203	0	0	0	17035	17035	-5168	0.
***TOTAL DIRECT COST	130072	0	0	130072	1571	9910	0	103456	113366	-16706	8.7
**TOTAL FIELD INDIRECTS	18797	0	0	18797	0	0	0	16146	16146	-2651	0.
**TOTAL PRO-SERVICES	12933	0	461	13394	551	2733	2733	12833	15566	2172	17.6
**TOTAL OTHER COSTS	6099	0	0	6099	0	0	0	5618	5618	-481	0.
***TOTAL INDIRECT COST	37829	0	461	38290	551	2733	2733	34597	37330	-960	7.3
**TOTAL OVERHEAD	5417	0	173	5590	271	1059	1059	5136	6195	650	17.1
***TOTAL	173835	0	643	174469	2339	13702	3792	143831	157533	-16936	8.7
**TOTAL ESCALATION	0	0	0	0	0	0	0	0	0	0	0.
**TOTAL CONTINGENCY	14001	0	0	14001	0	0	0	13194	13194	-807	0.
****GRAND TOTAL	187836	0	634	188470	2339	13702	3792	157025	170727	-17743	8.0

Figure 8.5 Cost progress report summary.

show the total costs. Figure 8.4 gives a summary of the major equipment and materials accounts. Each of the listed accounts is a roll-up of the detailed listing for all equipment items and further breakdowns of the bulk materials. If the summary numbers for any account show a cost-control problem, the detailed listing should be checked to find the source of the trouble.

The header section of the report is repeated on every page; it shows the title, project, location, and reporting period. The title of the report indicates that this is a summary sheet of all areas, which means that the body of the report lists all materials and equipment for each area of the project. Sectionalizing the report is handy when a project is subdivided into areas for individual supervisors, who can readily use their specific sections of the report to control costs in their assigned areas.

In the column on the far left side are the account code numbers, which are from the standard cost codes used on the project. They are the stem accounts used in the detailed breakdown for equipment. The first of the column heading to the right of the description column is the Original Budget. As was said earlier, this column should never be revised unless it has become necessary to reestimate the project. Any movement of equipment and materials among the accounts is handled with the column headed Trans(fers).

The next column reflects revision to the project in the form of approved change orders, which show up in the cost report only after they have been officially approved. The computer combines the values in the change columns with the Original Budget numbers, which then results in the Current Budget column. The subsequent calculations are based on the current budget number as the latest baseline. Because we are early in the project, no change orders have been approved.

The next series of columns contains the "action items," which report the financial activities as they occur on the project. The first two columns cover the *commitment activities* for the report period and the project to date. Commitments are the values of purchase orders and subcontracts that will spend project funds at future billing dates. No cash changes hands in either of the two columns.

The Cost to Date column shows the actual funds spent against each account. As invoices are received and checks are drawn against the project account, the amounts show up in the column representing the actual cashflow for the month and the project to date.

The Est(imate) to Complete column represents the cost of bringing that account to completion in accordance with the project scope. In equipment and material accounts it represents the money which has yet to be committed for unordered physical resources or subcontracts. After all project materials and equipment have been committed under fixed-price orders, a small sum should be held in the Est(imate) to Complete column to cover any unexpected costs that might arise before final delivery or installation.

Each estimated cost-to-complete value is added to the corresponding value in the Committ(ed) to Date column to get the revised expected total cost for the account. The resulting number is the most current Proj(ected Final) Cost for that account and, when all the accounts have been totaled, for the total project.

The new projected final cost is then compared by computer to the current budget figure, and the difference is shown in the Proj(ected) Var(iance) column as a positive or negative number. A positive number is an overrun and a negative one is an underrun. This is one occasion when being all negative is beautiful! That rarely happens, but a minus sign on the variance column total does reflect a favorable cost trend for the project.

Project and construction management would be beautifully simple if all we had to do was look at one number, but it doesn't work that way. Even if the total variance number is negative, we need to analyze all the accounts to see how it got that way. A negative value is definitely a good sign, but here it's no cause for complacency. Perhaps the number is not really factual; perhaps it should not be so low or high; or perhaps we could even improve on it. Proving to yourself that the number really is a *correct statement of the cost picture* is the only way to avoid nasty surprises at the end of the project.

Is there a cost-control problem in Figure 8.4?

Figure 8.4 is, in fact, a good example of how reading only the bottom line of a cost report can get one into trouble. The bottom line of the variance column shows an underrun of $353,000, which indicates that the total material and equipment account is under budget and therefore in good shape. In looking at the numbers in the Projected Variance column, however, it is obvious that something is wrong. The equipment account shows an overrun of $1,719,000 alone, certainly not a good showing. The bulk material account shows an underrun of − $2,072,000, which is fine. But how can that be, when we are overrunning above-ground piping by $2,400,000?

Account number 0199 for a Steam Pipeline shows an underrun of −$5,040,000, which turns out to be the culprit. The item was recently deleted from the project as being unnecessary, but it is still carried as an underrun instead of being shown as a negative change order in column 3. Carrying the account in the wrong column has created a false sense of well-being by turning a $4,687,000 overrun into a − $353,000 underrun. Although $4,687,000 represents only a 7 percent overrun of a $70,236,000 Total Direct Material account, the overrun does represent 33 percent of the Total Contingency account of $14,001,000 shown in Fig. 8.5.

It is still early in the project, with only 14.2 percent of materials committed, so there is cause for concern about cost control in the physical resources portion of the project. A detailed analysis of the materials account should be made to better evaluate the final projected outcome of the material accounts.

Cost-control summary all areas

Figure 8.5 shows a typical budget summary sheet for the same project. Let's examine the major accounts to see if we can spot any problem areas from the numbers shown. From the Total Pro-Services (home office costs) numbers, we can see that the design is about 17.6 percent completed, so the design is just starting to build a peak. The total home office services is the sum of Pro-Services, overhead, and fee, totaling $21,826,000. That number divided by the total installed cost (TIC) of $187,836,000 gives a design-cost-to-TIC ratio of 11.6 percent, which is in line for a petrochemical project.

On the other hand the Pro-Services account is already showing a projected overrun of $2,172,000, or a 17 percent increase. The higher value is still within the 11.6 percent rule-of-thumb number relating to the TIC, which is favorable. However, predicting a 17 percent increase in design cost that early in the project, without any change orders listed, does look suspicious. It could indicate that the design cost was underestimated or that change orders are not being processed properly, because the 17 percent overrun prediction has surfaced so early in the project. In any event, a thorough review of the home office services account seems to be in order.

We have already discussed the false indication of an underrun in direct materials caused by the deleted steam line. The next account in Fig. 8.5 is Construction Subcontracts, which looks peculiar at this stage of the project. The account shows a projected underrun of $11,185,000 on a base budget of $37,633,000, or 30 percent. Since no subcontracts have been committed as yet, it is difficult to imagine that such a prediction could be made without having a reduction in the scope of work. The same comment goes for the field labor account, showing an underrun of $5,168,000, or 23 percent, before one field labor hour has been expended. All of them add up to a nifty underrun of $16,706,000 when only one account has been 14.2 percent committed. If we drop down to the total line, only $13,702,000 has been committed, which is really only 8 percent of the funds budgeted.

Those are only a few points about the cost report that we developed from analyzing just two summary pages. I am sure that there are plenty more suspicious areas hidden away in the backup to the summary pages. Having wide swings of that sort in some of the accounts so early in the project indicates the possibility of faulty estimating or someone painting an overly rosy picture in the cost-control report. Analyzing a project cost report is not unlike reading a company's financial statements: Is it making money? Is it solvent? What are its cash reserves?

Most project and construction managers have technical backgrounds and very little accounting and financial training. Therefore it is vital for them to get the knack of using rule-of-thumb ratios to analyze and interpret a cost report. If you don't come by that knack naturally, you should take a few financial and accounting courses to develop your skills in that area.

Escalation and contingency

The project team elected to handle the escalation account differently than I had recommended. The escalation was distributed in lump sums to the major accounts as shown in the last lines of the Major Equipment and Bulk Materials accounts in Fig. 8.4. The escalation on equipment is only 5 percent, so it appears that we are in an era of low inflation. The Major Equipment Escalation account shows an underrun of $246,000 in the face of a projected overrun in that account of $1,719,000, which seems a bit optimistic at such an early stage of the buying program. Apparently the overrun is not due to price increases.

The Bulk Material Escalation allotment is slightly higher, at about 9 percent. The odd thing about the account is that it projects a $106,000 overrun in the face of a total account underrun of $2,072,000. That is a further indication that the underrun on bulk material is suspect, possibly due to the steam line's being eliminated.

The Total Contingency account is shown as a lump sum of $14,001,000 on the next-to-last line of Fig. 8.5. That's about 8 percent of the TIC, a fairly conservative figure even for a project of such large size. It indicates a high level of confidence in the estimate at this early phase of the project. In addition, the cost report is forecasting a reduction of $807,000 at this point—another sign of optimism.

In earlier chapters we discussed holding escalation and contingency in separate reserve accounts until needed. That is being done in the budget, even though the escalation has been parceled out to the major material accounts. Assigning fixed amounts to the major accounts is probably a good practice, since it allows more flexibility in putting the money where it is most needed. Contingency and escalation funds should never be assigned to individual account items, as we mentioned earlier, until a real need arises.

The contingency account could have been handled in the same way as the escalation account, with some slight advantage. There must be some parts of the estimate that are felt to be more accurate or less accurate than others. By breaking down the contingency fund, we make it possible for more money to be assigned to the weaker areas of the budget. In either case, the funds for accounts needing cost correction can be drawn from the main contingency account as required.

It is normal to reduce the contingency and escalation funds as the commitment percentage increases and the exposed portion of the unspent budget becomes smaller. Any changes in those accounts require the approval of the owner and the company management. As project and construction managers, it is best to manage those funds with a lot of thought before recommending any revisions. The complexion of a budget can change overnight, so maintaining a conservative balance for as long as you can makes good sense. You will be an even bigger hero if more healthy escalation and contingency balances show up at the end of the project!

Manual versus computerized budgets

The personal computer revolution has pretty well taken the argument out of the discussion about manual versus computerized budgets. I used to recommend using manual methods on smaller projects, because the cost of putting the figures in the mainframe computer was so great. Now, however, we can put a PC with commercial software on a small project at moderate cost. In fact, the PC and software often are already there. All that is needed is to assign a clerical person the job of inputting the data. The only caveat is to make sure that the person is trained and does the budget posting regularly enough to maintain current figures.

As we said earlier, if a PC is not available on the project and there is no chance of getting one, don't develop an inferiority complex. You can do a small job manually with equally good results, if the budget is good and you manage the funds properly. The format for the manual budget should be similar to the computerized sample seen in Figs. 8.4 and 8.5.

Specific Areas for Cost Control

We have been addressing the subject of cost control on capital projects in a general way; now it's time to get specific. The number of specific areas is considerable, and we will touch on only some of the more fertile areas for cost reduction. It is a general rule of cost control that the front end of a project is the area most likely to offer cost-reduction opportunities. After a project has passed the 60 percent design milestone, the likelihood of attaining significant cost reductions becomes much smaller. The same thing is true of the field work; it is the preconstruction planning and the setting up of the field operations that offer the best areas for cost reduction. As the construction proceeds, we go into a more *cost-control mode* to ensure that the budget does not get out of control.

The design phase

The overall cost parameters of the project are set during the basic design phase of any project. The initial project scope definition establishes the size and function of the facility. Many specific questions involving major impacts on project cost must be resolved. The feasibility of various approaches to the project must be tested—usually along economic lines.

In architectural projects, the basic uses of space and traffic flow are established. The function of the facility must be dissected, and each part must be questioned and studied to assure an optimized design. Architects call this formative stage of a project *programming*. William Peña has written a book, *Problem Seeking,*[7] which lays out the underlying principles of programming in these five steps:

1. Establish *goals.*

2. Collect and analyze *facts.*

3. Uncover and test *concepts.*

4. Determine *needs.*

5. State the *problem.*

The book goes on to delineate a complete programming system for establishing the basic design for any type of project. An organized approach to establishing the basic design parameters for the project can get the cost-containment program off on the right footing. Although I have not seen this concept used in construction work, it is something that might pay some dividends there as well.

In subsequent design steps, there are also many decisions affecting cost. Can expensive or redundant equipment be eliminated? Are we using the optimal construction materials? Can energy be conserved? Any of those design factors can yield cost savings in capital and operating costs if properly handled.

After the basic and schematic design phases have been completed and we have entered the construction document phase, design factors affecting cost continue to play major roles. Here we need to keep a sharp eye out for a natural tendency to gold-plate the project. This can come in the form of redundant systems, unnecessary controls, overdesign, and other aspects that do not improve the overall mission of the facility. Beware of the belt-and-suspenders syndrome!

Detailed design can yield more substantial cost reductions in such areas as more efficient plot planning, better building and equipment layout, and better systems integration. The civil, electrical, mechanical, and instrumentation areas also offer fertile fields for cost reductions. Specific areas to look at are unnecessary earth moving, paving, drainage, foundation design, main power supply, area classification, and excessive code requirements in general. These are all areas that can be improved through constructibility analyses made early in the project.

It is difficult for design and construction managers to ferret out all the design problem areas on their own, because they may not have the expertise to do so. What they can do, however, is alert their project engineers and design team leaders to search for cost-cutting ideas in those vital areas. The cost-cutting ideas are then referred to the PM/CM for screening and for deciding which ones should be considered for implementation. Because many of these ideas will represent major changes, the potential savings also must be weighed against their effect on project schedule. The time value of money must always be considered in implementing a major change that affects project schedule.

Home office construction costs

The home office budget usually represents only 2 to 5 percent of the total installed project cost. Although it is a relatively small part of the construction cost, it does rate scrutiny by the CM.

When design is done in the field, the CM becomes responsible for controlling that design budget. In some larger engineering-construction company operations, the design phase is sometimes offered as a "loss leader item" to get a highly profitable construction project. In that case, the design estimate will be on the low side and provide little chance to yield a profit. The design group's performance will still be judged as a separate budget center, so the design manager has an added challenge to eke out a positive result. As we discussed earlier in this chapter, the main items of cost in the design budgets are:

- Design personnel hours
- Average labor-hourly rate

- Computer costs (including CADD)
- Other out-of-pocket costs

The design hours must be monitored at least monthly on large projects and more often (i.e., biweekly) on smaller projects. Using some sort of design office cost report, similar to the Field Indirects Cost Report shown in Fig. 8.3, is vital to successful cost control. This is where a running estimate of the labor hours is essential to keeping tabs on how the time is being spent during the report period. Keep in touch with the design discipline leaders to get a feel for how they really think their budgets are faring. Often they will give you an indication of a problem area even before it turns up in the cost report.

Getting out into the trenches is *so* important in the area of design cost control. By reviewing the actual work to complete with the discipline leader, you will better be able to spot overly optimistic or pessimistic estimates to complete in the cost report. An early response to off-target results is vital to successful cost control. If you spot an adverse trend developing, get into the matter immediately so you can respond with an effective plan for corrective action.

Procurement cost control

Buying the large amount of physical resources and subcontracts required for capital projects is an important area of cost control that is often overlooked. Some project and construction managers tend to look on purchasing as a clerical or staff function without much bearing on the outcome of the project. Nothing could be further from the truth! A savings of 5 to 10 percent in the equipment, materials, and subcontract budget can represent a sizable savings to the project. Pick your procurement people with the same care as you do your technical staff.

Procurement is especially critical on construction work that is bid lump-sum to completed design documents. Fast action is required to place orders for equipment, material, and subcontracts within the validity dates on the bids used in the estimate. At the same time these purchase orders must be properly reviewed for scope, specifications, delivery, and price before the purchase order is issued. In many cases suppliers and subcontractors have been known to voluntarily reduce their prices after the bid has been awarded to the general contractor. Those potential savings shouldn't be overlooked in the urgency to place orders!

Spend some time developing a sound procurement plan to ensure timely and efficient delivery of vendor data for approval and the physical resources to the field. As we saw earlier, good productivity by the people in design and construction depends heavily on having the project resources available when they are needed. Sound cost containment results from an effective purchasing program.

Effective cost control in procurement is founded on good procurement procedures, which should comprise the following major factors:

- A good approved-vendors list
- Ethical bidding practices
- Sound negotiating techniques
- Change-order controls
- Control of open-ended orders and subcontracts
- Control of procurement, expediting, and inspection costs
- Maintenance of purchasing status reports

Procurement's goal is to buy the specified quality of goods and services for the project at the best possible prices. The first four items on the list contribute the most to maintaining cost control early in the project, the second three later.

The suppliers on the vendor list must be capable of supplying the right kind of goods at competitive prices to meet the schedule. Therefore they must be carefully screened to ensure that they make what you want to buy and have the capacity to deliver when you need it. A fully loaded shop will not be able to deliver and will not be as competitive as a hungry one.

Do not load up the approved-vendors list with more bidders than are needed to ensure competitive prices. Any marginal suppliers should be dropped, along with any having full order books. Having more bids than necessary adds to the time needed for technical and commercial bid-tabulating without improving the desired result. Good practice calls for having a minimum of three bidders and a maximum of six. It is well to go with the higher number during periods when "no bids" are prevalent, to ensure having at least three qualified bidders to choose from.

Ethical bidding practices have been established over the years in most western countries to get the best offer the first time from reliable bidders. If bidders know their prices are going to be peddled to the competition, they will not give their best price the first time around. The minor cost reductions that result from using so-called "auctioneering tactics" are more than likely to be offset by time lost in the schedule.

That theory on ethical bidding does not preclude the need for negotiation in the procurement process. Sound negotiating techniques can reduce costs, especially on large orders and subcontracts. The negotiating process often brings out some of the more obscure aspects of the procurement. Is the supplier really meeting scope and specifications? Are all requirements written into the specs really needed? Is it possible to get a better warranty or delivery at no additional cost? These are only a few hints as to the possible fruits of the negotiating process.

In most cases the negotiating team is made up of procurement and engineering personnel, to ensure that the technical and commercial points are covered as they arise. Team negotiation is always better, since the ball can be passed among the members to suit the occasion.

Minimizing change-order costs in the first place is the best line of defense in procurement. That means that the quality and definition of design must be stressed before the procurement process begins. Because change orders are inevitable, their cost is best controlled by obtaining unit prices for potential changes during the bidding process, while the sense of competition is still present. When you are already locked into a vendor or subcontractor, your position for negotiating the cost of extras is weakened.

A good example of protecting oneself on change orders is the practice of getting unit nozzle prices when bidding vessels, or volume price breaks when buying structural steel, rebar, and concrete. Having the prenegotiated prices for changes eliminates a lot of haggling over change-order costs that can consume a sizable piece of your procurement budget. Getting competitive unit prices for subcontractor extra work is also effective cost-control practice.

Open-ended purchase orders or subcontracts invite cost-control problems, so they should be used only when they are absolutely necessary. If you are forced into the position of using open-ended agreements, be sure to put a check-and-balance system in place for early warning of problems. The checks can be in the form of funding limitations, so flags will go up to indicate that more commitment is needed. At that time, progress and productivity can be reviewed to predict what the final number is likely to be. If something like that is not done, you will no doubt get a nasty surprise at the end of the job just when you thought that the financial goals had been attained. Make a point of having a checklist of all the open-ended commitments you have in order to ensure effective control.

A rule-of-thumb for estimating the overall procurement cost is about 5 percent of the total material, equipment, and subcontract cost. That breaks down to 2.5 to 3 percent for buying, 1.5 percent for expediting, and 1 percent for inspection.

Purchasing hours and costs are reported as part of the cost report, which was discussed earlier. The hours spent should be compared to the performance of the procurement commitment curve. The commitment curve is the S curve, resulting from plotting the projected value of orders placed against elapsed time. The procurement curve lags the project progress curve but leads the cashflow curve.

I have included controlling and reporting purchase-order status under cost control, because lack of that information can adversely affect the project. Keeping abreast of the procurement status of tagged equipment is relatively simple, but controlling bulk material orders in detail is something else. Proprietary computerized systems for tracking project physical resources have been developed by many firms. The systems are really spreadsheets that schedule each procurement activity for all the equipment and materials. As procurement progresses, the report is updated to keep the information current. The PM/CM should monitor the material status report, in conjunction with the procurement commitment curve and the cost report, to monitor the output and cost-effectiveness of the procurement program.

From this discussion of procurement activities, it again becomes obvious that the most money can be saved with good planning in the early stages of the project.

Construction costs

Since construction spends the lion's share of the project budget even in the design-build mode, we expect it to be a fertile area for cost reduction and control. As we look at the possible areas listed here, we can see that the best time to save is early in the project:

- Constructibility analysis
- Contracting strategy
- Planning temporary facilities and utilities
- Planning heavy lifts and associated equipment
- Assigning laydown and working areas on site
- Organizing the field supervisory staff
- Establishing the construction philosophy
- Field personnel mobilization and demobilization
- Field personnel leveling, and field productivity

As we said earlier, it's difficult to get the construction team into the project early enough to realize some of the potential cost savings. In split projects for which the design is completed before the construction is bid, it's virtually impossible. The fact that owners miss the potential cost reductions from early construction participation is one of the major drawbacks in that type of contracting. However, when it's possible to get the constructor's involvement early through a negotiated third-party constructor, the savings are still possible. Privately financed owners should remember that.

The constructibility analysis is relatively new; it has been developed and promoted in recent years by the Construction Industry Institute (CII), the Business Roundtable, and some of the large contracting firms. The advantage to the contractor is readily apparent, but the potential cost reduction to the owner also is quite real. Owners must consider the pluses and minuses of that approach in relation to their individual project needs.

The idea is to bring a few highly qualified construction specialists into the project during the design phase. These specialists review the design as it develops and offer suggestions that will make the facility easier to construct and thereby reduce construction costs. Although the people are fairly expensive, they do not have to be used full-time to reap the advantages. The best situation is have them carry right through to become part of the construction team. The knowledge gained during the design phase makes them especially valuable when it comes time to plan the early field operations.[5]

Temporary construction facilities are another important area in construction cost containment. The facilities should be adequate for the job and suitable for the local climate and labor conditions. Temporary utilities should have adequate capacity to reliably carry the expected construction loads until the permanent utilities are actuated. Frequent breakdowns of overloaded temporary utility systems during construction can be costly and are intolerable on a cost-conscious project. Pennies saved up front can cost big dollars through lost productivity during the peak of construction!

Gold-plating the field facilities also should be avoided, for obvious reasons. Make sure that the field facilities budget is spent in the right places. It may, however, be necessary to add some special touches if local conditions are poor and some morale-building is needed.

Early planning of the heavy lifts for the project is normally an offshoot of the constructibility analysis. The equipment for the heavy lifts is usually costly and hard to find, so it must be scheduled early and closely monitored for efficient use. On international projects, it may even have to be imported. Ensuring that the design group incorporates the proper lifting lugs into the plant equipment is equally important. Well-planned lifts are the safest lifts.[9]

Custom-designed construction equipment such as special form systems, lift-slab equipment, and concrete batch plants, must be planned early. They are key "management" contributors to improved field productivity, which is the number-one cost-reduction area in field work.

Laying out the construction site and assigning the laydown and work areas for the subcontractors is an important component in controlling field costs. The modern field site must be planned like an industrial plant to ensure smooth work flow and high productivity. An overly remote construction parking lot, for example, can cost thousands of dollars in direct and indirect labor costs.

Organizing the field supervisory staff is the largest single cost-sensitive item in the field indirect cost budget. Since many of these people travel from site to site on large projects, the salary, living, and relocation allowances tend to make the hourly rate higher than normal. My philosophy for any organization is to be slightly understaffed for best efficiency. Idle hands in the field organization can cause even more problems than they do in the design office.

Selecting the most effective contracting plan can affect project costs dramatically. Certain situations favor using an integrated contract approach over a negotiated third-party constructor arrangement, and vice versa. Other avenues to explore are going union or open-shop, direct-hire or construction management with subcontractors, or acting as your own general contractor. All those options should be investigated to determine the most feasible contracting approach. The decisions cannot be made solely on the basis of contract fees.

Monitoring and improving field labor productivity is the single most important source of cost improvement during the construction phase. If the contract is cost-plus, the owner has the greatest stake in maintaining above-average productivity. If it is lump-sum, the contractor has the greatest stake in field productivity. In either case, the CM should have the greatest interest of all!

The field cost engineers and the construction supervisors must evaluate the field productivity for all crafts at least weekly. Any unfavorable trends must be dealt with immediately to prevent erosion of the field labor budget. Field labor leveling can make a significant contribution to the overall field labor productivity by not overloading individual work areas. The leveling process must be carefully handled by using schedule float so that the end date will not be adversely affected.

Strict adherence to the field procedures and systems can also improve cost performance. This is based on the logical assumption that the procedures and systems were well thought out before they were put into use. If any procedures do turn out to be faulty and are adversely affecting project performance and cost, by all means change them for the better.

There are many free spirits in all areas of the capital projects business who find standard procedures and systems unsuited to their lifestyle. However, project disorganization is costly and a destroyer of budgets, schedules, and cost control. As CM, you cannot afford to let the construction project run amok!

In addition to reviewing the budget and cost reports, the CM should sign off on all bid tabulations for equipment, materials, and subcontracts. This puts an immediate screen in effect to catch any out-of-budget purchases before they are made. It is not advisable to delegate this important function to others who might not point out potential problem areas to you. Checking the values of the out-of-budget orders will give you a running mental tab of the winners and losers in your physical resources budget. You will need only the budget report to confirm what you already know.

Monitoring construction change orders is another province of cost control important to the CM. Your staff members must be taught to watch for change orders and to prepare the detailed forms for approval, but it is the CM who is responsible for getting the orders approved. A budget that is not adjusted for change orders is a budget that will soon be out of control.

The other items on the list have already been discussed in detail. They are listed in the construction team's list of responsibilities because *the buck stops at the CM*. It is impossible, however, for the CM to cover all areas of cost control alone. The field is just too broad and too specialized for one person to handle on a good-sized project. Use your staff as your eyes and ears and for advice, but the final decision on the corrective action has to be yours.

All cost-reduction areas in construction are the direct concern of the CM. If the CM is reporting to an overall project manager (or project director), that person also is responsible for seeing to it that the field is aware of the cost-reduction goals. Frequent PM site visits to review progress and costs are the only way to ensure that the field control goals are being met satisfactorily. Reading the field reports alone will not give the PM the necessary gut feeling of what is really going on in the field.

Controlling Cashflow

Forecasting and controlling the project cashflow is important to several project groups. These are mainly the financial people, but we are all interested in knowing that enough cash will be available when the bills come due.

The owner's needs

Information on the cashflow for a large project is vital to the owner's financial people. Capital projects often involve budgets in the millions and hundreds of millions of dollars. In either case, we are speaking of large sums of money in interest costs. Large construction loans are structured in a myriad of ways, but they generally follow a pattern of commitment, drawdown, and repayment phases.

During the commitment phase, the interest rate will be in the range of 1 to 1.5 percent. In that case the bankers have agreed to have the money ready only when it is needed by the owner. The money is actually loaned short-term somewhere else until the funds are drawn on. As the project progresses and cash is needed to make progress payments, the owner draws against the loan commitment funds. At that point the loan rate increases to the negotiated rate because the borrower now has full use of the money. If the owner draws down amounts greater than are needed for actual expenses, the surplus funds are banked at an interest rate as nearly equivalent to the loan rate as the owner can get.

The repayment phase starts when the project is completed and running and is generating cashflow to repay the loan. The interest charges on the reserve and drawdown funds have been accumulating during the project execution period. These interest costs are a significant part of the owner's project financial plan, which will be badly upset should the project finish late.

An accurate and current cashflow projection is extremely important to the project financial planners. Any wide swings from the plan should be brought to the owner's attention as soon as a problem is recognized. The project and construction managers should evaluate the cashflow, along with the other key project financial controls, every month.

The contractor's needs

Other important project participants with an interest in the cashflow are any contractors who may be involved. Like any good businessperson, the contractor wants to run the project with the owner's money. Although interest rates as of this writing (1993) are fairly low, the contracting company can't afford to have its cash tied up in the project. This is especially true if the contract is set up to run all of the project's materials and equipment procurement through the contractor's books, which happens when contractors write the orders on their purchase order forms and pay the vendors' invoices.

For example, let's assume that the contractor is carrying an average invoice balance of $3 million against the owner over a period of one year. Borrowing that money at 8 percent interest would cost $240,000 interest per year, or $20,000 per month. If contractors used their cash reserves to carry the invoice account, those interest expenses would represent lost income from their bank accounts.

The situation is easily solved on CPFF jobs by means of a zero-balance bank account set up just for the project. Each month the contractor tells the owner how much cash outflow is expected that month. The owner then deposits that amount in the zero-balance account. The contractor writes the checks for the payrolls and approved vendors' invoices out of that account, essentially drawing the account down to zero each month.

Zero-balance bank accounts do not pay the depositor any interest, so the bank makes its profit from the short-term interest it earns on the average balance in the account. If the account is not drawn down to zero in any month, the amount left over is deducted from the cash projection for the next month. The project cashflow curve is used as a guide for making the requests for funds each month. The actual amount transferred and spent is plotted on the cashflow curve each month to become the *actual* cashflow curve.

Cashflow curves

A typical cashflow curve follows the traditional S shape, with a slow start and finish and a straight center section. Cashflow is sometimes shown in the form of a table or as a vertical bar chart. A typical cashflow chart may be seen in Fig. 8.6.

The cashflow projection is made up of three major cost components as follows:

- Home office services progress payments
- Payments for goods and services made through procurement
- Direct and indirect construction costs such as labor, materials, subcontracts, and field indirect costs

The cashflow curve lags the design and construction progress curve, since the contractor normally finances the costs for up to 30 days. This may not be the case if the project is being done in-house, in which case design costs show up on the curve immediately. The cashflow also lags the procurement commitment curve, since most payments are made on delivery of the goods. On some international projects the delay may be shortened, due to the custom of making vendor-supplier progress payments starting with a down payment on order placement.

The cost engineers use the projected design, procurement, and construction S curves to draw the projected cashflow curve. Like most S curves, it is subject to some slippage to the right caused by schedule delays, late deliveries, and so on. Therefore the cashflow projection must be recycled monthly or quarterly and the forecast updated as required.

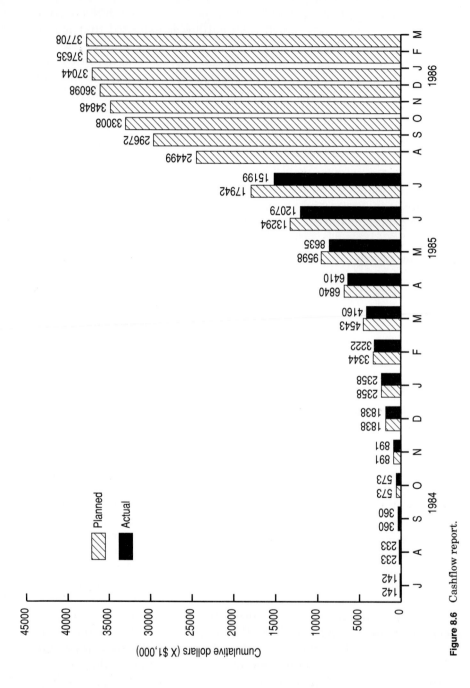

Figure 8.6 Cashflow report.

Cashflow on lump-sum projects

The cashflow on lump-sum projects requires the CM's special attention. It is vital that he or she ensures that the progress payments are keeping up with the cash outflow. On most lump-sum projects, progress payments are set up on a schedule geared to certain project-completion milestones. In that case the contractor will try to load more cashflow into the front end of the project to get a forward-loaded payment schedule.

One reason for doing this is to offset the payment losses due to the normal 10 percent retention on the contractor's billings. The owner's purpose for the retention is to keep the contractor interested in completing the project on schedule—especially the final cleanup items. Retention is another financial burden that contractors wish to avoid. It results in another case of lost interest on the funds retained by the owner.

One way to avoid the loss, and still satisfy the owner's needs, is to convert the retention funds into a performance bond until the project is completed. A performance bond costs the contractor only about 1 to 2 percent, as opposed to 8 percent in lost interest. Also, clauses can be introduced into the contract to reduce the amount retained as the project moves into its later stages and the owner's exposure lessens.

Project invoicing

It is normal practice for CMs to review all invoices before they are mailed to the client. That is not so much to check the arithmetic as to check for possible errors in the formatting that is required by contract and project procedures. A minor error can be a reason for not processing the invoice, which means another delay in the contractor's cashflow.

It is the job of the client project manager to review all incoming invoices before passing them on to accounting for payment. The review also ensures conformance to project procedures rather than to arithmetic. It is vital to good contractor-client relations to keep the invoice procedure moving on schedule. If minor errors are found, only the part in error should be held up for resolution or payment in the next invoice.

On most domestic projects the cash usually flows smoothly, so collecting the money is not a major problem, but on international projects it is often another matter. Most foreign work operates under an irrevocable letter of credit to the contractor's bank for the amount of the contract. This ensures that the funds to pay for the project are available in the contractor's country, which in turn makes it much more difficult for the owner to renege on legitimate progress payments. If the contractor can prove the work was done, the bank will release the payment.

Cashflow summary

Although cashflow is often overlooked in the plethora of other demands on the CM's time, it is too important to be passed over lightly. The top managements of all of the companies involved in the project are going to be watching the cashflow. The needs of each group must be satisfied if you are to be rated as a top-performing project or construction manager.

A realistic cashflow plan, followed by effective procedures for controlling and reporting it, will go a long way toward improving the project's cash management. The main thing to remember is not to surprise any of the cashflow participants with major changes at the last minute. If cashflow problems start to surface, get them out into the open quickly so they can be resolved.

Schedule Control

We sometimes hear schedule control referred to as *time control,* in relation to making the *time plan* for the project work. Actually time cannot be controlled, since it marches on relentlessly in fixed amounts and without any yet-known means of controlling it. Time is, in fact, quite inelastic.

The project schedule is a plan of work per unit time, so it is somewhat flexible. By speeding up the task (doing it in less time), we can make up for past or future lost time and still make the predetermined scheduled date. Therefore the work is *elastic* and must be conformed to the *inelastic* time scale. I realize that this discussion sounds very basic, but you would be surprised how many managers think they can create more time to make the schedule!

Monitoring the schedule

If we are to control the schedule, we must monitor progress against the time scale. We broke the work down into specific tasks (work activities) as a convenient way to check elements of our work against time. The average completion status of those work activities is our measure of overall physical progress on the total project.

We also must consider that each work activity has to be *weighted* to arrive at its percent of the total project. Each weight is determined by the value of the human and physical resources expended to accomplish the task. The weighted value of the activity multiplied by the physical percent completion tells us how much *earned value* that activity is contributing to our percent complete. Conversely, the earned value divided by the total budget for that activity gives us the physical percent complete for that activity.

Measuring physical progress

Again it is necessary to point out the fallacy of using labor-hours expended to calculate the physical percent of completion. If you are not operating at exactly 100 percent efficiency, the percent completion will be in error. That is why we stress the necessity of estimating physical progress and not expended progress.

The physical percent complete must be tied to the *earned value* of the work accomplished to date. The earned value of the work activity is measured by breaking the task down into a logical system of checkpoints and assigning a percent of completion up to that point.

To calculate earned value, it is necessary to break down the various construction tasks into each trade's contribution to the work. A ready example is the installation of foundations that is generally required on all capital projects. A typical breakdown can be made as follows:

Operation	Percent progress
Layout	5
Excavation	20
Forming	50
Set rebar and anchor bolts.	75
Pour concrete.	90
Strip forms and backfill.	100

The cost-effectiveness of any earned-value system must be evaluated in relation to the benefits derived. On a large project the amount of clerical work can become voluminous, even with a computer. On smaller projects the problem becomes even worse, because limited funds are allotted for such controls.

Much of the data gathering takes place during the monthly status evaluation as the group leader evaluates the percent complete of each task and the section's physical percent complete. The cost people then take those figures and review them for accuracy before converting them to the overall project physical percent complete data.

The field engineer or quantity surveyor evaluates the status of each operation being worked on during the reporting period and arrives at the physical percent complete. The field cost engineer then collects the input and calculates the physical percent complete for each task unit and eventually the whole field operation.

That system of cross-checking physical progress between the line and staff groups makes it more difficult to hide overly optimistic forecasts or to make errors in the estimate to complete. It is all done before the schedulers get into the act and plot the actual progress against that which was planned.

Planned vs. actual physical progress

Once the physical progress for the period has been calculated, it becomes a simple matter to plot the new value on the planned progress S curve as shown in Fig. 8.7. The curve is derived from the original project schedule by plotting planned percent complete and personnel hours over the elapsed time for the project. The resulting S curves are shown as the solid lines in Fig. 8.7. The actual physical progress and hours are then plotted monthly as dashed lines on the same graph (Fig. 8.7).

If the plot for the actual completion falls on the solid line, the project is on schedule. If the actual progress curve falls off below the

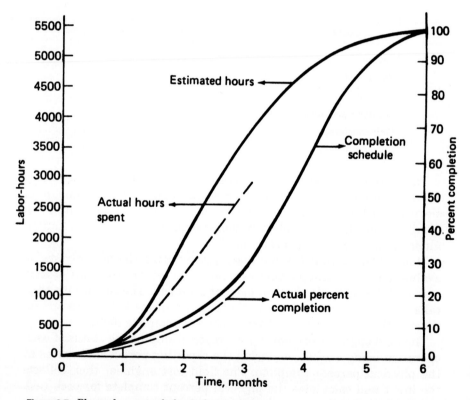

Figure 8.7 Planned vs. actual physical progress curve.

solid line, (as in Fig. 8.7), the project is running behind schedule. That could be due to an understaffed project, because the actual-hours-expended curve is also below the planned progress curve.

Should we be fortunate enough to have the actual progress line fall above the solid line, we are ahead of schedule. That's no cause for complacency, however, because progress can easily fall below the line in the following period!

If there are no extenuating circumstances such as unrecorded change orders, lack of owner's performance, force majeure forcing the actual progress curve below the planned progress curve, there must be another serious problem that is causing the adverse trend. The failure to maintain schedule then lies at the door of the construction team, and the CM, in particular, must take urgent steps to rectify the problem.

First, determine the cause of the problem. Is it due to such general causes as staffing problems, poor productivity, low morale, and lack of leadership? Or is it due to the failure of a specific group to perform up to standard? In most cases the problem will involve more than one cause, so the suspected areas must be prioritized in order to attack the worst ones first.

Although I stress immediate response to substandard performance, be careful not to overreact. Sometimes that only serves to exacerbate the problem rather than solve it. Discussing the problem with key staff members is a good way to get the pertinent input on the subject. After weighing this input from the staff, you, as PM/CM, should then make the decision as to the course of action to be taken. Hopefully you will be able to get the key staff people to buy into your decision, which is important if the selected solution is to have a good chance of success.

If we study the situation presented in Fig. 8.7, we can learn why it is so difficult to get a job back on schedule once it has slipped below the planned curve. After the project has passed the buildup part of the curve (the lower arc of the S), it enters the straight-line portion when the peak staff is on board.

On the straight part of the curve, most projects make about 7 to 10 percent progress per month. This period includes the time when we should be at peak productivity. Productivity is at its lowest during the curved portions at the beginning and end of the S curve. It is inherently difficult to get monthly progress above the 7 to 10 percent progress rate over the period needed to regain the lost time, and return the actual progress to the planned curve. Remember, if we expand the time, we slip the schedule!

I take that negative view just to let you know that getting back on schedule isn't easy. Nevertheless, we must make the effort to make up the lost time. We have to look to the unfinished work for the solution. The time lost in the past is history, and no amount of wishful thinking will bring it back.

Ways to Improve the Schedule

We can control the future, as we said, by speeding up those activities that are now expected to finish late. This discussion refers to all applicable phases of the project such as design, procurement, and construction. There are a few ways to improve your rate of progress when you are behind schedule in any of the project phases.

- Improve productivity
- Increase staff
- Work overtime
- Reduce the work
- Subcontract part of the work

Improving productivity is the best and cheapest way to increase the speed of doing the work, either by using better-qualified workers or by improving management and work methods. Because it takes time, improved productivity may not work when a quick fix is needed, but it is a good cure for the longer term.

Provided that there is ample room to work and that qualified people are available, adding more people is probably the next best way to make up time. If neither room nor work for the new people is available, the productivity of the larger work force is likely to fall off, giving a net gain of zero. Rarely do we find qualified people sitting around awaiting assignment to our project, so this solution is not suited to a quick fix either.

Overtime work is the next option, and it is the one most commonly used. The system can immediately increase the available staff hours on a project by 15 to 30 percent. The added hours come from people already trained in the work, so learning curve losses are nil. The disadvantage of overtime arises after its extended use. Productivity has been proven to fall off with prolonged overtime work. The people put in the time, but they do the same amount of work as in an eight-hour day.

Construction people usually work more overtime than do design people. In fact, overtime is often used as a ploy to attract workers during periods of skilled-labor shortage. I recommend that this practice be avoided because the high premium for construction pay and the resulting loss of efficiency can combine to ruin the field labor budget.[4]

The cost of premium-time payments also must be weighed against the value of the anticipated time-saving benefits. Many home office people fall into an overtime-exempt category, so their overtime does not carry a premium rate. Drafters and CAD operators usually fall into a premium-time category, as do all contract design people and construction craft laborers.

All things considered, using scheduled overtime to make up lost time and get back on schedule offers the most advantages. The increased cost due to premium time, and the potential efficiency loss over long periods of overtime, have to be taken into account. If the productivity is already poor, however, using overtime at premium rates is only throwing good money after bad!

Reducing the work load to regain schedule is another route we might consider taking. However, it is applicable only in those rare cases when some work operations can be eliminated to save time. If the eliminated items were easy to find, they shouldn't have been in the work in the first place, so beware of using this approach to the problem. Occasionally people may be found to be doing work beyond the scope given in the contract, which is definitely poor policy.

Sometimes we try to shift work from the design area to the field. For example, we could do the piping isometrics and material takeoff in the field to reduce the home office work. This is merely shifting the work to another location to make the home office schedule look good. However, the arrangement could be an improvement on remodeling jobs where doing the design work in the field is more efficient when field measurements are required.

The latter work-reduction ideas need to be well thought out before trying them. Usually they are recommended only as a last resort, and often they represent that short step from the frying pan into the fire.

Tips on Schedule Control

When monitoring the schedule, it is best to keep close tabs on the 20 percent or so of activities that are on the critical path. If the critical items slip, they can make some of the near-critical activities late. If the critical-path items rise much above 20 percent, the project will have to be rescheduled. That may or may not affect the strategic project completion date.

Be especially careful at the beginning of the project, because getting behind schedule then will probably dog your project performance the rest of the way. Quick detection and reaction are important at any time, as schedule problems hardly ever get better by themselves. However, be careful not to overreact and do something that proves counterproductive.

Monitoring Procurement Commitments

On integrated projects, especially in the early stages, procurement activity has a profound impact on the project schedule. The design and construction downstream activities are closely tied to the availability of vendor data and delivery of the physical resources.

That statement does not apply to straight architectural projects where the construction is bid lump-sum after completion of the design. In that case, the design is done in a broad fashion to suit a variety of equipment types that might satisfy the specifications. Also, the contractors usually are responsible for procuring all their materials and equipment as well as subcontractors, so they are in control of their own destiny on schedule. The strictly lump-sum, split-contract arrangement does lose the advantage of early ordering of long-delivery equipment to fast-track the schedule. In turn, that places greater strain on the contractor's procurement schedule.

The tool used to gauge the performance of your procurement team is the purchasing commitment curve. As with any activity with buildup, peak, and build-down phases, the result is the familiar S curve as shown in Fig. 8.8. The value of the materials and equipment to be purchased is plotted against the time planned for the procurement. The curve should show an early upturn due to the early placement of the long-delivery equipment orders, which are usually big-ticket items. Most bulk materials are purchased later in the project.

The commitment dollar value is posted to the curve only when the written purchase order actually is issued. It's bad practice to count verbal orders as commitments. Pressure must be maintained on procurement to get the written purchase orders out, because vendors don't perform nearly as well on verbal orders.

If the commitments fall below the line, the PM/CM must investigate the cause as soon as it becomes apparent. Procrastination now will eventually affect the schedule across the whole project. The problem could be due to delays on the technical side in terms of preparation of specifications and requisitions, making bid tabulations, and so on. Delays also are possible on the purchasing side, due to late issue of bids, poor vendor response, adverse market conditions, or a variety of other reasons.

After the buying activity has been completed, procurement activity will continue in the expediting and inspection mode until all equipment and materials are delivered to the field. Progress in that area is hard to chart and must be gauged through the expediting reports and reported late deliveries from the field. The PM/CM must regularly monitor the expediting reports to keep his or her finger on the pulse of deliveries, starting with the delivery of the vendors' data.

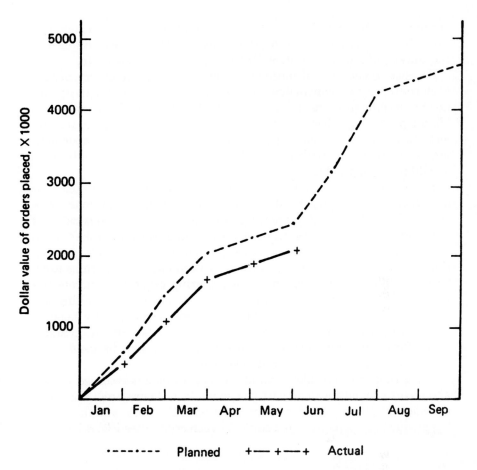

Figure 8.8 Procurement commitment curve.

When equipment and material deliveries especially are critical, I have often held the expediting review meetings in the field. That shortens to zero the communications lines between the field forces and the home office expediting people. That close proximity to the field work seems to give the expediters a real feel for the urgency of making the deliveries on time.

Monitoring Construction Progress

Keeping track of the field work on larger projects is the responsibility of the field cost engineers. They are responsible for setting up and monitoring the systems for measuring the performance of the field forces. Unfortunately, there is not a good nationwide productivity standard with which to compare individual project performance.[5,6] The CICE report listed in reference 5 gives a detailed account of why the Bureau of Labor Statistics (BLS) reports are not adequate.

Because of the lack of national (or even area) productivity norms, each contractor is responsible for developing its own productivity values for specific projects. That's not all bad, because CMs are responsible for producing effective productivity rates on their individual projects to meet their project goals. As long as the original labor estimates are produced with the firm's historical productivity data, that should not be a problem.

The earned-value system and the estimate to complete are used for each field work activity. Work activities are broken down into the specific crafts such as civil, piping, rigging, steel erection, electrical, and painting, by project areas. In turn, each item of work is broken down into its major components. As we said in our earlier example, foundations can be broken down into values for each operation such as excavation, building forms, setting rebar, pouring concrete, concrete finishing, and stripping forms. In some cases, however, breaking the work down into such detail is not considered economically feasible within the field indirect cost budget. In that case an all-inclusive value for each cubic yard of foundations placed can be used, with slightly less accurate results but lower cost.

The actual work completed on all field operations is recorded weekly and reported monthly along with the actual labor-hours used. The latter are compared to the standard hours allotted to the work, and the resulting productivity is plotted on a curve for each craft.[3] (See Fig. 8.7.)

Monitoring field productivity

There are about as many ways of monitoring field productivity as there are contractors.[6] The important thing is to select a process that works for your project environment. Basically all productivity measurements use some format that compares physical progress against labor-hours expended.

One system for calculating the productivity rate is represented by the following formula[1]:

$$\text{Percentage of productivity} = \frac{\text{budgeted hours}}{\text{actual hours}}$$

This calculation is made weekly for each craft in each area in order to monitor the productivity trends. Any unexplained adverse trends must be evaluated and acted on quickly to stave off time and money overruns. The productivity rate does vary over the life of the project as shown in Fig. 8.9, so don't be too upset when the productivity starts below 1.0 early in the job.

The productivity starts low because of job start-up problems and the learning curve for new people on the job. Typically, a productivity peak occurs during the peak loading period on the labor curve and then tapers off as the work winds down at the end of the project.

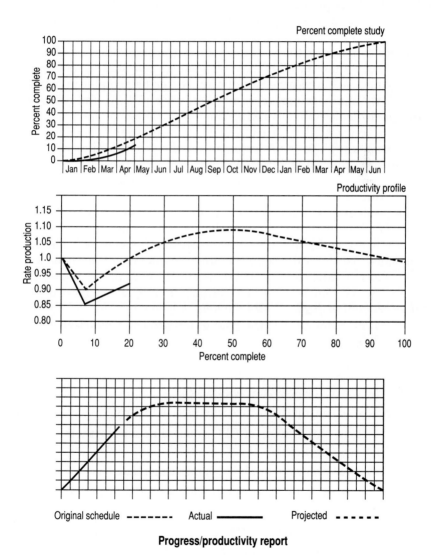

Figure 8.9 Productivity curves.

Finishing under 1.0 efficiency in the final stages of the project is virtually guaranteed due to the nature of the cleanup operations at the end of the job.

The objective of a good field performance is to finish the field work at a cumulative efficiency of 1.0 or better. This means that we have to have a period of better than 1.0 productivity during the periods of peak personnel loading to offset the losses at the start and finish. If the productivity never gets above 1.0, the budget and schedule are bound to be exceeded and the project expectations will not be met.

CMs must monitor the productivity of the individual field craft units to keep their productivity at least as planned. Some units will stay on track and others will slip. Never be complacent about productivity in any group—it can fall off very quickly, and recovery is always difficult. When walking the job site, keep a sharp eye out for productivity-killing practices. This is another key entrepreneurial skill a top CM must acquire over the years. Make sure that your construction superintendent and area engineers are aware that the productivity goals and weekly results are one of their vital responsibilities.

Good productivity is not solely a matter of being lucky in getting good labor. It is rather a combination of sound management, good work methods, and a qualified and motivated labor force.[6] The following checklist gives the combination of management methods and human factors that the CM must skillfully manage to maximize field productivity on every project level. Virtually every facet of the *total construction project management* approach espoused in this book must come into play if the goal of top project productivity is to be reached!

A schedule control checklist

- Cycle and check the schedule regularly.
- Look ahead to control the future; the past is history.
- Give the three major areas of EPC equal time during their critical periods.
- Check *physical* progress, not labor-hours expended.
- Check the schedule by exception, using the CPM critical-item and look-ahead sorts.
- Take early action to correct slippage.
- Make corrective decisions logically; don't overreact.
- Use overtime wisely.

Equipment and Material Control

On a major EPC project we may be handling from 500 to 1000 tagged equipment items, along with large quantities of bulk materials in the piping, electrical, civil, and other trades. The logistics of getting all that material to the construction site *on time* represents a major administrative undertaking.

Controlling major equipment

Keeping tabs on the tagged equipment is relatively straightforward, beginning with the equipment list that is developed as part of the basic design data package. Specifications for the procurement of these items are written and sent to procurement for purchase. Bids are taken and analyzed and a vendor is selected for each item. That initiates the vendor data cycle as well as the manufacture of the goods. The equipment is manufactured, inspected, and shipped to the field, where it is received and stored until it is installed. A field receiving report is prepared and sent to the home office so that the vendor's invoice can be paid.

To simplify that process, like items are grouped to make combined purchase orders. For example, all the centrifugal pumps can be gathered into one order, all pressure vessels into another, and so on. Each item can be listed on the status report, but grouping them greatly accelerates the procurement process.

The start and finish dates for each of the equipment-control activities have to be scheduled to ensure that the equipment will arrive on schedule for installation. Earlier, we discussed ways of monitoring the delivery of the equipment by use of spreadsheet programs to record the progress of each activity.

As each step is accomplished it is logged into the system, which is used to generate a weekly material status report. The report is used by the project and construction managers, the schedulers, and the expediters to monitor the progress of the procurement program. In the early stages, the accent is on delivery of vendor data; later, it is on the equipment itself.

Bulk material control

The bulk materials are handled similarly, except that a material takeoff (MTO) stage is required if we are to know how much material to order. On an integrated EPC project, a preliminary order of bulk material is made for up to 50 percent of the material required. The initial order is usually made for the estimated quantities for the whole job at the most favorable bulk order prices. Here we combine smart scheduling with thoughtful cost control! Another example of early bulk ordering is the early takeoff of the structural steel for placing a mill order before the detailed design is completed. It's another of the many time- and money-saving ideas to keep in mind when planning your project materials management program.

As the design progresses and detailed MTOs are done, the final order quantities are set and additional materials are ordered against the previously agreed-upon favorable bulk material prices. Generally, an excess of about 5 to 10 percent over the takeoff amount for bulk materials such as piping and electrical fittings is ordered to cover losses and rework in the field. Minor problems with disposing of surplus materials at the end of the project are better than having materials shortages damage our peak field productivity.

Naturally, some common sense is called for when ordering the surplus materials; much more care must be taken in ordering surplus exotic materials that are prohibitively expensive and are made only to special order. The situation also can set up a time trap, because of the long delivery time for most exotic materials. If you do come up short, waiting for delivery of items critical to project completion can ruin your schedule.

All of these points support my thesis that project and construction managers need to realize the importance of planning, organizing, and controlling their project material resources. A well-managed materials program can save time and money at every turn and make a valuable contribution to meeting the project goals.

Controlling Quality

International competition has made quality control a primary concern in industry worldwide, and capital projects are no exception. In addition to maintaining a company's reputation, the incidence of design- and construction-quality-related lawsuits has also been on the increase. There is certainly no cause for complacency about quality control in our business! The major areas of quality control in the capital projects arena are:

- The project design basis
- Design and engineering
- Equipment and materials
- Field construction
- Final inspection and acceptance

We will examine each of these areas in detail, along with the PM/CM's role in quality control.

The owner's design basis

Quality control cuts across many organizational lines of the project, but it begins with the owner's design basis for the facility. That is where the baseline for quality of the facility is established. Some of the many external and internal factors affecting the design basis are as follows:

- Market forces
- Expected market life of the product or service
- Product, service, and facility economics
- Financing capability
- Operational safety
- Environmental and zoning factors
- Government regulations
- Company or agency policies
- Maintenance, services, and operating costs

Each of those factors can have a profound effect on the cost and economics of the facility, which will in turn affect the quality that the owning entity establishes in its design basis. The owner must specify the economic life, which, along with the financing limitations, establishes the desired quality level for the facility. The design team then uses the applicable code requirements and designs to a quality standard within the limits of the project financing.

Design quality control

The design group, whether captive, contractor, design office, or field, develops the owner's design basis into the conceptual design within the limits set by the owner. In that phase the preliminary plans and specifications used to finalize the scope and quality of the facility are developed. Design and feasibility studies, along with cost estimates, are made to arrive at the optimum design within the economical design basis.

During this stage, an important job for the owner's PM/CM is to satisfy his or her management that the quality established in the design basis is being adhered to. Open communication between the owner and contractor PM/CMs during the developmental stage is vital to successful implementation of the design basis. When the conceptual design review meeting with the owner results in approval of the design basis, the detail design phase begins. From here on, control of quality passes to the design team, with only limited review and approval of key documents by the owner to assure conformance to the design basis and the contract.

The design team is responsible for its own design quality, which means that accuracy and consistency among the calculations, drawings, specifications, and all project documents is essential.

It is the design leader's obligation to ensure that the design quality-control systems are in place and functioning. The best reason for sound quality control during design is to minimize costly changes and corrections in the field. Correcting field work is expensive in and of itself and has a traumatic effect on the schedule.

Equipment and materials quality control

Quality assurance for the project's physical resources is generally left to the owner's or contractor's inspection department. That level of quality control prior to delivering the goods to the field is normally found only on process and high-technology types of projects that have a large amount of custom-designed equipment. To avoid costly rework during construction, it's vital to confirm the quality and measurements before specialized equipment is shipped.

That part of quality control often functions as part of the procurement or engineering department. Inspectors make shop visits at critical points during the fabrication cycle to monitor the quality of the fabricator's work. Final testing and inspection are made prior to acceptance for shipment. Conformance to the applicable drawings, codes, and material requirements are certified and made part of the equipment file. Inspection of the engineered equipment is important to prevent later more critical delays when errors are discovered in the field.

The PM/CM's role in the quality-control activity is to monitor the inspection reports and take action in those areas where quality control comes into conflict with meeting schedules. Conflicts require a delicate balance in terms of the compromises that have to be made to protect the best interests of the owner, project quality, and the project schedule.

In developed countries, it is not considered cost-effective to inspect such off-the-shelf items as pipe, valves, fittings, and conduit and wire. If those items are being produced by third world suppliers, a quality inspection may be necessary to ensure that standards of quality are being met. Material certificates certifying the chemical analysis and metallurgy of those types of goods are often required to satisfy code requirements. All that documentation is filed along with the final test reports and turned over to the owners for their plant records.

The project managers for the owner and the design-procurement team should work out the scope of the materials-testing program to assess the program's cost. This is usually established as part of the procurement services scope. The degree of material quality control is a function of the owner's policies, code regulations, and supplier reliability. Owners usually reserve the right to attend any inspections that have been delegated to others, but they rarely exercise the option.

Construction quality control

Construction quality control is perhaps the most involved and complex process on capital projects. That is to be expected, because the major portion of the budget is spent in this area. Also, working in the open with the diverse number of crafts involved adds to the complexity of field quality-control work.

The construction team establishes its quality-control program, to the degree that the finished work will finally be accepted by the owner without any costly rework. In A&E types of projects, the owner often hires the design firm as the final authority on acceptable quality. Most construction quality standards have been established in specifications such as those for concrete testing, steel erection tolerances, building code inspections, and equipment standards.

On larger A&E projects, construction quality inspectors will be on the site full-time to observe daily operations. On smaller projects the inspections may be intermittent, usually at critical stages of the work. Field inspection reports should be filed on a regular basis as a record of the quality-assurance program for the project.

On integrated EPC projects involving major engineering works and process projects, the quality of the field work is monitored by the contractor under the watchful eye of the owner's resident construction team. As you may remember, that group showed up on our field organization chart in Chapter 7.

The field engineer is responsible for field quality-control operations and reports directly to the construction manager. That arrangement raises a question: Is it sound quality-control practice to have the field organization inspect its own work? To avoid that situation, an owner often will insist on setting up its own quality-control team or on bringing in an outside inspection group to monitor the quality of the construction work.[8] The only real disadvantage of that approach is the additional cost.

In any case, the field engineer arranges for the services of local field testing laboratories to perform the soils, concrete, radiography, and other testing required by the specifications. The field engineer also is responsible for inspecting concrete pouring, steel erection, electrical grounding, and welder qualification, and for setting the lines and grades for the project. All test results and reports are maintained in a quality-control file and turned over to the owner at the end of the project.

The owner project or construction manager's role in the field quality-control process is to monitor overall performance to ensure that the program meets the owner's project goals. The owner and contractor's CMs coordinate their efforts in that regard through their respective field teams, and mutually resolve any problems that rise to their level for action.

The ultimate goal of any construction quality-control program, for all the parties involved in the project, is to have a smooth final inspection and acceptance of the facility as described in the next section. Outstanding examples of how *not* to ensure field quality are some of those nuclear projects that had to be virtually torn down and rebuilt in order to get them accepted.

Final facility checkout and acceptance

The final checkout of the completed facility actually starts somewhat before mechanical completion. As construction reaches the 90 percent complete stage, some units enter the "punchlist phase." The construction team makes its own inspections to determine what is left for it to finish before it can say that portion of the project is complete.

By that time, the owner is very anxious to take over the facility and start operations. To that end, owners will often accept the facility on the conditional basis that all remaining work will be finished before the contractor leaves the site. Depending on the type of project, this is known by a variety of names. In A&E work it may be known as "substantial completion," in other projects "mechanical completion." In any case, it is the point at which the owner takes over the *care, custody, and control* of the facility for its "beneficial occupancy." In other words, it is the point at which the owner moves in and starts to operate the facility. This point is often reached on a unitized basis in a process plant, where the owner accepts the utility systems first—followed by successive process units later.

The contractor may still have people in the area finishing up such details as painting, insulation, caulking, and other minor finishing work. At that time the construction and design team, in conjunction with the owner's representatives, starts a series of final inspections that delineate what has to be done to bring the project to completion. The result is a detailed check to see that the facility conforms completely to the plans and specifications as well as the contract.

The final check also applies to the design portion of the project, to ensure that all contractual requirements for as-built drawings, mechanical catalogs, code requirements, and so on have been turned over to the owner. It is the project and construction managers' responsibility to ensure that all contractual requirements have been complied with. All loose ends will have to be cleared up before the project or construction manager can get a final letter of acceptance for the facility. At that time all retained funds should be released and all bills paid, to ensure that the owner has a lien-free environment in which to operate the facility.

On industrial facilities the final checkout can be much more involved, because the quality check has to go all the way back to the engineering flow diagrams to ensure that the plant will operate safely and properly. Testing and adjusting the piping and instrumentation systems requires detailed inspection and calibration before the systems can be accepted as operational. At the same time, the owner brings in its operating personnel, who use the turnover period for orientation and on-the-job training.

If the process project involves a contractor's process design, final performance tests also will have to be run. These should be attempted only after the plant is completely operational and has reached a steady-state condition. When the tests have been properly completed and documented, the plant acceptance can be signed by the owner.

Project managers often are required to take up residence in the field for several months to handle the details of the complex acceptance process. Very often, everyone's nerves become frayed during this stressful time. The project and construction managers must exert strong leadership and instill a spirit of cooperation to bring the final checkout and acceptance process to a successful conclusion.

This is the time for the final accounting to determine whether all of the project's goals and expectations have been met. Have we really delivered the facility that was contracted for? Was it constructed *as specified, on time, and under budget?* If the answer is yes (or even close to yes), it is a happy time for all concerned. If the answer is no, it is up to the project and construction managers to put the best spin possible on a problem-plagued outcome. Special attention should be given to the project history to avoid making the same mistakes the next time around.

Summary of quality control

Quality control involves all project activities from the conceptual stage to project turnover. Substandard quality is not acceptable at any price. Everyone working on the project—owner, project team, design organization, procurement group, and construction forces—plays a major role in quality control. It is the owner's role to establish the quality desired and to ensure that it is delivered. The project and construction managers are responsible for implementing the quality-control processes for the project and for leading and monitoring the quality control effort.

Operating the Controls

As a means of summarizing the far-flung requirements of project control, I have prepared a list of recommended guidelines for handling them, each item of which is a heading in the next few pages. The best-designed control system in the world is not going to lead to a successful project unless we operate the controls correctly.

Establish priorities

The number 1 requirement for effective construction or project management is the establishment of priorities for your work activities. The most urgent activities should go to the head of your things-to-do list! Don't handle the activities in the order in which they arrive in your in-box. Remember, the project organization is functioning on the basis of what comes from your *out-box.*

Sort the whole contents of your in-box at least daily, in addition to handling the day-to-day operations of your project. Make sure those items related to project quality, cost, and schedule receive top priority. Also, items related to project and contractor/client relations should receive high priority. Be responsive to project people problems. To a slightly lesser degree, those items arising from your company management should also be treated as priority items.

All of the priority items have to be worked in with the routinely scheduled project activities such as project reviews, monthly progress reports, and trips, which also are making heavy demands on your valuable time. I have found that keeping a written priority or daily things-to-do list is a valuable tool when it comes to managing your time. It is a simple way to keep yourself from overlooking or forgetting a high-priority item now and then.

Control by exception

Using control-by-exception techniques is the only way to stay on top of the plethora of daily project activities. This is true even if you have properly delegated as many of them as possible to other project team members. Many control systems such as CPM schedules, control budgets, and quality- and material-control reports, are designed to highlight exceptional conditions. Make sure to check the exceptional items first, then go into the normal items as time permits.

Keep looking ahead

Running a project is like piloting a ship. It is nice to know the depth right under your keel, but it is more important to know what dangers lie immediately ahead. Keep your sonar turned on! Take advantage of the control systems that have look-ahead capability, such as periodic CPM look-ahead sorts and estimated-to-complete projections. Keep an eye on the applicable project performance curves for design, procurement, and productivity to note any trends away from the projected norms.

Check the actual work of those groups performing the work, whether it is design, procurement, or construction that is charted on the curves. Has there been a sudden drop in personnel just when the group should be at maximum production? How is productivity holding up? Checks should be made by visiting the groups performing the work and physically observing the work in progress. Also, talk to the group leaders; you will be amazed at what they will tell you verbally that they would not dream of putting into their written reports. Take a supervisor to lunch, visit the craft managers, keep tuned to the project grapevine to get a real feel for how things are going. Check all of the intelligence you have gathered to ensure that it is correct, then act on it accordingly.

Reaction time

The function of any control system is to have a rapid reaction time to off-target items. Corrective action should be initiated immediately. The longer you delay, the worse the condition is likely to become and the longer the time to get back on target! Once time has been lost from a schedule or money from a budget, it's very difficult to get even again. Also, you will have lost one of your opportunities to beat the schedule or budget. Don't postpone taking needed action to correct a situation just because it may result in unpleasantness to someone else or yourself. When you feel that the need for action is real, get right to it!

Single-mindedness

Single-mindedness is a long word, but it is in the dictionary. It is a noun and it means "the ability of a person to have or show a single aim or purpose." In the case of project and construction managers, their common aim or purpose is to have successful projects and attain their projects' goals.

A lot of people along the way will try to blunt your single-mindedness and divert you from your goals. Some people may not share your goals; they may even have a personal dislike for you. Others may be incompetent or just plain lazy. All of those factors must be brushed aside and not allowed to deter you from reaching your intended goals. There will be times when a suitable compromise may be the only solution, but never sell out completely!

If you have reached an impasse when going through your normal channels, don't be afraid to go over the opposition's head to meet your project goals. Remember, top management gave you a mission to accomplish the project, and you accepted it. I have found most top managements to be very supportive of their project and construction managers when an impasse arises on a project. The first place to start in a case such as that is the management sponsor for your project.

A final word of caution: Single-mindedness is essential, but the way you present it to get your project needs satisfied is equally important. We will talk about the fine points of avoiding the bull-in-the-china-shop syndrome later in the book.

Some common mistakes in operating the controls

A few mistakes that I have found to occur fairly often are these:

- Failure to plan, organize, and control the project.

- Failure to maintain good project and client/contractor relations. We will discuss this high-priority item in Chapter 12, Human Factors in Construction Management.

- Underestimating the construction cost and thereby making it impossible to meet the profit goals. The same can be said for overestimating the construction cost and thereby making the project uneconomical.

- Failure to get a good procurement program organized.

- Failure to properly document meetings, project changes, telephone calls, and other important project happenings. If the job runs into trouble later on, you will need the documentation to defend your position against a variety of possible criticisms.

- Failure to hold construction scheduling and review meetings with client and management, to properly check progress and coordinate the work. A corollary to this is to hold too many meetings, which wastes valuable time.

- Failure to control meetings, which makes them nonproductive. I will detail a procedure for avoiding this in Chapter 11, Project Communications.

■ Failure to exercise proper management control over field activities. More money and time are spent here than in any other part of the project.

Most of these tips start with the word "failure," and that is what you are going to have if you don't avoid the pitfalls listed in this section on project controls. Keep in mind that it takes about three or four good projects to wipe out the bad reputation acquired from one bad one!

References

1. Forrest D. Clark and A. B. Lorenzoni, *Applied Cost Engineering,* 2d ed., Marcel Dekker, Inc., New York, 1985.
2. H. W. Parker and C. H. Oglesby, *Methods Improvement for Construction Managers,* McGraw-Hill, New York, 1972.
3. Arthur E. Kerridge and Charles H. Vervalen, *Engineering and Construction Project Management,* Gulf Publishing Company, Houston, TX, 1986.
4. The Business Roundtable, *Scheduled Overtime Effect on Construction Projects,* CICE Report C-2, November 1980, 200 Park Avenue, New York, NY, 10166.
5. The Business Roundtable, *Measuring Productivity in Construction,* CICE Report A-1, September 1982, 200 Park Avenue, New York, NY 10166.
6. Louis Alfeld, *Construction Productivity,* McGraw-Hill, New York, 1988.
7. William Peña, *Problem Seeking: An Architectural Programming Primer,* 3d ed., AIA Press, Washington, DC, 1987.
8. Dan S. Brock and Lystrel L. Sutcliffe, Jr., *Field Inspection Handbook,* McGraw-Hill, New York, 1986.
9. Howard L. Shapiro, *Cranes and Derricks,* 2d ed., McGraw-Hill, New York, 1991.

Case Study Instructions

1. Prepare an outline of the cost-control program that you feel best suits all phases of your selected project. Also, list the goals of the program and how you would present your cost-control philosophy to the key members of your project team.

2. What are the most attractive areas of cost reduction on your project? How would you go about investigating and screening them for presentation to your client?

3. Assume that you have full procurement responsibility on your project, if the scope does not already include it. What key features would you look for in your project procurement agent's proposed procurement program?

4. Outline your proposed schedule-control program in conformance with the type of schedule selected earlier for your project. Assume the schedule S curve shows you are 5 percent behind schedule at the 50 percent complete point. What is your program for getting the work back on schedule? Would your solution be any different if

the procurement commitment curve were also 5 percent behind schedule?

5. Your labor survey report indicates a potential labor shortage during the peak construction effort. Your management wants to start out working five ten-hour days a week in order to attract the necessary craft labor. As construction manager, would you support or try to defeat the proposal? State your reasons either way.

6. Your construction productivity is below plan at the 50 percent complete point. What percent productivity should you be showing at that point? What do you propose to do about it? How are you measuring earned value and/or physical progress?

7. How do you plan to control and report the procurement and delivery status of equipment and bulk materials on your project? Sketch an outline format for the type of report you would use.

8. Outline a comprehensive quality-control program for all phases of your selected project. Make a detailed outline of the facility turnover sequence and procedure that you recommend for your project.

9. Make a list of areas in which you would be especially careful in operating the controls on your project. Expand on any areas that you think are particularly critical. Are there any areas in operating the controls where you have had problems in the past? How will you try to improve in those areas?

Construction Project Execution

In prior chapters we have been discussing how to meet our project goals by *planning, organizing,* and *controlling* a construction project as individual operations. Much of the material we covered was background information on how we got into the position of being the executing contractor. This chapter looks at the key operations one is likely to encounter while managing a construction project.

Using some imagination should allow readers of these discussions to place themselves in the role of either the owner's or contractor's construction manager. Owner's projects usually are handled with captive forces and subcontractors, or by letting whole or parts of a project to outside firms. In the second case, CMs are acting in the owner's behalf. The operating procedures are very similar in both cases except for the contractor's *profit motive.*

Construction Project Execution

In this chapter we will walk through the major construction management activities from the notice to proceed to the final turnover of the facility to the owner. I will try to make the project activities broad enough to suit readers serving the major segments of the construction industry. However, the variations among the four major contracting strategies (shown in Fig. 2.2) used in the execution of a construction project are great enough to require individual discussion in some cases.

The four basic contracting strategies we will cover are:

1. Owner lets lump-sum construction contract(s) bid on completed design by A&E.

2. Owner contracts for design-build with single contractor on CPFF or performance-incentive basis.

3. Owner lets separate CPFF or incentive contracts for design-procure and third-party self-perform construction contract.

4. Owner lets separate design contract and separate construction management contract. Construction management could be by the design firm or a third-party constructor.

Although the above four methods can be blended and modified in a number of ways to suit the owner's project execution needs, those variations should fall within the execution modes of the basic four. Naturally the project execution mode must always be modified to suit the owner's and CM's management goals for a given project.

We set our starting point for this chapter at the point of the receipt of a signed contract, letter of intent, or an official notice to proceed with the contract. The CM is in the starting blocks, the starter's gun has sounded, and the race is on! Some important things to remember are: the schedule clock is running, money is being spent, and it's now time to finalize the execution plan and build the project team.

Basic project parameters

Many of the basic matters affecting the execution should already have been at least partially resolved. The construction manager and/or the project director should have been selected and should be available for duty on the project. The construction labor posture and the strategic end date should have been set. The overall budget or cost target has been agreed to by the owner and contractors. Some sort of design basis ranging from a sound conceptual design basis to the complete design has been set. In most cases the site has been selected. The project scope and scope of services have been adequately defined in the contract. The project has been suitably financed to service the planned cashflow. The quality goals should have been set in the contract or the design basis. Some sort of construction cost estimate is in hand.

If any of the above matters is not in place, project execution problems resulting in unmet project expectations almost certainly will occur. At least the owner's and contractor's project management will have to take any missing factors into account in executing the work.

Construction project initiation

This phase of the project is especially critical because resources vital to meeting the project goals can be squandered on false starts, poor planning, and organizational mistakes. The project initiation covers the brief period from contract signing until the field office is up and running. During that period the CM is on the run between the home office and the field and is intensely involved with getting the project's administrative procedures into place. This can be a really chaotic period for the CM if sound management practices and procedures are lacking.

Our first example will be a project that has been bid lump-sum against a set of completed design documents. The ideal approach to the project would have the CM involved during the proposal and bidding stages. If that is not the case, the CM must be available *at least* 80 percent of the time to kick off the new project. The contractor's management must keep in mind that the *new* CM will be only about 50 percent efficient during the first few weeks until he or she gets up to project speed. Simple math tells us that the effectiveness ratio in those critical first weeks of the project could be as low as 40 percent (80 percent × 50 percent = 40 percent). Sound management cannot consider that an acceptable figure!

Project initiation assets

The construction firm has spent a considerable sum of money developing documentation for getting into the position of contract award. That preliminary project data should not be ignored, and the time used in its preparation wasted. Valuable items on hand at this point are:

1. The contract
2. The cost estimate
3. A preliminary schedule
4. The contracting plan and construction technology to be used
5. Firm proposals for materials and subcontracts
6. A preliminary project-execution plan
7. A preliminary project-organization chart
8. A pool of home office and field personnel
9. A complete package of project design documents
10. Preliminary design of temporary field facilities
11. Contacts with the owner's and designer's project people
12. The field site inspection and labor survey
13. Estimated field craft-labor hours and staffing requirements
14. A list of needed heavy construction equipment and small tools

15. Corporate operating standards and manuals

16. Other available corporate resources

17. The project proposal files

18. A slate of unsuccessful contractors' bids

The above list is quite comprehensive but still doesn't include all project specific data already available to the CM team. Its scope does give a good idea of what a new CM faces if he or she wasn't involved from day one in the proposal and preparation of the data.

There are a few other necessary project factors that must be in place before we can have a successful project:

1. An effective owner's project team dedicated to the project

2. Input from the design team for approvals, design interpretations, quality reviews, etc.

3. An existing infrastructure of business and government services serving the project site

4. A competitive marketplace for necessary project goods and services

5. Adequate project financing

These are key factors that can make or break the project and are outside the contractor's control. It's important to note that while any of these five factors can seriously affect the project schedule, they are *not* subject to force majeure under the contract.

The first thing the CM has to do is make a project priority list for the key project-initiation activities. Such a list might look like this:

1. Get up to speed on existing project documentation.

2. See that a contract number and accounting procedures are in place.

3. Order long-delivery materials and equipment.

4. Place the one or two basic subcontracts to open the site. (For example: site preparation, demolition, etc.)

5. Organize in-house and client project kickoff meetings.

6. Finalize a detailed project construction schedule.

7. Convert the project cost estimate to a control budget.

8. Finalize the organization chart and start bringing key people on board. (For example, construction superintendent, project scheduler, field engineer, cost engineer, etc.)

9. Finalize temporary site facility design and schedule installation of same.

10. Start preparation of Field Procedure Manual (FPM) and issue within three to four weeks.

11. Meet with personnel/labor relations manager(s) to assess field staff requirements and site-labor agreements.

12. Finalize the field overhead budget.

13. Initiate heavy-equipment and small-tools policies

14. Ensure that required field insurance, bonding coverage, and risk-management needs have been met.

15. Set up a contract administration tickle file for contractual requirements to be proactive in those matters.

These are a few of the normal project-initiation pressure points that are found on most construction projects. They are not necessarily listed in order of priority. Project goals unique to your specific project may also bring many more priority items into play.

Project initiation is an especially important time to get the work off on a solid footing. The owner's project and design staff will be evaluating your overall project performance during the difficult start-up period. In addition to the effects of Murphy's Law, there are a number of areas mitigating against a good performance during this critical time. Things like a new project environment, a newly organized team, adding to ongoing projects in the office, and availability of key people are a few that come to mind. The CM must be especially watchful for *people problems* during this formative stage to ensure a good foundation for the project's client and staff relations. Construction managers who can extend their abilities to the esoteric area of *human relations* will be the outstanding performers in today's difficult construction marketplace.

The Project Kickoff Meetings

Although many technically-oriented CMs consider them necessary evils, project kickoff meetings are valuable communications tools designed to get the project off the ground smoothly. Timing the meetings is important because the internal meeting needs to be before the client meeting, and the client usually sets the second meeting date a week or two after contract signing.

The in-house meeting is designed to acquaint the company management, key home office people, and the new members of the project team with the goals and needs for the new project. The agenda should allow space for top management input, a general scope review, a summary of the main contractual points, an overview of the project execution plan, a schedule and budget review, the proposed project organization, a description of the job site, and a review of the client's goals for the project. If possible the CM should act as the organizer for the meeting and moderate the agenda, calling on specialists as required. Naturally, attendees should be encouraged to ask questions to ensure as complete a transfer of project information as possible.

Spending a few hours with the in-house kickoff meeting saves valuable hours later on when the CM marshals the corporate resources from those groups represented at the meeting. The CM doesn't have to give each one a separate introduction to the project. Having the top management in the meeting to reinforce the importance of everyone's supporting the CM in the project is also vital to success. Again, this is a proactive effort in good *human relations* for the project! Methods for *planning, organizing, and controlling* all project meetings are covered in detail in Chapter 11, Project Communications.

The client's project kickoff meeting in the lump-sum, construct-only mode is not as critical or as time-consuming as in the other modes. By the time the project has been designed and bid, most of the problem areas and questions have been brought out and resolved. About the only thing left to do is cover any matters that have developed since the contract signing and to introduce the new people assigned to the project. Chief among those are the owner's and A&E field inspectors, the construction superintendent, the chief field engineer, and the like.

The agenda will most likely be set by the owner or the A&E, since this is their meeting. The person in charge of the agenda will assign those parts of the meeting to be covered by the contractor. The CM usually is responsible for developing presentations for the contractor's part of the program.

This is also a good time to review the channels of communications between the organizations involved, reviewing the details of the payment procedures and the project goals, and getting a general overview of the construction execution plan. The latest development of the final construction schedule, especially how key milestone dates will be met, is also a good topic to review. Key milestone dates can involve beating certain seasonal problem areas such as winter, monsoon rains, labor shortages, and the like. Be sure to include the criticality of any owner and A&E inputs to meeting the key milestone dates. Now that the contract has been signed, it's time to start focusing on those key areas critical to meeting the project goals.

The owner's kickoff meeting presents the contractor with an excellent opportunity to present an image-building performance. CMs always must remember that they represent a potent selling feature for the construction firm. Successful contracting firms never let the selling effort stop with the signing of the contract. This is Act 1, where you tell the client what you are going to do. Act 2 is doing the job as you had told them you would, and Act 3 is turning over the job with all the project goals met. Pulling off that sort of complete performance requires the CM's undivided attention to the management, technical, and human relations aspects of the project from start to finish. Putting it another way, the CM must learn to practice *total construction project management!*

Kickoff meetings in a design-build contract mode

The design team takes the lead role in this situation, with construction playing an important supporting role. The role played by the design project manager is fully described in my companion book *Total Engineering Project Management,* so I won't repeat it here. Because it will be some time before the CM is actively assigned to the project, the initial construction input usually is done by home office construction people. The designated CM may be used on a part-time basis, schedule permitting. The construction input at this early stage doesn't involve high personnel commitment, so a satisfactory arrangement usually can be made.

A separate construction kickoff meeting is held when it's time to open the field operations several months later. The CM takes the lead role in those kickoff meetings, with support from the design manager and the project director. The meetings are handled in much the same way as the bid-to-final-design meetings described above. In the latter case, however, the A&E and the construction contractor are the same firm, so only the owner and the contractor are involved.

In this mode, attention has to be given to the design document delivery dates. Once the construction has started on a fast-track project, it can't be delayed by late design information. The same applies to the delivery of any engineered equipment. Both of these factors must be considered when setting an ideal field-opening date.

Kickoff meetings with a third-party constructor

These meetings occur when the owner has let a design and equipment procurement contract followed by a construction and bulk procurement contract to construct the facility. In this mode the client construction kickoff meeting takes place shortly after the signing of the design contract. Remember, the big advantage of this contracting mode is that it fast-tracks the project while using separate design and construction contractors.

Because the design and construction contractors have no contractual connection except through the owner, it is important that they establish a detailed operating procedure, subject to approval by the owner. The owner lays out its requirements in the kickoff meeting and assigns the development of the key parts to the participants in accordance with the division of work for the project. Later meetings among the owner, designer, and constructor will lead to the detailed coordination procedure some weeks later.

In general, the client and internal kickoff meetings should follow the formats discussed above. Some variations can be expected depending on the client's needs, the size and complexity of the project, and other project factors. The general rule is to be responsive to the needs of the meeting attendees and the information to be transferred.

Kickoff meetings in the construction management mode

Project kickoff meetings in the construction management mode are somewhat less critical because the contractor is acting as an extension of the owner's organization and the field staff requirements are much lower than in the other modes. The main area to stress is the methodology for getting the construction subcontract work packages out to bid. Because we are trying to fast-track the construction, setting the scope of the bid packages and their issue schedule is especially critical.

The methods and procedures for controlling the project cost, schedule, and quality also are very important in meeting the owner's goals. The size and quality of the field organization are of special interest to the owner, along with the field indirect cost budget.

The CM needs to tailor the kickoff meetings to suit the above special areas and to select the applicable ideas presented for the other modes as described in this section. Owners are always interested in just how the contractor plans to meet its project goals.

Starting the Materials Management Program

Regardless of the construction contracting mode, the materials management program must be started early in the project. The design-build mode allows a more deliberate procurement start because we can't buy equipment or materials before the work is designed. Even so, materials management planning starts early during the constructibility review stage.

The mode of lump-sum bid to complete design requires the greatest urgency in getting the materials plan into action. The project materials plan actually starts during the estimate preparation stage, with the taking of firm bids for materials and subcontracts for the estimate. Also, the material takeoff for the estimate gives the quantities of materials, equipment, and services required to build the project. Many bidders now are making contractors list their selected subcontractors and equipment suppliers in the bid. Any decision to change these later must be approved.

The estimating bids also included the vendors' *expected delivery* of the goods and services offered. Those data are used by the scheduling group to evaluate the owner's desired construction schedule.

As part of the preliminary scheduling work, a list of long-delivery equipment was developed. That's a list of equipment furnished by the general or subcontractors that falls on or near the critical path. The materials and equipment on that list must be placed at the head of the materials management queue. All key players in the materials management plan must be made aware of the need to order early and expedite delivery of those items.

Of equal importance to getting the project initiated and the site opened is the placement of any subcontracted work in the site-development package. This can vary from some minor subcontracts for underground utilities or demolition work to the complete site-development package. Subcontracts or letters of intent should be ready to go for that type of work in the first days after contract signing. Paperwork for those subcontracts should have been started during the final contract negotiations.

A good materials management plan has to be comprehensive in scope, covering all necessary goods and services for the project. It should also cover the goods and services for takeoff, procurement, vendor documentation, delivery, and use in the field. To be effective, it must be a *total system*. The first thing we notice are the several *departmental* functions involved in materials management other than the CM. Estimating, procurement, engineering, subcontracts administration, transportation, field warehousing, and accounting, to name a few, all make a key input to getting the physical and service resources to the job when they are needed.

On smaller projects, the materials management coordinating effort falls to the CM. On larger projects, a materials manager can be set up to handle the task. In either case, the use of computers has gone a long way toward implementing the paperwork that organizes the data and executes the task. Simpler systems can use commercial spreadsheet software, while more complex systems may find available software systems tailored to the task to be more cost effective. *Cost-*

effective is the operative word here—operating the materials management system can not cost more than the funds available in the field
indirect budget.

Judging the cost-benefit ratio of the materials management plan
and control system also is very difficult. Estimating system cost is
easy enough, but estimating the cost savings or *benefit* is the problem. It's difficult to accurately estimate the cost of not having the necessary goods and services available at the site on time. Largely, the
type and operation of the material management system is a function
of general management's commitment to the system and of the CM's
judgment and ability in tailoring it to meet the needs of the individual project.

Implementing the Project Organization

This facet of the project initiation is again a function of the contracting mode, with design-build offering the least pressure and lump-
sum, bidding-to-complete design offering the most pressure to the
CM. Let's take the worst case first.

It's vital for the CM to get the most-needed assistance on board
first to help in setting up the project. People high on that list are the
chief field engineer, the scheduler, the cost engineer, procurement,
personnel, and the field office manager. The field superintendent may
also be needed early, but certainly no later than the start of field personnel buildup. Please note that I used the definite article *the* rather
than the indefinite *a* to indicate that these should be the *permanent
people* for the job. It is short-sighted to assign temporary people just
to get the job off the ground. The permanent replacements will only
change everything the temporary people did, making for total confusion later on.

With the above key people on board, the CM can start delegating
some of the high-priority project initiation activities to them as shown
in Table 9.1.

The number and type of early field staff buildup depends largely on
the size and complexity of the project, the schedule, site location, and
the field indirect cost budget. It's up to the CM to exercise sound judgment in tailoring the early staff assignments to suit the contracting
mode and to create a cost-effective management team during the vital
project-initiation phase.

Field-labor policy and staffing

The field-labor policy that was agreed to with the owner during the
project bidding and negotiation stages is now ready for implementa-

Table 9.1   Project Initiation Staff Assignments

Key person	Project initiation responsibility
Chief field engineer	Design and execution of temporary site facilities; set quality-control program; set up document distribution; establish design interface; set up technical files and procedures; set field monuments; monitor site-development work; implement applicable government regulations.
Field scheduler	Finalize construction schedule for approval and issue; prepare field scheduling procedures for FPM.
Field cost engineer	Prepare and issue approved project control budget; prepare field cost-control procedures for FPM.
Procurement manager	Start materials management program; ready first material and subcontract orders for approval and issue.
Field personnel manager	Assist with project staffing; organize site labor agreements; determine craft-labor sources; finalize site-labor agreements; implement and monitor government labor regulations; etc.
Field office manager	Start to organize field office staff, equipment, bank accounts, communications systems, and field office procedures. Publish approved FPM as required.
Field superintendent	Organize field staff; set construction equipment and tool requirements; study design documents; evaluate safety needs; interview and select area engineers.

tion. The CM, in conjunction with the personnel or labor relations department, sees to it that the proper site agreements for the planned field-labor sourcing are in effect and ready for use.

The field craft-labor requirements are finalized and checked against the available labor pool. The most important people in this group are the craft foremen, the craft supervisors, and the area engineers. This group forms the vital interface between labor and management that is critical to good craft-labor productivity. The foremen and craft supervisors often come up through the ranks, so they are more labor- than management-oriented. Anything the CM and personnel people can do to strengthen the craft supervisor's and foremen's company loyalty will improve project performance.

Physically bringing craft labor on board must be done carefully to protect the project budget. The crew mix and the quality of people required for each craft must be evaluated and established before bringing any craft people on board. A reasonable backlog of work for each craft also should exist before loading up the job.

In the construction management contracting mode, the craft labor is furnished by the subcontractors. However, it's still the managing contractor's CM who is responsible for seeing that the subcontractors have sufficient qualified labor on the site to perform the work. The managing contractor and the owner set up the site-labor posture and the site agreements under which the subcontractors will operate. The goal of that arrangement is to ensure a compatible site-labor situation to minimize labor disputes among all participants at the construction site. That constitutes a strong argument for going the construction management route.

CMs play a strong role in site-labor relations, even though there may be a resident labor or personnel manager assigned to the field staff to handle the day-to-day labor matters. Normally the role consists of resolving those labor problems that have become problem areas, such as work stoppages, working rules, or the like.

As with most construction problems, they should be solved at the lowest level possible, so CMs should try to solve them at the site level. If reinforcements are necessary, call in the home office labor experts. Labor relations generally respond best to a strong approach and to living up to the site-labor agreements.

The field procedure manual (FPM)

The FPM is the seminal document for organizing and controlling the field operations to meet the project goals. The contents are described in Chapter 8, so we will cover only timing and need here.

The main value of the FPM in the early stages is as an indoctrination tool for new people coming onto the project, so it must be issued early to be of value. Typically, an FPM is modeled on a previous version used on a similar project. The easiest way is to use the prior similar FPM and mark it up for present project conditions. That way a preliminary issue can be made rather quickly and then polished up as new project information becomes available. "Holds" can be put on those areas that are not yet firm in the preliminary issue. If the old model was produced on a word processor, the production of the new model is even quicker.

On cost-reimbursable jobs, the FPM is issued for client approval to ensure that the owner agrees with the procedures set up to control the project performance and cost. That presents no problem because CPFF work normally involves an open project-execution format, with owner's right to audit the contractor's books for the project.

The main goals of the FPM are to set the project ground rules and to reduce about 90 percent of the field administrative activities to routine tasks. That allows for about a 10 percent demand of routine activities on the CM's time. The other 90 percent can be devoted to the nonroutine activities that normally beset a CM during a project. CMs need to remember that concept when developing the FPM, in order to force the decision-making process down to the lowest possible level on the project.

The FPM should be completed and issued in final form within the first 30 to 60 days of the project start. The FPM is always subject to revision as the project progresses, and changes to the procedures may become necessary. Producing the document on a word processor and issuing it in looseleaf form makes the updating process relatively painless.

The contracting format has a considerable effect on the FPM content and structure. Those effects should be obvious to the CM practitioner, so we won't go into the details here.

Finalizing the project budget and schedule

These two essential control documents may have undergone some changes during the final contractor selection process and contract negotiations. In other words, the management might have had to do some price negotiating and schedule revision to nail down the job. If the CM was not present at those times, the changes may be startling because they rarely become more favorable to the contractor.

If the contractor's negotiators did their job right, they may also have negotiated some corresponding scope changes to suit the price and schedule changes. These revisions have to be incorporated into the project scope description, the budget, and the schedule to get everything in writing. The CM needs to see that these matters have been taken care of and are reflected in the final project documents.

The details of these matters should be investigated by the newly appointed project cost and scheduling people. Any positive or negative effects of these changes should be investigated in detail. After the final budget and schedule have been approved by management, they can officially be issued as project documents. These are the baseline documents for reaching the construction team's project financial and time goals.

Risk analysis

The CM must promptly analyze the project documents to ascertain those areas of risk that the firm has assumed in the contract. The chief and simplest of those risks are the ones covered by bonding and insurance. The CM must check the insurance and bonding requirements of the contract and see that the proper coverages are in effect for the prime and subcontracts and are maintained for the life of the contract.

Other less well-defined areas also must be evaluated so that a defense mechanism can be activated as required. Among those areas are the scope of work definition and the change-order procedure, the budget and cost-control procedures, any schedule or budget penalties or incentives, quality-control requirements, currency fluctuations, or the like. None of these is fully coverable by insurance policies or performance bonds. These risks can be controlled only by the project team's exercise of *total construction project management,* practiced under the watchful eye of the CM. The CM has to take a leadership role in this area to infuse his or her construction management philosophy into all areas throughout the project execution.

Temporary site facilities

The temporary site facilities were planned and estimated during the project bidding phase so their cost could be included in the contract price, and now it's time to put them in place. In Table 9.1, the CM delegates that task to the field engineer.

The CM oversees that work to ensure that the basic plan is adhered to, that the best temporary facilities are obtained within the budget, and that they are adequate for the service. Use of existing facilities should be maximized where economically justifiable. The temporary facilities must be ready to serve the needs of the field operations in a timely and economical manner.

The temporary site facilities are handled a little differently under the construction management contracting mode. The managing contractor is responsible for furnishing the common facilities such as access roads, temporary utilities, site fencing, first-aid facilities, and the like. This is done to eliminate expensive duplication by the various subcontractors. If the owner doesn't want the managing contractor to furnish any of these items in its scope, they can be apportioned to the most applicable subcontractor's scope of work. The underlying criterion is to see that the temporary site facilities are covered by the contractor who is in the best economic position to supply reliable temporary site services.

Summary of project initiation activities

This discussion has covered the major tasks involved in the hyperactive start-up period of the average construction project. There are probably hundreds of more minor ones you have experienced that were brought on by a particular project's circumstances. The main thing to remember is to put any important activities for starting that project in the right place on your priority list.

We have stressed certain areas involving human relations as being especially critical during project start-up. If you get them off to a good start, it makes it easier to continue them throughout the project, which is equally important.

Often productivity during project initiation is low because of the newness of the organization, so it's possible to get behind schedule in the early phases. If the time lost is minor, you may be able to make it up with improved productivity later during the field labor peak. We all know, however, that there are plenty of reasons for that not to happen without some good luck. Therefore, it is absolutely critical to do the project initiation right to avoid problems later.

A key factor in project initiation is to have your project look-ahead system in operation. Normally, CMs have the field scheduler give them a two- to four-week look-ahead schedule of construction milestones. That prevents the CMs from getting so entangled in current issues that they lose track of their longer-term goals. Don't let that happen to you during project initiation!

Project-Execution Phase

The move from project-initiation to project-execution mode is virtually undetectable except for the disappearance of initiation activities from your priority list. As time goes on, *planning and organizing* activities decrease and you move into the project *control* mode.

The FPM establishes the working rules for operating in the control mode by laying out the routine control activities and reports that are to be issued periodically. The FPM's goal of making 90 percent of the field activities take place routinely allows the CM and key staff members to devote most of their energies to handling nonroutine tasks and putting out fires. That gives the field staff the means to *manage by exception,* which means that on-target matters proceed routinely and off-target matters get immediate attention for corrective action.

Project buildup phase

Field activity has now passed through the lower part of the S curve as the craft labor begins to build toward peak loading. Cashflow is starting to increase with the growth of field payrolls, and project material and invoices are starting to arrive. The materials management system is in operation and the field warehouse is starting to fill up with incoming equipment and material. Subcontractors are moving in and setting up to start their work. The weekly scheduling, cost-control, and coordination meetings have been scheduled. The technical and administrative functions are in operation and the job has fallen into its routine operating mode. The CM is now at the site full-time, monitoring and coordinating the field activities.

The happy scenario described above can happen only if the CM has done a superior job of *planning and organizing* the work in the project initiation phase. The keys to success in that regard are the selection of proven supervisors and *effective delegation* of their duties in a *written* job description for that project. The job description should spell out the goals and duties of each supervisor in simple terms that form the basis for routine performance monitoring.

Handling the paperwork

Proper application of those management tools places the CM in a position of monitoring the routine project reports for problem areas combined with trips out into the trenches to observe the actual work in progress. Part of the CM's routine should include regular attendance at weekly meetings run by others. Examples of these are the weekly scheduling and field superintendents' coordination meetings. When sitting in as an observer, the CM must be careful not to usurp control of the meeting, which sometimes happens by default. Remember, you are there to gather information and to observe the performance of the participants. If the occasion demands your input, provide it carefully and in a low-key fashion.

A checklist of the routine items CMs should monitor to assure themselves that administrative matters are proceeding according to plan is as follows:

1. Weekly schedule evaluation and report
2. Weekly craft productivity and loading report
3. Weekly staffing report
4. Material status and expediting reports
5. Warehouse activity report
6. Change-order log
7. Quality-control reports
8. Cashflow and payroll reports
9. Safety, accident, and security reports
10. Purchase orders
11. Invoice payment report
12. Subcontractor status reports
13. Heavy-equipment operating report

14. Minutes of meetings

15. Copies of incoming transmittals and correspondence

16. Field office cost report

17. Daily log

The above list of key items plus the day-to-day incoming and outgoing correspondence creates a mountain of paperwork to review and act on. Effective CMs need to be able to speed-read in order to keep current. They should learn to scan documents, picking out the essence of the document and focusing on the hot-button areas requiring prompt action. To properly manage a fast-moving construction project, they have to learn to read to suit the *management by exception* mode.

This list mainly covers routine reports, but good CMs also alert their staff members to raise problem areas as they arise without waiting for the routine report. This means that the CM maintains an open-door policy for key staff members. Receiving the problems in a positive, upbeat manner encourages early-warning practices in key staff members, while "shooting the messenger" creates the opposite effect. If your staff people are saving the surprises for their routine reports, they are not being properly supervised.

To counteract the office-binding effects of monitoring the routine activities on larger jobs, CMs need to set aside time each day to walk the job. It shouldn't be the same time each day so the people in the field get accustomed to the tour time. In addition to the supervisors, spend time talking to the foremen and craft workers to find out what is happening at the lowest possible level of the work. Your interest will raise morale and you may find out some surprising things about the job. This is a direct connection into the job grapevine that is a good source of intelligence on the inner workings of the project.

It is also a good idea to walk the job with the field superintendent or the field engineer at least once a week to get their feel for how the work is going. That way they won't think you are always walking the job alone to find fault with the work. The solo days will give you enough opportunity to form an independent opinion. The joint site inspections are also a good time to see how well your key people are learning to conduct proactive site tours for visiting dignitaries. Remember, one of your duties is to train future CMs and to develop a trained assistant to leave in charge when you are away from the site.

In addition to the human factors mentioned above, CMs also need to note the physical progress and craft productivity in all areas, comparing it to that claimed in the reports. This calls for that entrepreneurial sixth sense that outstanding CMs develop with experience.

That sixth sense may not come naturally, so you may have to work at it to achieve it.

Discovering the job's critical areas

Each type of construction project has its typical list of critical areas. It's up to the CM to look for those likely problem areas before they go critical in accord with Murphy's Law. Fortunately, the CM has some management tools and many years of construction experience to draw on. Evaluating the project for critical areas forms the basis for building the CM's ongoing project priority list.

The first place to look is the CPM schedule, if one is available. The list of items and milestones that are sorted with the lowest total float are the ones to watch. As we said earlier, the critical-path items will thread through the whole job. Naturally, the contracting plan also affects the critical path. If the project is being fast-tracked, for example, the design, procurement, and construction are intimately dependent on each other. As the CM is the manager of the last party in the project parade, that mode places the greatest stress on him or her. The CM's successful performance is at the mercy of the design schedule and quality, along with the delivery of the equipment and materials. That's in addition to the normal day-to-day construction problem areas.

For the lump-sum, bid-to-complete design project, the CM faces the problem of early ordering of critical equipment, materials, and services to meet the schedule. There is usually a great deal of pressure on the construction schedule because the earlier design and bidding procedures have already eaten up most or all of the planned float. The lump-sum mode places most of the contractual risks for cost and profit onto the construction contractor, with heavy emphasis on the cost-control side of the project. Close attention must be given to project scope, quality, change orders, and the like.

The construction management contracting mode places most emphasis on management practice because most of the contractual cost and schedule risks are passed on to the subcontractors. The CM must be mindful of the qualifications and capabilities of the winning subcontractors to ensure a good overall project performance. The coordinating role also emphasizes the CM's strengths in human relations, problem solving, and leadership in administering the subcontracts. The CM's priority list starts with the overall project goals, followed by the individual problem areas of the subcontractors.

The CM's intimate role in developing the construction schedule has the benefit of highlighting the project's high-priority problem areas.

Running a CPM schedule as a project-execution model on a computer gives the most detailed analysis of impending problem areas. Doing some "what-if" analyses on the CPM before and during the project brings the problem areas into even sharper focus.

Discovering the job's problem areas is only half the battle. Effectively solving them is the other half. Problem solving and decision making is much too broad a subject for us to cover in these few paragraphs, but we will talk about them later in our discussion of human factors in Chapter 12. Solving problems calls for a single-minded approach while keeping the overall project goals in mind. The CM must evaluate the risk-versus-reward ratio and a fallback position when evaluating the solution. Also, CMs must draw heavily on their own and their staff's experience and know-how in the decision-making process. After considering the time factor and all the data available, the CM must make the decision.

Operating the construction quality-control systems

Quality-control procedures are vital to the success of any construction project. Although quality control is delegated to the field engineer, it's the CM's responsibility to see that effective procedures are in place, are documented, and function correctly throughout the project.

Each type of construction project has its specific quality-control hot buttons that must be recognized by CMs for close attention. A couple of outstanding examples of how this matter was not properly handled are numerous nuclear power plants and the famous Trans-Alaskan Pipeline project. No doubt there are still many CMs with bruised reputations left over from those quality-plagued projects. No owner, construction firm, or its CMs can do less than their best when it comes to maintaining quality control and assurance on their construction projects!

Documenting the project

Properly documenting a construction project involves virtually every aspect of the work. Most documentation already is required by the contract, the FPM, company procedures, government regulations, departmental standards, and the like. The project documentation is collected and organized in the various project files, and is tied together by the CM's project file. CMs should evaluate the existing systems as applied to the needs of their present project. Minor modifications or additions to the existing documentation system may be required to ensure meeting the present project's goals. Wholesale changes to the existing systems may not be well received by your firm's management and coworkers, so care is required in that regard.

Why bother going to the expense of documenting a project? Could a project run without documentation? Perhaps, but even if the project ran well, we still would need documentation for efficient day-to-day operations. Project operations feed on a continuous flow of accurate and timely information. If there isn't a smooth flow of labor, materials, tools, and equipment to the field construction forces, productivity suffers badly, and we cannot meet the project goals. It would be virtually impossible to communicate the necessary information orally because nobody could remember who said what to whom, and who had the authority to say it.

In addition to all the operational benefits of paperwork, it's needed to form a project history or paper trail for past and future needs. When things go wrong, as they often do (schedules are not met, budgets overrun, quality is substandard), we need the paper trail to figure out the solution and the proper corrective action to solve the problem.

It is the CM's responsibility to see that the field activities are properly documented. People sometimes become lazy and forget to write minutes of meetings, telephone call confirmations, change-order submittals, each one representing a potential information gap in the project documentation. Usually those are the documents that are key to winning or defending a claim or similar financial obligation. Meticulous and well-organized project documentation is just another of the hallmarks of the superlative CM.

Project documentation culminates in the monthly project progress reports. CMs are responsible for preparing or at least editing these documents. The quality and appearance of the monthly reports tend to set the tone for construction project operations in general, so a good set of reports is essential. CMs must be sure that their reports will stand up to thorough scrutiny and project an exemplary project image.

The Owner, Constructor, and Designer Interfaces

The interfaces among the owner, designer, and the constructor are the most critical communication channels on the project. The whole body of design information flows through these channels. Therefore, it is vital that these interfaces are handled effectively to further everyone's project goals.

In my forty-five years in the engineering-construction industry, I have worked on all three facets of this interface. During that time I have seen the interface function in varying degrees of success, from a spirit of responsive cooperation to open warfare. As we will find later in the discussion of human factors in Chapter 12, we just can't afford to operate at the latter end of that spectrum if we have any hope of having a successful project. None of the three key players (in our case the CM), can allow his or her personal attitudes toward the other players to adversely affect these key project relationships.

The owner, designer, and constructor all have vital stakes in making the key interfaces work. However, it's the owner who has the most to gain in making them operate effectively to meet the project goals. Any discord among the key parties must be investigated by the responsible people and resolved before any permanent damage can occur.

A major factor contributing to the age-old antagonism between design and construction is often the absence of contractual relations between the designer and constructor. An obvious exception is in the design-build mode where the designer and constructor are the same firm, and even that doesn't completely eliminate the problem. There probably will be a natural lack of mutual respect among the three groups over the next centuries as there has been in the past. Owners have a natural suspicion of contractors (design and construction) placing their financial goals ahead of their project goals. The owner, after all, does underwrite most of the project cost in the end. Constructors usually question the designer's ability to design the project the way the constructor thought it should be done. Countering that thought are the designers' doubts about the constructor's being able or wanting to build the project the way it was designed.

In the heat of project execution, relations between the field organization and the owner frequently can become chafed. A large measure of the stress surfaces when the contractor's profit goals are placed in jeopardy for any number of reasons. Other causes can be ingrained mutual dislike, personality problems among the team members, overbearing attitudes, abnormal construction problems, and the like.

As the leader of the field forces, the CM is responsible for maintaining good client and project relations throughout project execution. We will investigate this factor in more detail in Chapter 12. CMs must maintain a proactive owner and project relations stance, calling forth a positive attitude from all members on both sides of the unified construction team. Personal disharmony can only hurt the success of any construction project and result in dashed expectations on all sides.[1]

Obviously, none of these preconceived notions about owners, contractors, and designers is entirely true, or the construction industry wouldn't enjoy the success that it does. However, some individuals do have difficulty getting beyond their preconceptions, which lies at the root of the problem. CMs must show leadership by submerging any negative impressions and encouraging the most cooperative feeling possible among their team and the other players. Some may find that an *acquired trait* in their personalities, but it *can* be acquired!

Projects with good preconstruction planning input usually find that the design interface runs much more smoothly. The construction, design, and owner groups have effectively investigated and resolved any differences as to the various construction materials and constructibility factors by the time the project gets into the field. That leaves the probability of design and construction errors as the remaining area of contention.

Controlling design errors of omission and commission will always be with us, although computer-aided design (CAD) has eliminated some of them. Because there are about five to seven hours of field labor for each hour of design, design errors can be very expensive. Owners should be sure that their design team has an effective quality-control system in place before drawings are issued for construction. If many design errors are found early in the job, the CM should discuss the matter with the owner and designer to resolve the problem. If the design documents aren't thoroughly clear, the field team should request clarification before proceeding. Late notice of design problems can be costly and is likely to upset the flow of the construction work.

Even the contracting method can create an environment that fosters design deficiencies. The most common one occurs in the lump-sum, bid-to-complete design contracting mode. Many of the more obvious design inconsistencies are discovered and corrected during the bidding stage, but there are usually some left over to be discovered and resolved during construction. These errors are in addition to the systemic problems described next.

In the interest of increasing competition among system suppliers, and to make it possible for designers to accept the design variations of multiple suppliers, designers often design to a general standard. For example, a building design with a large window-wall requirement may not be completely detailed or specified to keep the application open for several window-wall suppliers. After the most competitive system has been selected, it is up to the general contractor or the system supplier to complete the final installation details. That in turn may involve some additional cost that wasn't made clear in the bidding documents. This is a fertile area for problems in the design-construct interface, even though a statement in the general conditions may mention that possibility in general terms.

Other possible problem areas are cost and specification matters arising during the shop drawing approval process. When alternate sources and designs of materials and equipment are submitted, there is often disagreement on the quality and equivalence of the alternate products used in the lump-sum bids.

Another case occurs when the fast-track scheduling approach is being used and the field is opened too soon to suit the design schedule or the design schedule falls behind. In either case, the schedule and field-delay costs place additional pressure to get the design documents issued for construction *without proper checking and quality control.* That practice can lead to some horrendous problems when major field installations are installed to the incorrect drawings. That situation will eventually devolve into an often devastating breakdown in all facets of the design, owner, and construction interface, resulting in serious unmet project goals and additional costs.

All three players have a stake in avoiding this sort of situation. The CM is placed in a difficult position in trying to rectify the resulting technical and financial problems in the field. The owner faces the bottom line of having to pay for the nonproductive extra costs involved. The design firm faces financial as well as reputation losses. There is very little chance of anyone coming out clean in this no-win situation, so avoid it at all costs.

Looking at the positive side of the field-design interface, it's important to have its operation well organized. A channel between the field technical representative (usually the field engineer) and a designated person in the design office must be arranged to handle routine design document interpretation and to deal with unforeseen problem areas. As we said earlier, there is a normal flow between these groups in the approval process for vendors' data, design document revisions, and other technical matters. Problem areas not fitting into the routine flow must be handled separately as they arise. Change orders are a case that falls into the latter category. They must be handled as called for in the contract and well documented, including official approval or disapproval. Don't let unresolved change orders accumulate because they cost money and they must be resolved one way or the other. CMs cannot ignore the change-order process and the need for keeping the change-order log current!

In the construction management mode, the CM and the field engineer handle the interface between the construction subcontractors and the design organization. In this case the design organization may be part of the construction management team or a separate organization. In either case the legal status and responsibilities of each group must be defined to make the system work effectively. Because the construction and design contracts usually are made directly with the owner, the owner's project representative plays a major role in resolving conflicts.

On large and/or complex projects, it's common for the owner and design organizations to put teams in the field for quick resolution of questions and to supply an ongoing review of the construction work.

That arrangement has proven itself cost-effective by minimizing corrective work during the late stages of project completion and start-up. It also shortens the lines of communication and allows the key players in the design-construction-owner interfaces to deal on a face-to-face basis in the field. CMs are well advised to promote that sort of arrangement and to work hard to ensure that it's working effectively toward the meeting of project goals.

Project Safety and Security

The background and needs for sound construction safety and security programs will be discussed later in a separate chapter. Here, we will address the CM's role in integrating those two important functions into the ongoing site operations.

Site safety is an important function regardless of project size. Certainly, larger and more complex construction sites do require more sophisticated safety programs. On small projects, the CM may even wear the site safety engineer's hat. On megaprojects, the site safety group must be more sophisticated to meet the special needs of that large-project environment. Owners also play a role by setting the tone for site safety in the contract documents. The latest and perhaps largest players in today's site-safety programs are government agencies such as OSHA and EPA and their mandatory construction safety regulations. Any site-safety program must at least comply with those regulations at a minimum. Many larger construction firms exceed the minimum standards based on safety considerations and accident experience in their particular type of work. They have found that it pays dividends in containing workers' compensation costs and losses due to lost-time accidents.

CMs ultimately are responsible for site safety and accidents that occur on their jobs, even though a safety engineer or department handles the details of the site-safety program. Normally it's the duty of the CM to notify the next of kin in the case of serious accidents or fatalities. Having handled that emotional experience a few times generally commits a CM to spearheading an accident-free safety program. As part of the CM's ongoing management, he or she must constantly be evaluating the field operations and the efficacy of the site-safety program. Revisions to safety procedures may have to be made to suit some unanticipated hazardous conditions. This can be especially true when working in operating facilities engaged in ongoing hazardous operations.

Just as the firm's general safety programs are founded on the support of top management, the site-safety program is based on the CM's full commitment. The CM must be proactive by word and deed to continuously promote safety on the site. CMs should lead the way in setting up work-safety goals and in organizing an award system to reward their successful achievement. Following through to see that the incentives are carried through when the goals have been met also is important. That can be difficult to do on smaller projects when the craft-labor force on the job contracts quickly near the end. Usually the reward functions are held near the end of the job, when high-risk operations have been completed.

The site-safety engineer usually handles the details for the governmental safety inspections, but the CM should participate in resolving any shortcomings or dealings with citations raised by the inspectors. It's usually not cost-effective to fight the inspectors on claims unless you feel strongly that you can win the case on appeal. Also, those items involving high additional cost must be examined closely and resolved by negotiation if possible. That situation sometimes arises when a costly safety or environmental procedure has been overlooked during the estimating phase. The example illustrates the point that the safety program actually starts during the estimating phase of the project.

A successful safety program relies heavily on a good on-site first-aid facility, backed up by an effective ambulance and hospital facility for more serious injuries. Dry-run testing of the first-aid and other emergency procedures also is effective in evaluating and maintaining their reliability.

Fire safety is another major part of overall site safety, particularly when flammables are present during construction. A high welding input to the construction also requires special fire-prevention systems. Keeping close contact with the local fire department is essential for those fires exceeding the capacity of on-site fire-protective capability. Remote sites and megaprojects may even require a self-sufficient fire department capable of handling any construction-site fire emergency.

Site security

Although security and safety aren't directly related and have separate responsibilities, they seem to be grouped together in practice. Security of the site is necessary for a lot of reasons, such as:

- General site safety
- Unauthorized personnel access control
- Protection of physical assets—tools, equipment, materials, personal effects, etc.—against theft and vandalism

- Site control during labor unrest
- Site monitoring during nonworking hours
- Contacts with local security forces
- Control of proprietary and classified information and equipment

Site security starts, where feasible, with a perimeter fence to control public access. If the design includes a security fence, get the finished fence in place early enough that it may be used for security during construction. There are many projects such as individual houses and housing tracts where fences are not feasible or even thought to be necessary. I was owner's representative on a new school for the mentally retarded where we were not allowed to fence even during construction. The board of mental retardation did not fence their properties and wouldn't allow one even during construction because of *image* considerations. Being in a vandalism prone area and not being especially welcomed by the local residents, we ran into unanticipated vandalism problems that led to unforeseen extra charges for night security.

Fences won't solve all the problems either, given the enterprising nature of thieves. We came up short 2 km of double conductor romex electrical cable on a construction site in Germany. It seems this was just the cable size suitable for wiring houses. We figured that the material had been passed through the chain link fence to an accomplice on the outside, either during or after working hours. We had the empty cable drums to prove that we had correctly ordered the necessary material!

Security can be a costly item on large and complex sites where the security budget plays a major role in the site indirect budget. On that type of project, CMs must include that item in their cost-control monitoring plans. Security costs are maximized on projects involving national security where classified documentation is handled. Even with the end of the Cold War, there are still some classified projects going on around the world.

Project Completion Phase

The project completion phase starts when the project approaches the point of *mechanical completion* or *substantial completion*, which is the time that the owner wants to take over the facility and start moving in. The owner's take-over action is sometimes referred to as taking *beneficial occupancy* of the facilities. Establishing the date of mechanical or substantial completion is especially critical because that is normally the date that the contractor's warrantee clauses for the facility start. If those dates are allowed to drag out, it results in extra time that the contractor is held responsible for repairs under the warrantee clause.

Mechanical completion occurs in industrial facilities, where it means that the buildings and process systems are *mechanically complete* but there still is minor insulation, painting, and cleanup work to be done. That phase is used to give some overlap between finishing the contract work and commencing the plant commissioning and start-up phases. Getting every last minor construction detail cleared up delays the plant commissioning phase, which in turn delays start-up. Using those intermediate completion terms and taking beneficial occupancy is another way to fast-track the plant start-up or owner move-in date.

Substantial completion is the term used for commercial projects because there is really no start-up phase involved there. The final punchlist of cleanup items is usually shorter in the commercial project case. It's also much more difficult to move in with a business operation with a lot of unfinished construction operations going on. Industrial clients are much more eager to move into uncompleted facilities so that they can use the constructor's craft personnel for maintenance until their own people are ready to take over.

Process project completion phase

Winding up a process facility construction project can be a hectic time if the closeout operations are not well organized. A great deal of coordination among the construction contractor, owner, design group, and the plant operating team is vital to the checkout, start-up, and turnover of the facility. Planning for these final events actually starts during the initial project planning stage with the scheduling of the unit-completion sequence. The first units scheduled for completion are usually the utility systems, be they simple tie-ins or complex stand-alone units.

Utility tie-ins usually are simple activities that are done fairly early in the job so that the permanent utilities can be used for construction purposes. This reduces the cost for temporary construction utilities and allows the systems to be debugged during construction. The same thinking applies to utility units, except that they come on-stream later in the job. The other offsites and process units are scheduled in a follow-on sequence that suits the start-up plan.

The complexity of phasing out a process construction project is shown by the following key construction procedures required to ensure project completion and a smooth start-up. These activities are all carried as punchlist items that must be closely coordinated with the start-up schedule. The CM is responsible for timely completion of these key items, either directly or by subcontractor:

- Pressure-testing and sign-off of piping systems
- Final P&ID check of all systems
- Electrical system checkout and energizing
- Calibration and checkout of instrumentation and control systems
- Flush piping of systems and equipment
- Installation of oils and lubricants
- Conducting rotational and mechanical tests of mechanical equipment and systems (often in conjunction with vendor's reps.)
- Installation catalysts and chemicals
- Curing field-installed refractory linings
- Turning over manufacturers' operating and maintenance manuals
- Turning units over to operations for plant commissioning and start-up

It is readily apparent that the work listed above can be quite costly and labor-intensive, involving personnel from the owner, contractor, subcontractors, and design staff. CMs must pay close attention to the scope of work in this complex and detailed area.

ANSI power piping and boiler codes require that all piping systems, vessels, tanks, heat exchangers, and the like be hydrostatically tested to ensure leak-tight and safe operating systems.

Shop-built equipment usually is tested in the shop and doesn't require retesting in the field unless the item has been field-altered. By far the largest field-testing load is in the area of field-installed piping systems, which can number in the thousands on a large process project.

Each piping system or field-erected equipment item must be tested before insulating or painting it, but only after all other fieldwork has been completed. Each test item is numbered and defined in a field-test log, including the test data and sign-off by the designated field engineers or inspectors. The owner's inspection team typically participates in or observes the tests and also signs off on each system. The systems then can be released for insulation or painting as required. The most convenient way to handle the large volume of test records is by computerized spreadsheets. The spreadsheet record is used as a punchlist to control the testing program and then becomes part of the record documents turned over to the owner at final completion.

The final activity of the design-construction interface is for the design team to check out the field installation against the final piping and instrumentation diagrams (P&IDs). This is a final quality check to ensure that all pipelines, valves, instruments, heat tracing, and the like have been properly routed and installed before start-up begins. System rework during start-up operations is time-consuming and causes safety problems.

Checking out the facility electrical system frequently involves high voltages that introduce safety problems if not handled by qualified personnel. Most often that activity is handled by the electrical subcontractor, so be sure to include it in its scope of work. If the subcontractor installed the system correctly, it should be able to bring the facility on-line electrically.

Calibration and checkout of the instrumentation systems falls into the same category as electrical systems. The instrumentation work involves more time, but is less hazardous to personnel. The main items to watch here are good record keeping, schedule, and progress of the work. Start-up operations can't begin until the control systems are operational and signed by the owner.

Flushing piping systems and connected equipment is a pretty routine operation that comes after final testing and acceptance of the system. It's amazing what comes out of piping systems and equipment during this process. The purpose is to keep construction debris such as dirt, rocks, welding rod, etc. out of close-tolerance mechanical equipment, valves, and instrumentation. In addition, temporary protective in-line screens are installed ahead of critical equipment items for the start-up period. The screens are removed later after the lines are completely clean.

The final checkout of rotating and special equipment is done by the millwrights along with the installation of lubricants. The equipment often is run for a short period to check out rotation, lube systems, speed, discharge pressures, and the like. Conveyor systems should be checked out for operability, safety switches, and interlock systems. All testing that can be done without actually handling process materials should be done to clear up potential operating problems before start-up begins. The field engineer is responsible for scheduling those vendor representatives required for installing and turning over complex mechanical equipment and/or systems.

Chemical operations often require the installation of catalyst materials, resins, filter media, and the like as part of the construction scope of work. This is another area prone to requiring supervision from vendor specialists to get it right. The timing of those installations is important to avoid damaging those costly and specialized materials.

An important requirement for start-up is the mechanical catalog consisting of the operating and maintenance instructions from equipment and system suppliers. This information is critical for the operating and maintenance support people in starting up the facility. Most process-project owners want that documentation early, so they can use it for operating and maintenance personnel training. The documentation itself normally is assembled in the home office, but its turnover to the owner is a contractual obligation so the CM also is involved to see that the material is ready on time.

Process project commissioning and start-up phase

In commissioning, construction activity passes from the lead role into a supporting mode as the facility enters the start-up phase. Final cleanup of punchlist items may still be going on to complete the project scope. There may also be some make-good work items discovered during the commissioning and start-up activities. All this detail is in addition to the contractor supplying maintenance support for the start-up operations.

Balancing the needs of these variable demands makes control of construction personnel levels a problem during this transitional period. Fine-tuning the field staff to keep enough craft and supervisory field personnel on hand to effectively support the start-up activities can be tricky. Remember, all this activity must be done according to the budget and schedule, and the CM can't afford to lose the budget during this frenetic period.

Nonprocess project completion phase

Each type of nonprocess project has its own key areas to be handled during the final checkout phase. All projects eventually pass into the final inspection and punchlist period. The contractor calls for the final inspection with the owner as a prelude to turning over the facility to the owner. Setting the final inspection date is a judgment call for the CM. Pressures from schedule penalties often force the inspection too early, which leads to lengthy, unmanageable punchlists. On the other hand, nobody closes out a project with one final inspection. The CM needs to get the owner's detailed input on the punchlist to discover any hidden problem areas and to prune the list down as quickly as possible.

Construction project general phase-out activities

Most construction punchlists include as-built drawings as a requirement in the contract. This is an item that field engineering should have been doing as any field changes occurred, so usually it's not a major problem.

Assembling and turning over to the owner the project documentation as defined in the contract is frequently a sore point in closing out the project. Problems occur if the responsible engineering group has let this area fall behind or if vendors and suppliers are late with their information. The owner won't release final payment and retention funds until the contractual requirements for documentation have been fulfilled.

On larger, multiunit facilities, most owners allow contractors to turn over units on a piecemeal basis with a final wrap-up of the whole project. CMs should take advantage of that arrangement to expedite the turnover process and to get units off their books as soon as possible. That process helps to bring the remaining construction resources to bear on completing the residual units.

While winding down the technical aspects of the construction job, CMs can't forget about closing out the administrative side of the work. The final goal of finishing the project *as specified, on time, and within budget* can slip away in the last days of the project. Many of your field staff are anxious to get away to new assignments (as are CMs) so they may slight the job closeout procedures. Remember Yogi Berra's cliché that "It ain't over till it's over," or you may experience "déjà vu all over again"!

It's very important at this stage to make sure that all outstanding subcontractor and supplier invoices and claims have been submitted, resolved, and paid to ensure a lien-free project. Hopefully, no open-ended contracts or purchase orders surfaced to ambush your job-ending cost report. The cost engineers and the accountants need to finalize the financial activities and close the books on the project in a manner suitable for an audit, should one ensue.

Scheduling and procurement should be among the early finishers on the administrative side. Their files should be condensed and made ready for storage per the applicable job closeout procedures.

The field warehouse should be ready to get out of business after the spare parts have been turned over to the owner and any surplus construction materials have been disposed of. Surplus materials often are a problem on CPFF process jobs because bulk materials on those projects are intentionally overbought to ensure a good supply on site. Surplus materials normally are turned over to owner's stock or sold off to local buyers. Although disposing of surplus materials is a one-time chore at the end of a job, it beats facing a costly and continuing shortage of materials during the project.

The temporary field services, facilities, and equipment should be phased out as the work force is reduced. Construction equipment should be removed from the project's books as soon as it is no longer needed. Company-owned small tools must be inventoried, packed, and shipped to the next project or to the home office. As the last personnel are phased off the job, the personnel files must be organized and consolidated in accordance with company procedures.

Project Closeout Procedures

The key to closing out the project is the receipt of the owner's final acceptance letter indicating that all construction and contractual matters have been properly completed. The contract typically describes in some detail how the acceptance will be handled. CMs should make sure the turnover procedure is being followed in a proactive manner to ensure a smooth transition with a minimum of revision.

Consolidating the voluminous project files seems to be the most nagging of the project closeout procedures, especially when the applicable procedures are vague or nonexistent. It's truly amazing how the volume of paper accumulates on a construction project. A general rule to follow is to save those personnel and financial data required by law and those files used to develop historical data for improving performance on following projects. Another good general rule is to eliminate duplicate files where possible.

Most firms now have the CM write a job history covering the positive and negative factors that were encountered in executing the project. Any special features in dealing with that owner or specific industry should be discussed. What went wrong and why? What or how could we have done better? What were the labor problems? Include anything you feel will help the firm or a colleague improve the next project performance. These points apply even if your job was a huge success; we must always strive to improve our project performance.

After the project has been closed out, you may still get a callback from the client during the warrantee period. This is often viewed as an irritating problem, but it can also be viewed as an *opportunity*. CMs must take the time to investigate the follow-up matter with interest to assist the owner and solve the problem if it is part of the contractual warrantee. If the firm is not legally liable, any offer of assistance probably will be appreciated. The *opportunity* part of the problem is to do some company and personal image-building while answering the inquiry. The worst thing you can do is to fail to handle the request professionally.

Any construction firm with an active sales or marketing group will follow up with clients on completed projects to see how the firm's performance was viewed. Any positive or negative feedback should be discussed with the CM who handled the job. It also doesn't hurt for

CMs to build a working business network by keeping up their personal contacts with past clients with phone calls or visits.

References

1. Robert D. Gilbreath, *Winning at Project Management,* John Wiley and Sons, New York, 1986.

Case Study Instructions

1. Prepare agendas for the proposed in-house and owner project kick-off meetings on your selected project.
2. Prepare a brief description of the temporary site facilities and services (include safety and security) you anticipate using on your selected project.
3. Describe the labor posture and labor sources for your project. How do you plan to load the craft labor onto the project? Describe briefly how you propose to handle labor relations on your project.
4. List the critical areas peculiar to your project that will require your special attention, and tell how you expect to deal with them.
5. How will you organize your project's design and owner interfaces? What steps will you take to enhance those interfacial relationships on your project?
6. How will you implement the project completion and turnover phases on your project? Include your ideas to expedite the process.
7. Briefly describe your project closeout procedures and how you propose to handle them on your project.

Chapter

10

Construction Safety and Health

Anyone working in construction has been ingrained with the concept of construction safety for as long as he or she can remember, but construction is still ranked among the highest accident-risk industries in the world. There are a number of reasons that these high numbers are endemic to the construction industry that are not found in most other businesses.

First, the industry is highly splintered into a formidable number of very diverse operating entities. There are a plethora of individual owners, contractors, subcontractors, A&E firms, and the like directly involved in the business of building facilities. This diverse group is underlain by an impressive lineup of government agencies, trade unions, trade associations, equipment manufacturers, insurers, and universities, all playing a supporting role to the main players. Collectively, these organizations have some sort of responsibility or are contributing to the construction safety picture, but not necessarily with the same safety goals in mind.

The high rate of employee turnover endemic to the construction industry is another problem area when it comes to executing an effective safety program. A high percentage of construction firms serving a widely distributed market rarely see the same craftspeople on successive projects. This makes it difficult to properly train the craft labor in the firm's safety programs and standards. Turnover also leads to unknowingly hiring a percentage of accident-prone workers who have to be retrained or weeded out. Small local contractors with long-term supervisory and craft-labor employees don't have the same problems. On the negative side, smaller contractors inherently place less emphasis on safety standards and on programs that tend to adversely affect their accident statistics.

However, we will be looking at how construction managers can overcome the fractured nature of the industry by bringing a sharper focus on effective safety programs for their individual projects. As with any key management function on the project, the safety program doesn't just happen, it must be *planned, organized, and controlled.* As the leader of the field activities, the CM must bring every available resources to bear in executing a superior project safety performance every time. CMs play a vital role in their firm's overall safety performance chain.

How Did Construction Safety Evolve?

Going back again 4000 years to the Egyptians and their monument-building, we can evaluate some of the accident and health problems they must have encountered. They, after all, represent the first organized heavy construction activity in our recorded history.

Because the Egyptian nobility were so obsessed with the afterlife and how to get there, they apparently made no written record of their construction safety problems and how they coped with them. We do know they didn't have what we today consider basic safety equipment such as hard hats, safety shoes, eye protection, respirators, and the like. As described in the *This Old Pyramid* show,[1] we know that they quarried and moved stones weighing up to 15 to 20 tons to build the pyramids. Their quarrying, tunneling, and chamber excavations in solid limestone were done by manually chipping 18-inch-wide channels around three sides to a depth equal to the stone's height. They then split out the resulting stone piece along the base by driving in bronze wedges.

The resulting stones were taken out and finished by stone-cutters using pointed copper or bronze adzes and chisels. All this stone-quarrying and excavation produced tons of rubble, flying chips, and clouds of limestone dust. It was a perfect setting for stone-cutters to contract silicosis and all sorts of respiratory diseases as we know them today. Assuming that the Pharaohs couldn't afford to eradicate their farmers, who worked on the tombs during the off-season, the workers must have developed some primitive methods to cope with the safety and health problems.

In the absence of modern safety programs as we know them today, the safety measures had to be simple *self-preservation* methods developed by the laborers themselves. Unfortunately, that self-preservation attitude seems not to have been passed down to many of our present-day construction craftspeople.

Considering worldwide construction activities for the next 4000 years, not much happened on the safety front even with the onset of the industrial revolution. In fact, the factory industries spawned in the industrial revolution had an even worse safety record than the construction industry. Finally, with the organization of trade unions, labor legislation, and workers compensation, which arrived in the early 1900s, owners were convinced that having good employee on-the-job safety and welfare programs was a sound business practice after all.

The heavy-industry portion of the industrial business then developed safety programs and basic safety equipment such as hard hats, safety shoes, protective clothing, and the like. Contractors working in those industries were then forced to adapt those heavy-industry safety rules and equipment to apply to their own employees. In response to their clients' wishes and the new legislation, contractors soon started to take a proactive stance toward on-the-job safety.

Considering the longevity of the construction industry, we can see that modern safety practices have developed only over the past five or six decades, so they are still relatively new. There is still plenty of room for improvement in the construction safety field, given the parallel improvements in technology, training, and management practices.

How Effective Is Construction Safety Today?

Despite titanic efforts on the part of government agencies, the insurance industry, trade associations, universities, owners, unions, and contractors, the overall accident and severity rate seems to stay about the same. The Bureau of Labor Statistics (BLS) has been keeping and reporting accident data for all industries for many years. Despite wide swings in the economy, U.S. construction accounts for about 7 percent of the work force and has contributed about 25 percent of the job-related fatalities.

It would seem that the present construction safety programs have brought the accident statistics down to an irreducible minimum that must be accepted as part of an inherently risky business. However, some notable exceptions to the general safety performance figures have shown up recently to prove otherwise. The Business Roundtable (BRT) has started a Construction Industry Safety Excellence (CISE) awards program, open to BRT members and outside construction firms that are nominated by a national contractors association. The CISE-nominated safety program must incorporate the recommendations BRT makes in their CICE Report A-3, "Improving Construction Safety Performance"[2] and have demonstrated improved accident figures.

As listed in Appendix A, BRT has published the results from three CISE Award winners: Air Products and Chemicals, Inc. (1988), Monsanto Chemical Company (1989), and Gulf States Inc. (1989), a contractor. These are three good examples of what can be done to improve a firm's safety program and results when sound management practices are applied to the problem.

The latest (1992) winner of the BRT CISE award was KCI Constructors, the U.S. construction subsidiary of M. W. Kellogg. KCI also earned several international safety honors for passing one three million safe-work-hour goal in the U.S. and Australia. In the past three years they have passed the two-million mark twice and the one-million mark six times. That is indeed living proof that a well-implemented safety program can reduce accidents.

Why Are Construction Safety Programs Important?

There are many mitigating factors supporting the need for effective construction safety programs today. Generally they fall into *humanitarian* and *economic* categories. The hard-nosed economic factors of safety have forced even the most *inhumanitarian* managements into taking a more humanitarian stance on construction safety. I say this because of the construction industry's widespread macho image, which tends to play down present-day management's general trend toward a *humanistic* approach toward employees.

Present-day construction safety programs must also stress *accountability* for safety throughout the organization. Owners' and contractors' managements must initiate the need for safety, and the resulting system must be clear as to who is accountable for carrying out the program. Because CMs have sole responsibility for delivering the project goals, they are held accountable for the success or failure of the site-safety performance.

Humanitarian factors in safety

The humanitarian factors in safety are quite straightforward. No one connected with the industry wants to see coworkers killed or injured on the job. Everyone working on the job must make a personal commitment to perform in a manner that doesn't endanger the lives and property of others. This is especially true for the various management groups involved in the construction project because they are responsible for managing the safety environment.

CMs are especially sensitive to safety's humanitarian side because they are customarily responsible for delivering the sad news of an accident to the next-of-kin. As leaders of their field organizations, most CMs feel a personal responsibility for an accident that happened on their watch.

Although the accident rate among CMs is relatively low, I have had a CM experience a serious lost-time accident on one of my projects. Considering a CM's overall project-performance rating, having a personal lost-time accident on one's record is probably the worst thing that can happen. No one on the project is outside the safety program's umbrella or immune from accidents.

Economic factors in safety

Construction-accident costs have been estimated in a variety of ways.[2,3] It's been estimated by several sources that accidents cost the industry 6.5 percent of the $300 billion spent on construction, or about $20 billion per year in the 1980s. There is no indication that it will be much lower in the present decade. Those numbers give ample proof that construction safety and accident-reduction offer one of the best cost-reduction routes available to meet the construction industry's vital goal of reducing overall costs.

In the final analysis, the above economic factors forced owners and contractors into taking a more pragmatic approach to construction safety. It's always easier to sell a humanitarian approach if one can offer an economic advantage as well. When one considers the direct and indirect costs for accidents, an effective safety program wins hands-down on the factory floor and the construction site.

The direct costs of a high accident rate are fairly obvious and easy to evaluate. Some of the key direct costs attributable to high accident rates are:

1. Higher workers' compensation insurance rates

2. Higher liability insurance rates

3. Losses not covered by insurance policies

4. Government agency fines

5. Depressed craft-labor productivity rates

6. Cost of investigation and filing accident reports

Indirect costs are a bit harder to evaluate but typically add up to more than the direct costs. Some of the more obvious indirect costs are:

1. Increased employee turnover

2. Lost time of injured workers

3. Training cost of replacement workers

4. Time lost on schedule (possible liquidated damages)

5. Lowered worker morale

6. Loss of worker efficiency

7. Lowered morale and efficiency of supervisors

8. Damage to owner's property

9. Damage to tools and equipment

10. Litigation support costs not covered by insurance

11. Loss of new business and damage to corporate image

Direct costs of poor safety

Everyone in construction knows that the more dangerous the industry (or even the craft), the higher the workers' compensation (WC) rate per hour. The published WC premiums usually are the national or statewide average rates to cover the costs of operating the program. Individual employers are assessed an experience modification rate (EMR) based on that employer's safety record.[3] The EMR can run as high as 50 to 100 percent of the base WC rate. WC charges are assessed at a rate of from $4 to $12 per hour worked, depending on the craft. Substantial project cost increases result when those rates are doubled by the EMR. One doesn't have to be a rocket scientist to figure out that a high EMR can destroy the contractor's competitive position in the marketplace!

Higher liability insurance premiums result when the insured's claims history reveals a high rate of accidents and accident-related litigation. Liability insurance claims can involve people and physical property such as tools, construction equipment, and physical plant. Again, these rates start from a pretty elevated rate due to the nature of the construction industry, so a percentage-rating increase can be significant. It's another *adder* to your firm's cost to do business in a keenly competitive world.

Uninsured losses represent another direct-job cost, covering deductible charges and uncovered losses. These losses can be substantial, especially on those projects where the project risk analysis has not been well thought out.

Government fines (primarily OSHA) for site violations have been relatively light up to now. However, when one adds in the stigma attached and the resulting indirect effect on the firm's performance rating with potential clients, the cumulative effect becomes worthy of consideration. Also, present indications are that the fines may become more significant in the near future.

The effect of high accident rates on field productivity is probably the most significant direct cost listed. Some may list this item as an indirect cost, but it can be estimated by assigning an overall percentage rate such as 1 to 5 percent to the overall field labor-hours. Also, remember to include the field supervisory hours in this cost because their productivity is significantly lowered by poor safety performance.

The time spent on researching the cause and effect of site accidents can be significant and is an extra nonreimbursed cost to the project budget. Typically this is high-priced, nonproductive time spent by supervisors who can ill afford it. The time requirement also percolates through the entire reporting process into home office and government agencies, setting off an ongoing, time-consuming, investigatory process. In addition to the project charges, accident-reporting increases the firm's overhead, which eats away at its competitive position in the industry.

The indirect costs of accidents

Because indirect costs are not listed separately in the project accounts, they are difficult to assess accurately. The BRT CICE report A-3[2] estimates that they run from 4 to 17 times as much as direct costs. Even a multiplier from the lower end of that scale can force your total accident costs into the stratosphere. Some of the indirect costs of accidents are so insidious that they can occur on your project almost unnoticed.

Increased employee turnover can result on projects with inferior safety records. High accident and near-miss rates go right to the core of poor morale on any construction site. That in turn makes the better, more-concerned employees move to another job, leaving the less competent, accident-prone workers behind. Such a turn of events can exacerbate an already dangerous and unsafe working environment. When that happens, indirect site costs can increase geometrically with disastrous results in project productivity and profitability. If immediate corrective action on the safety program isn't taken, the CM often faces a personally untenable position and can expect to be replaced.

The lost time of injured workers can upset the routine of work crews while the people are away, in addition to any direct salary or wages that may be paid. If the workers are killed or do not return, the training and learning curve costs of their replacements is an indirect cost to the job. Both of these items also reflect adversely on the field-labor productivity.

Time charged to injuries and lost time has an adverse effect on the project schedule, jeopardizing one of our major project goals. If the contract schedule clause specifies liquidated damages, the adverse affect translates directly into poor profitability. Most contracts specify the contractor's safety performance, so it's difficult to blame the owner or outside interests for poor safety performance as a way of evading the liquidated damages.

We have already mentioned the insidious effects of poor safety on the morale, efficiency, and productivity of the craftspeople, but the same is true of the field superintendent's supervisory staff. Although these people are relatively few in number, their inefficiency gets multiplied throughout the working crews they supervise.

Although we have already discussed the direct costs resulting from damage to construction equipment and owners' property, there are indirect costs involved as well. Damage to tools and construction equipment results in lost time while the units are repaired or replaced. If the works themselves are damaged from an accident, delays while the causes and effects are assessed, along with the time spent on actual repair, are significant. Because accidents and repairs are not classed as force majeure, the contract grants no relief for resulting schedule delays.

When litigation results from an accident, some of the direct cost for defending it is covered by insurance. However, there are expenses contributed by the defendant firm in the form of defense preparation costs. The people involved must contribute specialist support in the form of depositions, technical advice, and the like, which become part of the firm's overhead.

Perhaps the least-recognized indirect cost is the damage to the contractor's corporate image in the eyes of the marketplace. Many owners have become intensely safety-conscious in recent years, and now make special efforts to select only *safe contractors*. A poor safety record is often grounds for contractors not making the bidders list despite having an otherwise excellent track record.[2]

Naturally, the "no free lunch rule" also applies to construction safety. We are not going to reduce all of these direct and indirect costs without spending some money. It costs about 2.5 percent of direct field-labor costs to establish and operate a basic safety program.[2,3] That figure includes those costs mandated by government regulations that must be spent on safety as a minimum, and are often considered as direct job costs.

The cost of maintaining staff safety engineers, developing and maintaining safety standards and procedures, safety training programs, site-safety inspection trips, purchased safety equipment, and the like is directly proportional to the size and depth of the firm's safety program. Most owners and contractors have gone through the initial setup expense and are now in the maintaining mode. Those that have few or no safety procedures in place will have to take the initial financial plunge. However, safety programs can be built up gradually to spread the cost over time.

In the final analysis, all safety program expenses eventually are borne by the owner in both lump-sum and CPFF contract settings.

Because lump-sum contractors absorb safety costs in their overhead, safety practice is sometimes diminished by the competitive nature of that sector of the construction business. Contractors should pay attention to the owner's safety requirements in the contract when they are bidding the job, regardless of the contracting basis.

The owner and contractor upper managements are responsible for setting or approving the home office safety department's operating budget. This is their opportunity to put their money where their mouth is in supporting construction safety. Some managements are willing to *talk* effective site safety but aren't willing to back it financially.

Who Are the Main Players in Construction Safety?

The contributors to a successful construction safety program fall into two groups divided into line and staff functions as follows:

Line functions	Staff functions
Owner managements	Field safety engineers
Contractor managements	Home office safety groups
Construction managers	Government (OSHA/MSHA/EPA)
Contractor field supervisors	Trade unions
Field foremen	Trade associations
Craft labor	Academia

The line functions are responsible for implementing and applying the safety regulations that are developed by the staff organizations. The line people are directly responsible for the execution of and results from the safety program at the job site.

The corporate staff functions are responsible for formulating, maintaining, and supporting the safety programs at the site. The corporate safety groups (CSGs) act as intermediaries between upper management and the field organization as well as outside safety-oriented organizations. The outside groups are the government, trade unions, trade associations, and academia. We will look in more detail at the typical responsibilities and contributions of each group.

The owner's contribution to site safety

Because owners eventually pay most of the safety bill, they have historically been prime movers in the safety process. Like any sharp buyer, they want to see a good bang for their buck. Industrial owners were natural leaders in that regard as an extension of their in-house safety programs. If safety was seen to pay off on the shop floor, why wouldn't it be applicable to their construction projects? After all, they were picking up the tab in both places.

In my opinion, government and commercial owners are not as safety-conscious in their operations, so they tend to have a lower profile in the construction-safety arena. The starting point for this opinion in that area is the contract language devoted to the contractor's requirements to meet the owner's safety requirements. Rarely do commercial owners follow up with a detailed field safety manual as some industrial owners do. The type of far-reaching safety programs I am talking about are those shown in the Business Roundtable's CISE award programs listed in Appendix A. Contractors can then take the owner's detailed safety program and adapt it for their use in the field. That arrangement gets everyone playing from the same music.

A strong construction-safety program has to start with the owner's top management. They must issue a vigorous charter to the corporate safety department setting the goals and tone of the program. The charter must be followed with *continuing and meaningful reinforcement* every step along the way. The CSG can then develop and implement an effective safety program down through the contractor's management and into the field. At that point the construction manager empowers the safety program down to the craft level by ensuring that the field safety group is delivering an effective program. If there is a single weak link in the safety chain, the humanitarian and cost-reduction goals for the safety program will not be met.

In accepting management's charter, the owner's corporate safety department makes itself responsible for implementing a humanitarian and cost-effective corporate safety program. Generally this group is responsible for in-house as well as construction safety. Therefore, the construction safety branch is typically a specialized group specifically oriented to the field construction environment. This arrangement works out well on those expansion or revamp projects that are done while the plant is in operation. Doing construction in an already hazardous operations environment poses the utmost problems for safety-program execution.

The role of contractor's management in safety

The contractor's management must *at least accept* the safe-operating challenge originated by the owner and infuse it into their own organization. I say "at least accept" because many construction firms are even more proactive on safety than some owners. As we said earlier, an effective safety program is a strong sales weapon in promoting the contractor's services in the marketplace.

The contractor's management must formulate the firm's safety posture in writing and organize a safety group to effectively carry out the corporate policies. Accountability for safety throughout the organization is a vital part of any corporate safety program. The CSG must be made to perform their duties against established standards to ensure that the corporate, humanitarian, and economic safety goals are in fact achieved. The safety group is obligated to commit the corporate resources, support, and leadership necessary to meet the goals.

Both contractor and owner CSGs draw information from the various outside staff organizations (listed on p. 343) to update their safety practices. Like everything else in the construction area, safety practices must be kept abreast of the latest technology and operating experience as it develops in the construction and related industries.

Effective training techniques for all levels of the field organization lie at the heart of any successful field-safety program. The CM and the field-safety engineer usually analyze the proposed field operation methods and the hazards expected to be encountered on that specific construction site. They will then work with the CSG to formulate and implement the safety program for the site.

The respective corporate safety staffs handle the collection and reporting of the company's accident records from the ongoing projects. They also interface with various government inspection bodies to keep their firms in compliance with the applicable safety and health laws and regulations. The department also maintains a staff of field-safety engineers for assignment to the firm's construction projects. The field-safety engineers report administratively to the corporate safety department and functionally to the construction manager.

The safety program for each project typically is worked out among the CM, the owner's field representative, and the owner and contractor field-safety engineers, all within the guidelines laid down by the respective corporate safety departments. It is important to have the field-safety program carefully planned, with specific duties and responsibilities assigned to the participants to meet the proposed project safety goals.

The construction manager's role in site safety

We have already touched on some of the CM's responsibilities in the execution of the site-safety program. However, the number-one duty of any CM is to bring to all phases of his or her specific project a strong commitment to top management's safety philosophy. The four major phases for safety are the preconstruction activities, initial site planning, execution of the plan, and review of the final results. Safety planning should start during the constructibility analyses done during the design phase if possible. On construct-only projects, safety analysis must start during the bidding process.

During the project-execution phase, it is easy to lose sight of the safety program in the crush of other high-priority activities going on at the same time. CMs cannot relax when it comes to monitoring and enforcing safe operating methods on the job, no matter how tempting less-than-safe shortcuts may appear. They are committed to making safety a continuing high-priority matter in all their work activities; especially when walking the job. Any unsafe practices should be noted and rectified immediately. If a CM has a relaxed attitude toward safety, it will gut the safety program and lead to an accident-prone construction site.

One of the simplest enhancers of site safety is good housekeeping at the site, because sloppy site conditions breed accidents by themselves. Poor housekeeping also is a sign of a poorly organized and operated job and an ineffective CM in all areas including safety. Good housekeeping includes maintaining all construction equipment and tools in good working order, as well as a clean and orderly site.

Professional CMs can't succeed in the construction business with poor site-safety records. They will not get past their annual performance evaluation for their next promotion. Neither will they attract the top-quality supervisors and labor that are so essential to the execution of successful projects. These are only a few personal factors, without even considering the human factors of providing a safe workplace for your fellow workers and top management!

The role of construction supervisors in site safety

This supervisory level is on the cutting edge of the site-safety programs. If these people don't make the safety program work, all the management backing, sophisticated safety procedures, and good intentions will have been wasted. These people must be properly trained in delivering the correct safety message to the craftspeople working under their direction. They also need to continuously monitor their areas to ensure that proper safety procedures are practiced during construction.

When they are planning the execution of their work, they have to keep safety in mind to ensure that trained workers using safe equipment are available to safely perform the tasks as planned. Field-labor productivity can be seriously hampered when the construction process must be stopped for safety reasons after work has begun. The field safety engineer, the superintendent, or the CM each has the authority to stop the work if unsafe construction practices are observed.

The field construction superintendent and area supervisors are responsible for delivering the site-safety procedures to the craftspeople through the general and craft foremen. Because safety procedures are a two-way street, the supervisors also get feedback from the craftspeople about unsafe working conditions. The feedback may be in the form of worker complaints about unsafe practices, which should be investigated and resolved to all parties' satisfaction.

Much of the information transfer takes place through the venerable "toolbox meetings" (TBM). TBMs have been with us within everyone's living memory, and have served the industry reasonably well over those years. Their downside, however, is that they often become boring, trite, and forgotten in the press of getting on with the job. In a union-shop setting, the union site representatives usually ensure that the meetings are regularly held. In an open-shop setting, it is up to field management to ensure that the TBMs or their equivalent are regularly held.

With the recent advances in space-age training tools such as videotapes and multimedia presentations, the construction industry needs to update its training methods to improve information transfer in safety and other areas. Any training and information changes shouldn't lose sight of the need for two-way information flow between supervision and craft labor.

The role of craft labor in the site-safety process

We have now reached the level with the greatest exposure to danger and where most of the accidents actually happen. This is the level where 100 percent of the people should be most concerned about working safely. Although most craft labor is self-interested enough to be safety-conscious, there will always be a certain percentage who are not all that interested in safety on the job. These are the people who willingly bend the safety rules or who may be just plain accident-prone.

The personnel people should give specific attention to each worker's safety record during the hiring process. Those with poor safety performance should not be hired. Some poor-risk employees will still slip through in the initial hiring process, so everyone's safety performance must undergo continuous monitoring. If the problem workers don't respond to training, they should be weeded out quickly before they cause an accident. This may seem a hard-hearted attitude, but unsafe workers can injure themselves just as readily as their safety-minded coworkers. The humanitarian side of this philosophy is worth the effort alone, without considering the economic havoc unsafe workers can wreak.

The best-developed project safety plan in the world won't be effective if the information is not transmitted to those doing the work. It's vital that all employees be indoctrinated with what they need to know about the safety program that was developed for that specific project. This is especially true for those contractors working over a large geographical area that don't usually get the same workers on successive projects.

The key workers in this part of the safety chain are the foremen, for they have the most detailed picture of what and how the work crews are performing their tasks. They should know the goals of the safety plan and should be experienced in the way the tasks should be safely performed. Making a comparison to a military organization, the foremen are equivalent to the noncommissioned officers who actually carry out the battle plan at the combat level.

The role of the site-safety engineer

The resident site-safety engineer (RSE) is the only field supervisor who devotes 100 percent of his/her time to safety issues. To everyone else on site, safety is a secondary consideration. This presents RSEs with the challenge of promoting site-safety programs to the meaningful levels where they are fully effective. After all, the project isn't being run for the benefit of the safety program, so to get full cooperation the RSE is obliged to win the site people's confidence in a proactive and professional manner.

Larger projects typically have a full-time RSE assigned to the field staff from the corporate safety department. RSEs may be assigned full-time on a large project or divide their time among several smaller projects in a localized area. Whenever they are on the site, RSEs report functionally to the CM.

The role of RSEs is to lead the effort in preparing the safety plan for their specific project assignments and then to follow them into the field for execution. They handle the liaison between the CSG and the construction sites and report directly to the CM. RSEs also are responsible for approving and monitoring the safety programs for subcontractors operating on the site.

Their primary duty is to ensure that the work performed on the site is done within the safety program laid out for that project. RSEs must be proactive in reviewing the field operations to ensure that the proposed construction methods and technologies have been analyzed from the safety and health standpoint. They should be knowledgeable in OSHA/MSHA/EPA standards to ensure that the site is in compliance with government regulations. RSEs also handle the government inspections and follow up with the necessary reports of violations and accidents.

RSEs are responsible for all on-site safety training for supervisors and craft labor. They also monitor the training and safety follow-up done by the site supervisors. For example, RSEs should spot-monitor the toolbox meetings to see that the meetings are maintained in a relevant and effective manner. Safety training should be simple and uncluttered in its approach. After teaching the general site-safety regulations, it should focus on what each trade needs to know to perform its work safely.

Administratively, RSEs are responsible for ordering the necessary safety supplies and equipment for the project. They also are responsible for controlling the inventory of those supplies to match the project work load and personnel loading. RSEs should conduct periodic field inspections and emergency drills to see that the safety equipment and systems are being used and are functioning properly. They also are the field representatives for liaison with the local fire, EMS, and law enforcement groups serving the site.

RSEs should attend the weekly construction planning meeting and any other safety-related meetings on the project. They should be in regular contact with the CM to review safety plans and problems as well as to bring up emergency matters as they arise. A weekly safety report should be prepared and issued, along with a monthly summary for inclusion in the CM's monthly progress report.

An important function of the RSE is to handle the site-safety statistics for submission to the CM, CSG, insurers, and government agencies. An important aspect of that duty is to monitor the safety incentive and awards programs set up to improve the site's safety record. The incentives and awards program is one of the best promoters of site safety when it is properly planned and organized.

As a key member of the field staff, the RSE and the CM must work as a team to effectively execute the safety program. These two people bringing a professional attitude and approach to the safety program will go a long way toward assuring a successful safety record on the project. Neither one of them can do it alone!

The role of the corporate safety department

The CSG's primary role is to implement and unify the corporate safety strategy on all the firm's construction projects. They also play a leading role in cultivating the firm's corporate safety image with employees, clients, trade associations, and the public. Larger contractor CSGs play an important role in working with owner's CSGs in developing mutually agreeable safety programs for each major project. Details of such an arrangement are given in BRT's CICE Report A-3.[2]

CSGs start by formulating detailed safety standards and procedures and organizing a staff to meet the corporate safety goals. The CSG disseminates its information and promotes its image throughout the firm and other interested industry-related organizations. Another key duty of the CGS is to set an annual operating budget and monitor it. They also prepare figures of projected savings to justify the economic operation of the department.

Good ideas for improving safety and health come from many diverse sources, but especially from the firm's ongoing projects. The ideas and examples developed from the field are most likely to relate to the type of work in which the firm is engaged. Contractors working in chemical plants, for example, have an entirely different set of safety considerations than a commercial builder. Many trade associations who generate safety information do not always address a specialty contractor's needs.

It is the CSG's job to evaluate safety ideas from all sources such as labor unions, government, trade associations, insurance companies, and academia and incorporate the best ones into the firm's safety programs. Analyzing and comparing the firm's accident experience with industry and national statistics is a good way to locate the most serious problem areas that need attention.

Government's role in the safety process

Actually, the Federal government was rather a late arrival on the construction-safety scene when OSHA was created by the Williams-Steiger Occupational and Health Act of 1970. As required by the Act, OSHA promulgated safety regulations for general industry (book 2206) and the construction industry in particular (book 2207).[5] Regulations 2207 are basic safety regulations applicable to all contractors and subcontractors operating in the construction industry. The law provides for random, unannounced inspections at any construction site to ensure compliance with the regulations. Official citations are issued for noncompliance, and fines against the offending contractor may result.

The OSHA regulations require all contractors to maintain employee work-related accident and illness records on various OSHA forms. They also must post the required OSHA information and accident records on a project bulletin board available to all employees. Normally, ensuring conformance to OSHA regulations is the safety engineer's responsibility, but the CM must be knowledgeable about them to ensure a proactive stance for OSHA compliance. On those projects without a safety engineer, the CM has to handle the implementation of the OSHA regulations.

These regulations and inspections, together with the resulting citations, were quite bothersome in the early years, but have now come to be fairly routine. Any construction firm these days that receives OSHA citations on a routine basis needs to pay some serious attention to developing a better safety program. The fines themselves are fairly moderate and could stand to be more stringent to promote good safety on those sites that have substandard safety programs. Any firm that accepts the OSHA fines as an acceptable nuisance while continuing loose safety practices should also remember the potential for litigation from injured employees or their estates.

OSHA has recently published a monograph, reporting their survey of the 100 most frequently cited standards for 1993[6], along with a guide for the abatement of the top 25 associated physical hazards. You may be surprised to learn that virtually all of the 100 most-cited safety hazards are of the everyday, routine variety. For example, item 1 is fall protection, and item 10 is general housekeeping. A breakdown of the categories involved is shown in Fig. 10.1.

The second half of the book uses photographic illustrations giving OSHA's recommended abatement of the 25 most-cited violations in the list. This book is available at $5.50 from the Superintendent of

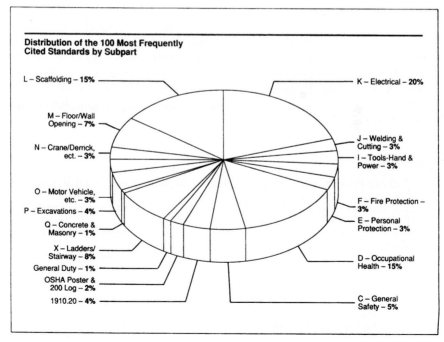

Distribution of the 100 Most Frequently Cited Standards by Subpart

L – Scaffolding – 15%
M – Floor/Wall Opening – 7%
N – Crane/Derrick, ect. – 3%
O – Motor Vehicle, etc. – 3%
P – Excavations – 4%
Q – Concrete & Masonry – 1%
X – Ladders/Stairway – 8%
General Duty – 1%
OSHA Poster & 200 Log – 2%
1910.20 – 4%

K – Electrical – 20%
J – Welding & Cutting – 3%
I – Tools-Hand & Power – 3%
F – Fire Protection – 3%
E – Personal Protection – 3%
D – Occupational Health – 15%
C – General Safety – 5%

Figure 10.1 Distribution of the 100 most-cited OSHA Standards (From *The Most Frequently Cited OSHA Construction Standards in 1991,* U.S. Department of Labor, February 1993.)

Documents (202) 783-3238, and will be worth many times that amount in your technical library.

CMs need to be aware of those sections of the regulations that pertain to their type of work. If reading the OSHA construction regulations is too onerous a task, Sidney Levy gives an overview of the more common regulations in his book, *Project Management in Construction.*[5] Remember, the OSHA regulations should be considered only a minimum requirement for a proactive and effective safety program.

Each phase of the construction industry has special conditions and technologies that have to be considered in greater depth. For example, process contractors working in existing chemical plants concurrently with ongoing plant operations are often exposed to hazardous chemicals not ordinarily encountered on a green-field construction site. The risks of illness and accidents to contractor's employees from exposure to or explosion of hazardous chemicals is multiplied.

Summary

In considering the CM's input to the site-safety program, I have stressed the management side along with what they must know to make safety succeed on their projects. Construction safety doesn't just happen; it has to be handled conscientiously from project start to finish. There are hundreds of ways to excuse a bad safety record, but none of them is really acceptable. CMs are the ones who must accept the blame for a poor safety performance, just as they are the only ones in a position to manage the safety program for a good performance. Safety is the fourth leg on the chair of *safely* completing the project *as specified, on time, and within budget.*

References

1. *This Old Pyramid,* NOVA show No. 1915, WGBH Educational Foundation, Boston, MA. Air date: November 4, 1992.
2. The Business Roundtable, CICE Report A-3, *Improving Construction Safety Performance,* 200 Park Ave., New York, NY, 1982.
3. James J. O'Brien and Robert G. Zilly, *Contractor's Management Handbook,* McGraw-Hill, New York, 1991.
4. Henry Parker and C. H. Ogelsby, *Methods Improvement for Construction Managers,* McGraw-Hill, New York, 1972.
5. Sidney M. Levy, *Project Management in Construction,* McGraw-Hill, New York, 1987.
6. *The 100 Most Frequently Cited OSHA Construction Standards in 1991: A Guide for the Abatement of the Top 25 Associated Physical Hazards,* U.S. Department of Labor (OSHA), February 1993.

Case Study Instructions

1. Describe how you have included the safety function in your field organization, assuming that your firm has proactive field-safety policies.

2. Evaluate your project from the safety standpoint. What standard and special safety considerations will you be giving specific attention to?

3. You have had a fatal accident on your project. How will you handle this situation with OSHA, your management, the local media, the insurance company, and the next of kin?

4. OSHA has issued a number of serious citations on your project. How will you handle the situation?

5. Assume that your project has a resident safety engineer on site. Describe how you will ensure a proactive safety stance and a favorable safety record for your project.

11

Project Communications

Successful construction project execution is virtually impossible unless you have an effective communication system. Good communication skills are basic to becoming a successful construction manager. Effective communication is the lubricant that keeps the successful construction project machinery running smoothly.

Figure 1.2 highlights the construction manager's key role in project communications. It shows the many communication channels that exist in a complex project/corporate organization. If communications break down, each of those contacts offers a potential barrier to information flow. The only way to control potential information barriers is through effective communication.

Much of construction management communication has to do with conflict resolution, which means that CMs must maximize their communicating abilities. If you are lacking in this critical area, you should enhance your writing, public speaking, visual presentation, and listening skills through training and practice. I will offer specific suggestions in this area as we discuss the various types of project communications.

Why Do Communications Fail?

There is a myriad of reasons why communications fail in all parts of our life. A few of the more common reasons business communications fail are the following:

- Not having a clear goal in mind
- Staying in a negative mode
- Concentrating on your own thoughts to the exclusion of the other person's ideas

- Not establishing rapport
- Assuming that others have the same information on the subject that you have
- Being impatient—not hearing the other party out
- Mistaking interpretations for facts
- Failure to analyze and to handle resistance
- Being ashamed to admit you don't understand
- Overabundance of ego

Communication is defined as: "1. a transmitting, 2. a giving or exchanging of information, etc. as by talk or writing." By comparing the common reasons for failure with the definition, it's easy to analyze the failures. I am sure that each of us can go down the list and pick out three or four failures that apply to us. They are the ones we must improve on if we are to become successful communicators.

One of my techniques is to mentally transpose myself into the intended receivers of my communications. Then I read the message while using their knowledge of the situation. If I can still understand what I said, the communication just might work. If I can't understand it, I rework the message. If you practice this technique, you can use it even in oral communications as well as written ones.

If we don't establish rapport in communications, anything we say is likely to fall on fallow ground. The easiest way to effect rapport in your project communications is to build your credibility as a professional manager. People will receive your messages more readily if they respect your opinion. Then, all you have to do is keep an open mind to build *full rapport.*

My personal number-one communications problem is being patient while receiving communications. After many years of trying to improve, I still have not perfected the art. If you have the same problem, don't give up easily. It can be corrected!

Handling resistance in the mind of the message-receiver also requires patience. If putting your idea across is important to you, try to analyze the resistance and cancel it. You must remember this factor when *selling* your idea *upward* or *downward* in your work setting.

The last two items on the problem list often go together. Never let your ego get in the way of admitting that you don't understand something. None of us will ever learn all there is to know about managing construction. Showing that you are big enough to admit ignorance now and then will increase your stature, not decrease it. Another way to handle this problem is to play the message back to the sender the way you understood it. That avoids the open admission of not having understood the message.

Those are a few basic points for improving your communications. Do take advantage of any communications training courses that will help to overcome any weak areas. Above all, remember to practice, practice, practice!

Communication Systems

In addition to improving their communicating abilities, CMs need to keep abreast of the latest communications hardware. In the past couple of years an explosion of new communications tools has hit the market. These new tools multiply your communications abilities through hardware alone. Fax now allows us to send documents anywhere in the world with the speed of placing a telephone call. The new computer links, portable computers, and paging systems will soon afford us very few places in the world to hide.

Multicity telephone conferences are now commonplace. Video-conferencing is also becoming cost-effective as its costs come down and travel costs rise. New, more sophisticated multimedia computer presentation systems are coming onto the market every day. CMs should check out and evaluate these new technologies and take full advantage of those suited to the project site. Calculate the cost/benefit ratio for each of these new tools and include it in the construction cost estimate if you feel it will pay.

Project Communications

CMs are the prime spokespersons for their respective organizations. That duty places them in the position of either creating or monitoring most project communications. To set a good example, CMs should make sure that all project communications are handled professionally. They should review all project correspondence and at least edit it to a consistent professional standard.

Typically, project communications fall into the following key categories:

1. Project correspondence
2. Audiovisual presentations
3. Project reporting
4. Meetings
5. Training
6. Listening

As we discuss each of these areas, we will find that the list covers almost every form of communications there is.

Project correspondence

The primary project communication link is that between the client and the contractor CMs. All major project correspondence should pass through that channel. That is the only effective way to control the project design, scope, and other contractual obligations. The contract usually sets up that communication channel by calling for official notices to pass between the respective managers. Remember, this applies whether you are contracting within the company or externally.

CMs often delegate some specialized outside correspondence to other key staff members and managers to speed up communications. Some examples of this are procurement, personnel, field engineering, and subcontract administration. However, any matters concerning scope, cost, schedule, and so on, must pass through the contract-defined channel, over the CM's signature, to be official. This is the only way to avoid conflicting instructions and project chaos.

Project correspondence serves an important function in addition to the sending of messages. Copies of all correspondence should find their way into the project files to document the project. When problems arise later in the job, it often becomes necessary to build a paper trail to find out what happened. The site document distribution chart should designate who receives copies of specific correspondence for their information.

The originating secretary should assign a number to and record each document in a correspondence log. The log makes it easier to locate correspondence without thumbing through the bulky files directly. This applies to incoming and outgoing letters, memos, document transmittals, and minutes of meetings. Maintaining those logs on a spreadsheet is a good PC computer application.

Audiovisual presentations

CMs often make audiovisual presentations in the course of the project. As we said in Chapter 2, the first one could be the "dog-and-pony show" used to sell the project. Later presentations occur during project and management review meetings, feasibility reports, PR talks to citizens groups, among many others.

In preparing for a presentation it is vital to select the single most important idea or thought you want to convey to your audience. Marshall all of your arguments, visual aids, and statements in support of your main theme. Start your preparation with an outline of your theme. Base it on the most positive way of presenting your arguments, from the introduction to the closing punch line. Work the outline over to refine it into a clear, concise, and effective presentation of your thoughts that supports your ideas. If you are working with a group, bounce your ideas off the other members as the outline develops. Write the copy, then prepare the visual aids supporting the idea. Remember, the clearly spoken word, reinforced with a strong visual image, increases the listener's understanding and retention of the idea.

Rehearse the presentation to get your timing worked out and to polish off any rough edges. Make the dry run in front of a nonsympathetic group playing a devil's advocate role to critique your effort. Encourage the dry-run participants to ask any difficult questions they can think of, so you will be ready to handle them should they arise in the course of the actual presentation.

Be sure that your visual aids are of good quality and are easily legible from any point in the presentation room. Check out the presentation facilities to be sure that the layout and the equipment will work well for your materials. *How to Run Better Business Meetings,*[1] published by the 3M Company, is an excellent guide to arranging the physical facilities for effective presentations.

Management consultants in presentation training have developed such training into a fine art over the years. Usually, courses take a management group into seclusion for several days to indoctrinate its members in the art. They teach the attendees the basic techniques of creating such visual media as slides, flip charts, overhead viewgraphs, videos, and multimedia presentations. Next, they develop storyboards incorporating the spoken and visual parts of the presentation. Through the benefits of videotape and a critique by the rest of the attendees, the presenter receives pointers on how to improve. This is an excellent way to gain basic skills in the presentation arts. Later, you can fine-tune them in actual practice. You should take advantage of available presentation training at the first opportunity.

Handling VIP Site Visits

A special case for handling presentations occurs whenever you have VIP visitors to the site. They may be high corporate management people from the client's or your own firm, or high government officials. Nonresident visitors like the V.P. construction, the home office project manager/director, and the like do not usually rate the full treatment. However, it is important that you show off your project and job site to all "outsiders" in a positive manner. Be sure that your human relations skills are well honed and in good working order for the occasion.

Getting prepared for the VIP site visits falls into three main areas. Area one is offering to make local housing, transportation, and social arrangements for the visitors. These matters will vary greatly from visit to visit, but make the offer anyway. Area two is to make a presentation in the office of what you are building and its present status before you tour the construction work. Area three is the tour of the work in progress.

Usually there is a lot of interest on the visitor's part in making a site visit, because of the many exciting things going on. For example, it's much more glamorous than making a visit to a design office. It's important for CMs to remember that the key goal here is to have the VIPs leave with a warm feeling about the site operations.

A fundamental item for this type of visit is to ensure that your housekeeping is in good order. Don't leave this one to chance, but instead make a personal inspection tour after you have put out the spruce-up order and before the guests arrive. A fundamental point of preparing for the visit is to have the site housekeeping in especially good order, even though it normally is (as it should be!). A sloppy site can offset all the effort a CM has put into the visit.

Have an orientation meeting with your key staff people that you plan to participate in the visit and tell them who's coming, what they will be looking for, and what part you expect them to play. I always believe in showcasing the capabilities of the field staff in their areas of expertise as long as they have a meaningful contribution to make. However, it pays to be careful with this one on foreign jobs where cultural differences may exist. I had this approach backfire on me with a Korean client. In Korean (and possibly Asian) culture, the head man is expected to handle the whole show. Bringing in one's key lieutenants is viewed as a sign of weakness. Fortunately, my boss understood the situation and reassured me after the client had left town that I had done the right thing. We just had to change the format a bit in later visits.

During the office briefing is where you will make the presentation to review the project scope, the present status, schedule, and special features of the project. This presentation usually doesn't require a lot of detail and should be tailored to the type of visitor. Remember, the focus here is on selling the capabilities of your firm or department, your project staff, and your management abilities, more or less in that order. In effect, you will be telling the visitors what they are going to see before you show them the site. The briefing should demonstrate the excellent spirit of cooperation of the client-contractor team in meeting the project goals.

The site tour should be done in conjunction with the field superintendent, who should be most familiar with the work. Ensure that you have thoroughly briefed him/her in your philosophy of describing the work before you start. Again, you don't want any surprises along the way.

Special site milestone celebrations such as ground-breakings, topping out steel, plant dedications, and the like may require the organization of special events after the usual speeches. These should be done in good taste and within the agreed-upon budget. Every effort should be made to promote the client, company, and field team's image as an extension of the CM's public relations duties.

Project Reporting

Another powerful project communications tool is project reporting. Creating project reports forces the project staff to make a thorough review of their project activities at least once a month. Setting the high standard for quality project reports is an important role for CMs.

Producing a concise and well-presented monthly progress report is the best way for CMs to set the standard for reports on the project. It is also a good time to review the quality of the other project-control reports and to check on how the project is going. The monthly progress report is a summary of how well the contraction team is moving toward meeting its project goals.

The main purpose of the progress report is to inform the key interest groups as to how the work is progressing. An important secondary purpose is to keep a running history of vital project activities. The key groups interested in the progress report are:

- The client's organization, local and main offices
- The contractor's top management
- Key staff people on the project

The operative words in this discussion are *to inform* key people of important project activities. A rambling, poorly written progress report will not inform the reader. The progress report must also be factual and results-oriented and report any problem areas to present a true picture of project status. It should be written in a positive, direct style to instill confidence in the reader that the construction team is in control of the project.

The writing style must be forceful and direct. Stay away from the passive voice and weak verbs. Keep your sentences short, and use a new paragraph when the subject changes. There are several new style manuals and PC software programs on the market today aimed at improving one's writing skills. One that I used in writing this book is *RightWriter*, which works with most word processors. Software won't write the report for you, but it can train you to write in a more forceful style.

The monthly report format should be consistent from month to month so readers don't waste time hunting for the key indicators each month. Most well-managed companies today have developed standard report formats to suit their type of projects. Using a standard format saves CMs from having to develop a new format for each project.

Even with standard formats, some latitude is possible to meet particular client requirements in the report.

Figure 11.1 shows a typical progress report format that I have developed over the years. The sample format is intentionally general, but you can readily adapt it to your type of project. Compare it with your company's present format to see if it can contribute to doing a better job of reporting project performance.

A major report-format decision is whether or not to include complete copies of the project-control reports in the monthly progress report. That may be the way to go on small projects, but it makes for a bulky and unwieldy report on large projects. In the latter case, a summary section for the major detailed reports highlighting any accomplishments and problem areas produces a more readable progress report. The client's control people are the ones really interested in perusing the detailed project-control reports. It's their duty to pass on any unusual conditions found in the detailed reports to their management for action.

The client and contractor managements receive and read the monthly progress report, so it should not contain any confidential contractor information. For example, a client should not expect to see the contractor's detailed cost information on a lump-sum project. Also, don't discuss any confidential personnel matters in the report. Prepare an addendum covering any of the confidential company matters as an attachment to your company management's copy.

Construction project reports should be factual. Painting a rosy picture by ignoring problems only leads to more serious problems later. Bringing problems out into the open as they arise, along with possible solutions, is the best way to handle the situation. This approach shows a positive attitude in the face of an adverse situation. Problem areas require immediate attention from both owner and contractor managements to mount a quick and concerted attack to develop the best possible solution.

Don't include a lot of dull and tedious statistics in the report. Graphs and charts present data in a more interesting and readable format, and break up long pages of text. For example, issuing a long list of rebar tonnage set, forms built, and embedded steel items set last month will not mean much to report readers. Only the people working with those detailed numbers will know what they mean— surely management readers will not. A chart or graph showing how many foundations have been put in place against the planned amount will be much more effective. The cost engineers can cover the detailed individual quantities in their monthly report.

Preparing the monthly progress report offers an excellent opportunity to practice my favorite communication tool of placing myself in the intended reader's shoes. Read the draft report from the intended

0.0 TITLE PAGE OR COVER SHEET

Project name, number, and location
Client and contractor names and logos
Report number, issue date, and period covered

1.0 TABLE OF CONTENTS

2.0 MANAGEMENT SUMMARY

A brief abstract of monthly project activities

3.0 OVERVIEW OF CONSTRUCTION OPERATIONS

Narrative Reports by Construction Activity

3.1 Construction management

3.1.1 Organization and major accomplishments
3.1.2 Site-safety review and report

3.2 Field construction activities by area

3.2.1 Site development and utilities
3.2.2 Foundations and underground
3.2.3 Buildings and structures
3.2.4 Architectural trades

3.3 Major subcontractors

3.3.1 Mechanical contractors
3.3.2 Electrical contractor
3.3.3 Other subcontractors

3.4 Project controls activity

3.4.1 Construction scheduling
3.4.2 Field cost-engineering and -control
3.4.3 Material control
3.4.4 Field engineering and inspection
3.4.5 Design-construction interface

3.5 Procurement Activities

3.5.1 Order placement
3.5.2 Material control

3.6 Administrative activities

3.6.1 Personnel matters
3.6.2 Labor relations
3.6.3 Others

3.7 Design engineering (if applicable)

3.7.1 Status report
3.7.2 Needs list

4.0 SUMMARY OF CONSTRUCTION CONTROL REPORTS

4.1 Cost report summary
4.2 Schedule and progress report summary
4.3 Cashflow report
4.4 Change-order register

Figure 11.1 Table of contents for construction progress report.

5.0 PROGRESS CURVES OR CHARTS

5.1 Construction progress S curves by major activity
5.2 Procurement commitments
5.3 Design activities (if applicable)
5.4 Personnel loading curves
5.5 Cashflow curve

6.0 WORK TO BE DONE NEXT MONTH

6.1 Construction work by major area or activity
6.2 Procurement
6.3 Design work (if applicable)

7.0 PROBLEM AREAS

7.1 Information and materials needs list
7.2 Decisions required
7.3 Problem areas with proposed solution
7.4 Review unresolved items from last month

8.0 CONSTRUCTION PHOTOS OR VIDEOTAPES

9.0 DESIGN DOCUMENTATION LISTS (Optional)

Figure 11.1 *(Continued)*

reader's viewpoint and knowledge of the project. Polish up any areas that are unclear from that viewpoint and you will have a *better-informed* reader. Well-presented progress reports can be real point-winners toward meeting your career goals. This opportunity comes around every month, so don't pass it up!

Progress report contents

The monthly progress report is an important document, so reviewing its contents section by section here is worthwhile. The title page or cover sheet (page 1.0) is self-explanatory. Using custom-printed covers is not that expensive now that the new low-cost printing processes are available. Also, using a decimal numbering system makes the table of contents and the report itself easier to follow (Fig. 11.1).

Section 2.0, the Management Summary, is a key section of the report, and it's perhaps the most difficult to write. This section should give upper management a thorough review of the project in no more than one or two pages. If upper management people are reviewing many projects, the summary section may be the only part of the report they have time to read. Ideally, executive summaries should be no longer than one page and *never* longer than two. If it runs beyond two pages it's no longer a summary, and busy readers will not read it all the way through.

Creating this section for a large project should tax your ability to write clearly and concisely and still cover the ground. You will find it difficult to include details in the executive summary and still keep it short. If details are necessary to clarify a situation, refer to the detailed section of the report. If the readers are interested in more information, they can pursue the referenced section.

The summary need not follow the same format as the complete report. Present the higher-priority items up front, in case the reader does not finish the summary. Using a news-reporting style is perhaps the closest analogy to describe the writing style needed here.

Section 3.0, Overview of Construction Operations, is a narrative report of continuing operations in the various construction activities. Activity leaders should prepare those sections for the CM in a format suitable for use in the final report. As CM (and editor-in-chief), you should issue the section leaders a standard format for submitting their input to the monthly progress report. Using standardized formats saves a lot of editing time to get the various sections into a consistent format for your report.

Collecting data from each construction activity gives you an opportunity to review each group's performance for the month. Go over each report with the activity leader as part of your MBO goals-review system before incorporating them in the progress report. Concurrently, you may collect a few news items for the executive summary.

Section 3.0 is where we go into more detail about each discipline's actual progress and what is needed to carry out the mission. Remember not to overload the section with boring statistics, just accent the positive and discuss the problem areas. The tone of this section will evolve over the life cycle of the project, starting with project initiation and site work and then shifting into facility construction. Don't devote a lot of space to those areas that have moved into a passive follow-up mode.

Section 4.0, Summary of Project Control Reports, summarizes the detailed control reports that were too bulky to include in full. The schedule report will be discussed in Section 5.0, so this section should highlight cost, material, and quality-control matters. Most of the information for this section comes from the executive summary sections of the control reports. The change-order log and the project cashflow projections are other good items to include in this section. Also, this section should relate to how well we are achieving our overall and specific project goals. Are we still likely to meet our goal of finishing the project *as specified, on time, and within budget?*

Section 5.0, Progress Curves, discusses project progress using the monthly updating of the bell and S curves. The cost and schedule curves come from the detailed control reports for use in this section of the monthly progress report. The curves shown in Fig. 11.2 and 11.3 represent a typical example of a set of construction personnel loading and progress curves.

Discuss the monthly and total progress to date for the curves in the narrative portion of Section 5.0. Analyze and discuss any deviation of actual versus planned performance. Give particular attention to clarifying the areas that fall below the planned performance curve. Explain the reasons for not meeting the plan and the corrective actions taken or recommended. Also, discuss any serious performance lapses later in Section 7.0, Problem Areas.

Overall construction progress is covered in this section. The sudden burst of construction personnel in the past month, as seen in the personnel loading curve in Fig. 11.2, calls for some comment. The heavy increase in personnel last month should have the desired effect of getting construction progress above the planned curve shown in Fig. 11.3. However, the huge increase in personnel also will create some other operating problems that need to be discussed.

The procurement commitments shown in Fig. 11.4 are starting to fall below the planned progress at about the 50 percent complete point. That adverse trend must be addressed in the report. The purchasing is now in the bulk materials stage, and that is critical to maintaining construction progress.

The cashflow shown in Fig. 11.5 has been tracking the projected curve fairly well up to now. This month it appears to be heading off the chart. An explanation of that condition and a short-term idea of what to expect next month are now in order.

In Section 6.0, Work to Be Done Next Month, each construction activity lays out its work activities and goals for the succeeding month. Discuss the key milestone dates to be met for next month. Here are some examples of milestone dates: complete the ordering of long-delivery equipment; finish 50 percent of building foundations; start steel erection; open pipe fab shop; and the like. Be sure that you can meet most of the goals called for, because the items not completed will carry over to the next month. Missing monthly goals will require an explanation.

Section 7.0, Problem Areas, is another key section of the report. It requires careful preparation to avoid arousing panic in the reader. Summarize the problem areas mentioned earlier in the main body of the report. For example, we usually have a form called a Needs List showing those information and hardware items needed to complete certain construction tasks. If the needs are slow in coming, discuss how the lack of information or hardware is delaying construction progress. The adverse effects of this condition will in turn jeopardize

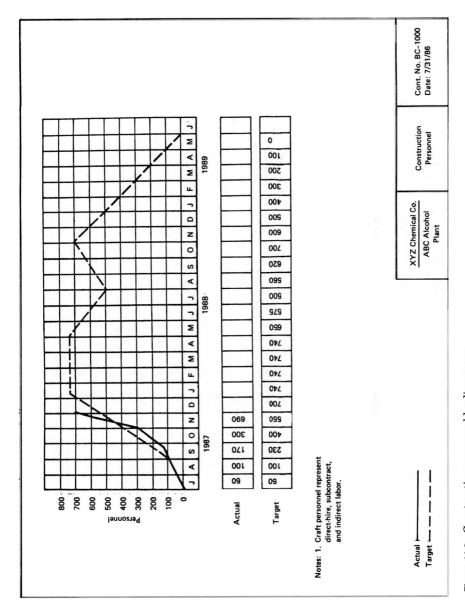

Figure 11.2 Construction personnel loading curve.

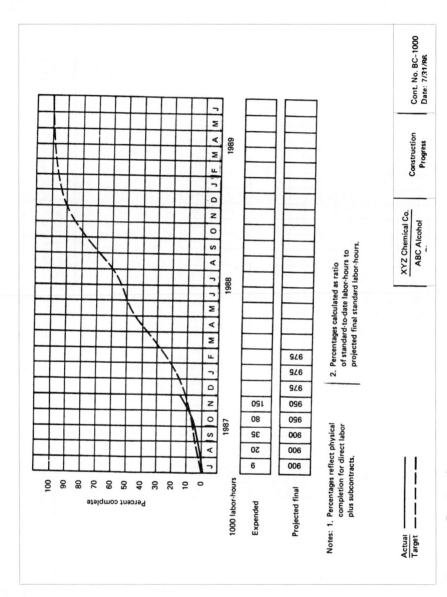

Figure 11.3 Construction progress curve.

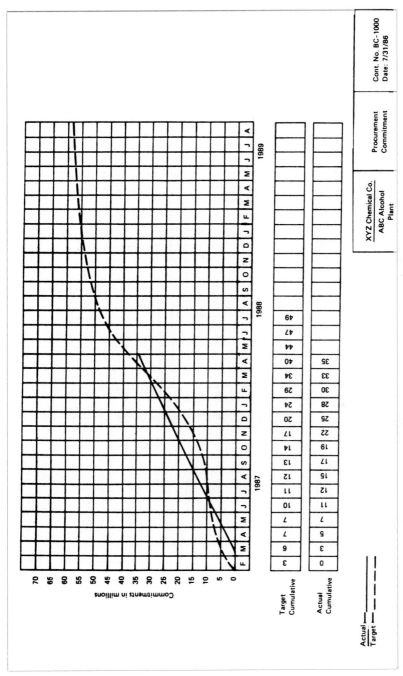

Figure 11.4 Procurement commitment curve.

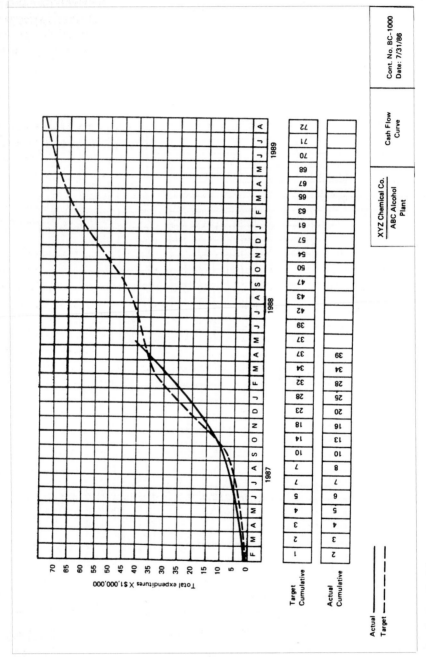

Figure 11.5 Cashflow curve.

the completion schedule. Go into the ramifications that this delay will have on construction progress, meeting the strategic end date, and project costs.

Above all, remember to propose some solutions to the problems you have raised. The top-management people reading the report are not going to have the answers. They expect the CM to produce the solutions. Leadership, conflict resolution, and problem solving are the CM's forte. This is a fertile area in which to practice the fine art of construction management!

Sections 8.0, Construction Photos, and 9.0, Drawing and Specification Lists, are optional. Construction photos are good if the project site is out of town or difficult to get to. Home office people are interested in seeing how their work turned out. Construction photos can serve as a good project-morale booster for those not able to visit the site.

Drawing and specification lists tend to be bulky and dull, so I wouldn't include them unless you feel that the delivery of design documents is going to be a problem. Bulky reports cause busy readers to lose interest. If any document serves no useful purpose, leave it out.

You should consider my suggested table of contents in Fig. 11.1 as a minimal requirement for a good progress report. Certain client or business conditions may require you to alter or add to it. That's no problem as long as you maintain the overall goals of good organization, standardization, and readability. Some government contracts may include a highly detailed progress report format that counters my suggestions. If the contract calls for it, you'll have to do it.

Project Meetings

Ineffective meetings are major time-wasters in project operations. The construction team spends a large part of its time in meetings. The range is from working meetings, planning sessions, problem solving, information gathering, and management review meetings to major presentations.

My goal in this section is to make the large blocks of time spent in meetings as productive as possible. As CMs, you are directly responsible for handling or attending most of the meetings on your projects. Here are some specific goals vis-à-vis improving your meeting results:

- Increase productivity
- Improve quality of results
- Cut wasted time and money
- Build your personal and company images

Our goals in improving meeting results are certainly in line with our total construction management commitment of *completing the project as specified, on time, and within budget.*

Establish need

"Is a meeting really necessary?" should be the first question asked when considering a meeting. Unnecessary meetings are a form of overcommunication that is a wasteful practice. If you can attain your goals with a conference call instead of a meeting, you will save time and money.

When group action is needed to exchange information, resolve conflict, develop ideas, or for organizational purposes, a meeting is useful. If a meeting becomes necessary, keep the group as small and compact as possible. Six to eight is a reasonable number for a working meeting, and twelve is an absolute maximum. Larger groups in information-transfer meetings can be effective because the role of the participants is more passive.

I have worked on projects where weekly meetings of 30 or more were commonplace. As expected, the results were minimal. Meeting costs approached $1.00 per minute per participant. That meant the cost was $30 per minute, $1800 per hour, and $3600 per normal two-hour meeting. That adds up to $15,600 per month or $93,600 over the course of a six-month project. Negative morale resulting from those ineffective meetings easily offsets any minor meeting accomplishments, so money spent on the meetings was completely wasted.

Estimating the cost per minute is a good way to keep your meeting size under control. It also allows you to estimate the actual cost/benefit ratio and productivity of your meetings.

Meeting execution philosophy

Since meetings are similar to small projects, they are subject to my Golden Rule of Construction Management. *Plan, organize, and control* your meetings to make them productive. That philosophy applies to all meetings, large and small. Obviously, larger, more complex meetings require a more formalized approach. The meeting checklist shown in the next section also applies to small meetings, but on a less formal basis.

I recommend that you follow the meeting checklist for all types of meetings, whether they are routine or one of a kind. The easiest *wasted meeting trap* to fall into is the routine weekly meeting, which can become ineffective through slovenly practices. Don't let familiarity breed contempt in routine meetings.

The most difficult-to-handle meeting on any construction project is the weekly planning meeting. In addition to the meeting being repeated weekly, it usually has a large working attendance on larger projects. There is no way to arbitrarily limit the number of attendees to this vital meeting because everyone responsible for meeting the construction schedule must attend and participate when their work is discussed. Each attendee has interfacing contact with many other attendees. All those conditions increase the difficulty of controlling the meeting and getting effective results. The CM must ensure that the project scheduler is properly prepared each week, and must attend the meeting each week to support the meeting leader in the difficult task of effectively running the meeting. If this meeting gets out of control, the schedule goals on the project will be placed in jeopardy.

A condensed meeting checklist

I have developed the following meeting checklist to serve as a guide for meeting preparations.

Before the meeting:
- Explore alternatives to having a meeting.
- Set your meeting goals.
- Prepare a detailed meeting plan.
- Make an agenda and prepare visual aids.
- Control attendance (calculate a cost per meeting minute).
- Check key attendee availability.
- Set the meeting starting time and time limit.
- Distribute agenda and reference materials before meeting.
- Select a proper location and facilities.
- Arrange seating and refreshments if required.
- Assign a recorder and timekeeper.

During the meeting:
- Start on time.
- Follow the agenda.
- Control interruptions.
- Keep your meeting goals in mind.
- Record decisions reached and actions needed.
- Keep participants on the subject.
- Don't waste time.

- Be tactful.
- Rotate part-timer contributors in and out.
- Adjourn on time.

After the meeting:
- Issue minutes of meeting on time.
- Critique your performance.
- Clear up any unfinished items.
- Follow up on action items.

Reviewing that condensed checklist will give you a good idea of just how involved planning, organizing, and controlling a meeting really is.

Planning the meeting

Thorough planning is essential if you are to meet your goal of a successful, productive meeting. Proper planning is second in importance only to the ability of the meeting manager to steer the meeting.

Meeting plans should be in writing. The amount of writing will be directly proportional to the size and complexity of the meeting. Even small meetings should rate a written outline.

The first step in the meeting plan is to set the goals you wish to accomplish. What problems are to be solved? What information is to be exchanged? In how much detail? Setting the goals should go a long way toward getting the desired results and having a productive meeting.

The backbone of any meeting plan is the agenda. As prime mover for most project meetings, the construction manager must insist that there be an agenda for every project meeting. Failure to have a written agenda virtually guarantees that a meeting will be unproductive even before it starts.

Assign a priority and a time allotment to each agenda item to tailor the agenda to the meeting plan. The agenda should follow a logical flow-pattern. For example, in a problem-solving meeting, state the problem. Next, give the needed background material before starting on the possible solutions. The agenda logic for project-review meetings can be more flexible in presenting the subject matter. One universal rule for agendas is to place less-critical items at the end, in case time runs out before completing the agenda.

State each agenda item as briefly as possible within the limits of clarity. Each participant in the agenda should have a clear understanding of the matters for discussion under that topic. That is especially applicable to the person who is presenting the subject in the meeting. Actually, the physical size of a properly drawn agenda has no bearing on the time needed for discussion. A well-prepared agenda serves to shorten the meeting time by giving the chairperson an effective tool for controlling the discussion.

After you have properly prepared and approved the agenda, issue it to the meeting attendees before the meeting. Allow attendees enough time to plan and prepare their contributions to the meeting. Be sure to include any written backup material for attendees to read before the meeting. Passing out reading material during the meeting destroys the tempo of the meeting and wastes valuable meeting time.

With the agenda in hand, you can now complete the list of attendees. Limit attendance of working meetings to only those involved or contributing to the meeting. Attendance is limited because each attendee adds communication channels to the group and increases the meeting-control problems. For example, 2 people have 2 channels, 4 have 12, 8 have 56, and so on. Controlling the additional communications channels adds to the difficulty of controlling the meeting effectively.

The meeting plan also must include any visual aids needed to communicate the message you want to deliver in the meeting. These may be flip charts, overhead transparencies, slides, or handouts. Use a medium that meets the specific needs of presenting the subject matter to the group. Don't go overboard on visual aids just for showmanship. A chalkboard for simple messages can be as effective as an overhead transparency and at less cost.

The meeting starting time and duration are important elements in the plan. Calculate the total elapsed time for the meeting from the time values assigned to the agenda items. Add some contingency time based on the type of meeting and prior experience, while keeping in mind the productivity goals. Using the elapsed time, back off from a natural cutoff event like lunch or quitting time as a time to start the meeting. A natural time barrier makes it easier for the meeting manager to keep the discussion closer to the meeting goals and agenda. Attendees, who are inclined to wander from the meeting subject, seem to be especially sensitive to natural cutoff times.

The window of time for the meeting must also be accessible by the major participants if you are to have a quorum. A brief telephone poll of key participants by your secretary can ensure schedule compatibility. Having a couple of options for time and date may be necessary to suit the most critical attendees. Of course, qualified alternate attendees are also usually acceptable.

Certain situations may require a dry run for fine-tuning the agenda and presentation. Such a situation occurs regularly in project-review meetings and major management presentations. Include that factor in the meeting plan to allow time for final polishing of the agenda before issuing it.

Organizing the meeting

Organizing the meeting involves making the necessary physical and administrative arrangements. The key action items are:

- Arrange a suitable meeting site
- Prepare visual aids, handouts, and so on as required
- Issue agenda and venue information
- Assign a recorder and timekeeper as required
- Arrange for refreshments*
- Make seating arrangements, if necessary
- Provide note pads and pencils*
- Hold dry run, if required*

The starred items are used for more formal, management-oriented meetings only.

Select the meeting site based on size, creature comforts, visual-aid equipment, proper lighting, and so on. The physical requirements are important to maintain the attendees' undivided attention throughout the meeting. Sometimes for critical meetings it is even necessary to move the meeting offsite to insulate the people from outside interruptions.

Visual aids are important communication tools for information-transfer meetings. Make sure that the visual aids are clear and legible to all participants, or they will do more harm than good. If the visual aids are complex, bring in a graphics consultant to prepare them and to operate the equipment. B. Y. Auger's book *How to Run Better Business Meetings*[1] is an excellent reference book for the preparing of visual aids.

A working meeting requires a recorder to take and publish the minutes. Issuing meeting minutes is essential to confirm the results for the participants. Publishing the minutes also informs those nonattending project team members of the meeting results. The recorder can also be the timekeeper who advises the meeting leader as to the timing of agenda items. The meeting leader should not handle the recording and timekeeping duties, because they detract from the main duty of leading the meeting. The recorder should be a good communicator and technically qualified in the subject matter. This often precludes using the project secretary in technically-oriented meetings.

It is pointless to record everything said during the discussions leading up to decisions. That would make the meeting minutes too wordy and lead people to ignore them. The meeting leader should sum up the decisions or action required as briefly as possible for the recorder.

It is vital to assign the necessary follow-up actions on matters decided in the meeting as part of the recording process. This can be done by providing an action column at the right-hand margin of the minutes of meeting format. This makes it easy for the meeting leader to follow up on the status of the action items after the meeting.

It may be desirable to serve refreshments for larger management-oriented meetings lasting longer than two hours. The serving should be at the scheduled break in the agenda or at the start of an early-morning meeting. I don't feel that serving refreshments adds anything to working project meetings unless the meetings are unusually long.

Prearranged seating assignments usually are used only in major management-oriented meetings. If necessary, group people from the same company or department together for easy communication. It's a good idea to arrange management people in the normal pecking order to prevent any hurt feelings. That factor is especially important when dealing with international groups having strict cultural customs in that regard. Place upper-management attendees near the meeting leader.

Sometimes, the meeting may require only part-time attendance by certain contributors. Schedule the part-timers in and out as required, to save their time and to keep the meeting within reasonable limits. This gives the meeting leader another incentive to keep the meeting on schedule.

The organizational part of meeting preparation is one area the meeting leader can effectively delegate to others. A good project secretary or administrative assistant can arrange most of the details based on the meeting plan.

Controlling the meeting

The meeting leader's ability to control the meeting is the most important requirement for the meeting's success. Even a well-planned meeting will fall apart if it is not properly controlled from start to finish. The meeting leader must always keep the meeting goals in mind while controlling the meeting. Writing the meeting goals makes them clearly understood and available for reference during the meeting.

Loss of meeting control is a prime cause of failed meetings. Although steering the meeting requires a firm hand, it also must be done with tact and diplomacy to permit valuable input from all attendees. Autocratic control often causes potential contributors to withdraw mentally from the meeting.

As the first act of meeting control, the leader must start the meeting on schedule. This is a good habit to set up, since it encourages perennial latecomers to improve their schedules. Starting late also places the meeting goals in jeopardy even before the meeting starts.

After a few opening remarks and any necessary introductions, the leader should get right to the agenda and follow it closely. Check a few of the milestone targets as the meeting proceeds to ensure that you are maintaining schedule. Some agenda items will finish early and some late, so the schedule should even out. Make agenda changes only if there is a good chance of improving the schedule.

Occasionally, some part of the meeting plan may break down and force revisions to the agenda. Handle the changes very carefully to protect the overall meeting goals. Quick thinking here can obviate the need to schedule a follow-up meeting.

Meeting leaders spend most of their steering efforts controlling the "problem people" in the meeting. Those people fall into these four general groups:

- Those who talk too much
- Those who talk too little
- Those who hold side conversations
- Those who want to usurp the leadership of the meeting

Fortunately, most people attending construction project meetings are technical types who are not overly garrulous by nature. When a participant becomes too wordy, the meeting leader must restate the sense of urgency required to keep the meeting on track. If that doesn't work, the chair must find another means to cut off the harangue. You may have to double-team the offender with the cooperation of the recorder or other participants.

Drawing out those who say too little can be even more difficult than controlling those who are wordy. Quite often the silent types can make a strong contribution to a successful meeting. Encourage active participation by the reticent through positive recognition when they do open up. Build active contact by drawing out their suggestions and opinions. Encouraging the clash of differing opinions also opens the way for those who are careful about expressing an opinion.

Control of side conversation is a must to avoid having two meetings at the same time. Side conversations are upsetting to those who are trying to make a contribution, and to the meeting manager. If there are any worthwhile ideas in the side conversations, bring them into the full meeting.

Squelching those who would usurp your meeting is the only way to prevent anarchy. Naturally, you should use extra tact if the usurper happens to be your boss. Occasionally, visiting executives do take over meetings, but normally they will back off when gently reminded of their actions.

Meeting control involves many human relations factors in a live performance. A good way to build your skills in this area is to study the human relations involved and practice various ideas in real situations. Use those that work and discard those that do not. Rate your performance after every meeting to see where you can improve. Antony Jay's article from the *Harvard Business Review*[2] presents some original ideas for that type of information.

Another useful meeting-control device is the tension-breaker. Quite often tensions will develop during meetings when conflicts are being resolved or sticky points negotiated. Perhaps the most effective tension-breaker is humor. It can relax tensions, but use it carefully. It can backfire if it embarrasses someone.

A prompt adjournment should occur when you have cleared the agenda or the allowed meeting time has expired. If some minor agenda items remain, try to settle them without convening another meeting. If they are important enough, however, schedule a later meeting.

Post-meeting activities

Issue the meeting minutes promptly, to remind people to get started on their "action items." Don't give them an excuse to forget about their assignments. Make every effort to issue the minutes no later than 24 hours after the meeting.

Some international meetings require signing a *protocol* before the participants can leave the meeting site. That puts extreme pressure on those people preparing and issuing the minutes to avoid tying up the meeting participants. In that case the production of the minutes must begin before the meeting is over.

In addition to listing the action items, the minutes serve as a record of the actions taken during the meeting. They are major contributors to project documentation. Place a copy of them in the project file as a part of the project record.

Shortly after the meeting, the leader should review the meeting results with key staff members to measure overall performance. Some searching questions to ask *and answer* are the following:

- Did we meet our meeting goals?

- Were we properly prepared?
- Did we control the meeting?
- Did we finish the agenda?
- What were the good points?
- What were the bad points?
- Where can we improve next time?

If you answered the first four questions "no," you need to work on those areas to improve your meeting performance. Also, remember the main purpose of the meeting. Follow up on the action items to be sure that the meeting time was spent effectively.

Approaching meetings with a proactive stance, as I have put forward here, should improve your meeting performance. Naturally, you will have to shape those ideas into your own operating environment. Effective meetings are too important a communications tool to waste, so the quicker you improve performance in this area the better!

Be a good listener

Most of our discussion to this point has involved outward messages, which comprise only half of the communication equation. Receiving the messages is equally important, but often it doesn't get the same emphasis.

Much construction-management communication deals with both business and personal conflict-resolution. Resolving conflicts effectively requires CMs to listen to both sides of the conflict, especially in the case of personal conflicts. Because field team morale is a key factor in successful project execution, CMs often play the project chaplain role. The most difficult disputes and decisions are the ones that wind up in the CM's office for final settlement. I don't mean to imply that CMs have unlimited time to spend on these matters. They must gain good listening habits that allow them to separate the wheat from the chaff to give timely and sound decisions.

To be a good listener, one must be patient. Many CMs lack patience because of the continual sense of urgency in construction management. Therefore, both listening ability and patience are traits most CMs need to develop.

I know of no training given in the area of patience, so this must be self-taught. However, there are several good courses for improving listening ability. I think the speed of improved listening comprehension from such training courses will pleasantly surprise you. A refresher course every few years is also effective in keeping this skill sharp.

Summary

Effective communication in every form is a potent weapon in the CM's arsenal of skills. This vital ingredient is essential for successful construction project execution as well as good client and human relations. CMs must be effective speakers, writers, presenters, meeting managers, and listeners if they are to present themselves, their ideas, and their jobs in the best possible light. CMs can use the best construction technology and have a spotless site, but poor communications skills may cause it to be perceived otherwise. Therefore, CMs must train and practice to become proficient in all areas of communication.

References

1. B. Y. Auger, *How to Run Better Business Meetings,* Minnesota Mining and Manufacturing Company, Minneapolis, MN, 1979.
2. Antony Jay, "How to Run a Meeting," *Harvard Business Review,* Cambridge, MA, March–April, 1976.

Case Study Instructions

1. As CM of your selected project, prepare the communications section of your Field Procedure Manual. Include all items such as correspondence (and logs), project reports, meetings, minutes of meetings, and telephone conversations.

2. Write a construction progress report for your selected project at the 50 percent completion point. Assume that field productivity is below standard and that you are three weeks behind schedule. The delay is starting to affect the meeting of the strategic completion date.

3. The owner's president and chief construction manager are touring the company's major capital projects. They are due to review your project in two weeks. As CM, you are in charge of planning and executing the field portion of the visit. Funding for the next project similar to yours depends on how well your project is going. Present your detailed plan for managing this major presentation.

4. Assume your project is being done on a construction-management basis. The owner (or client) has requested an in-depth project-review meeting before starting lump-sum construction bids. The design includes a model and design drawings, along with full construction specifications and an engineer's estimate. The owner's team is available for one day from 9 a.m. to 5 p.m. Prepare your detailed meeting plan and agenda for the meeting.

Include all physical arrangements, visual aids, and so on that you plan to use.

5. As construction manager, give a detailed plan of how you propose to operate the weekly field planning meeting on your selected project. Assume that the work is about 60 percent direct-hire and 40 percent subcontracted. The home office has provided a PC-generated CPM schedule. At peak workload there will be about 20 people attending the meeting every week.

12

Human Factors in Construction Management

If there were no people involved, construction management would be duck soup! Whenever I talk to construction managers about problem areas, *people problems* always head the list. Perhaps it's because we spend so much time working on construction technology, methods, and procedures and so little time on human factors.

In-depth investigation of human factors in the business world has taken place only in the past several decades. An example of that situation shows up in Fredrick Herzberg's basic study on satisfying human needs cited in earlier chapters.[1] Since the middle 1960s, considerable research and data based on Herzberg's theories have been published in management journals.

The general management community has eagerly accepted the work on human factors, and have sought to apply it. Also, literature on human factors in project and construction management has increased somewhat, but it doesn't seem to have affected actual practice that much. If it had, project and construction managers would not continue to rate human factors as their number-one problem area.

I believe proper handling of the human factors is the critical third leg of the stool leading to successful projects. It rates right up there with my Golden Rule of Construction Management and project communications as a requirement for successful projects. Even though we *plan, organize, and control* our projects perfectly, the projects can still fail if we mishandle the human factors. The human element in construction management involves human relations, personality traits, leadership, and career development. Strength in those areas can help us do a better job of managing projects, as well as improve the chances of meeting our personal goals. Years ago, good construction managers with a natural flair for human relations consistently had better projects than those without it. With the research and literature in the humanities available now, we no longer have to depend so much on chance. All we need is the right attitude, study, training, and practice in human relations.

Obviously, I am not a professional in human relations. I base my ideas in this chapter on practical observations made in managing capital projects for many years. This chapter cannot possibly cover all of the human factors involved in project work, because the field is so broad. The areas I have selected for discussion are the more important ones relating to management performance.

Qualities of a Successful Construction Manager

Let's start by taking inventory of the most important personal traits necessary to manage construction projects successfully. You can then study the lists to see how to shape your own qualifications to meet the requirements. A periodic review of your performance in these key areas can be a valuable tool for keeping the necessary skills up to date. The successful manager of construction projects must be

- An effective manager/administrator
- People-oriented
- Decisive
- A strong communicator
- Resourceful
- A problem solver
- Responsive
- Knowledgeable in the business
- Creative and imaginative
- Patient

A successful CM must possess

- Strong leadership and motivational capability
- High standards of ethics and integrity
- Personal drive
- Physical stamina and mental toughness

- Multidiscipline capability
- Common sense

Those two lists taken as one, present an impressive array of characteristics. It further confirms how broadly based construction management work really is. Most of the traits involve human factors in one form or another. We have talked about some of those traits in earlier sections of the book. This chapter concentrates on the traits not previously covered.

Personality

In looking over the lists, it becomes readily apparent that a person matching the requirements should have a dominant personality. Clearly, a person meeting the qualifications ought to be a self-starter with enough drive to stay the total course.

We also know that capital projects are not successfully completed by brute force alone. Because of the people-oriented nature of project work, a lot of persuasion also is necessary. Combining the two personality traits results in the dominant-persuasive personality that is ideally suited to project and construction management.

The largest number of construction management professionals come from the technical side of the business. Understanding the technical nature of the work is a prerequisite for managing capital projects. Technology seems to appeal to people with introverted personalities because complete concentration and hard work are needed to meet the college degree requirements. The intensity of the technical course work virtually excludes study in the humanities, which could broaden the technical person's personality. Given that type of background, most of us are still at the introverted end of the spectrum by the time we reach the construction management level. If your personality tests on the introverted side, it is time to do something about it.

Much psychological research shows that we are products of our forebears and environment. That research also shows that our personalities are formed by the time we reach age seven. I find it difficult to accept those theories at face value because of my personal situation. My memories of age seven are no longer very vivid, but I do remember being very shy, introverted, and lacking in leadership capabilities all the way through high school. Without the benefit of formal humanities training, I expanded my personality, became more outgoing and dominant, and built a successful career in project and construction management. If I could do that, anyone can do it! All it takes is desire, training, and practice, practice, practice.

Training for curing introversion is not so difficult. I recommend one of the personality-expanding courses such as that offered by the Dale Carnegie group. That course has proved itself over the past 50 years with millions of graduates all over the world. Following the course with some work in the Toastmasters' Club speaking experience should start you on the road to generating an enthusiastic, more outgoing personality. These basics also will open the way for further study and practice in the human element of construction management.

There are a few people on the extroverted end of the personality spectrum who make it to the construction management level. Extroversion is a bigger problem, because I know of no training for lowering high levels of it except in the school of life. If you have a problem with extroversion, work on controlling it gradually without killing off your enthusiasm in the process.

Ethics and integrity

As in any profession, high standards of ethics and integrity are important in project and construction management. Capital project managers often control large sums of money that don't belong to them. Usually accountants thoroughly audit the funds, but there is still a lot of room for possible conflict of interest. A conflict of interest can arise on either side of the client-contractor equation. Both parties must conduct project activities on a businesslike basis and show no favoritism to vendors, suppliers, or subcontractors. That includes accepting favors, kickbacks, and gifts. Today most companies have a statement of policy governing the behavior of those employees who handle company business. Be certain that you stay within those guidelines in all your dealings. You cannot build a professional reputation in this business if you don't adhere to the rules. There are a lot of insidious ways to circumvent the rules, so be alert for any traps. Some firms vying for the project business may not be as scrupulous as you are about bending the conflict-of-interest rules. As leader, maintaining the ethical standards of the construction team rests on you. Set a good example yourself, and make it known to the others where you stand on ethics and personal integrity.

Controlling that situation is fairly easy in the United States, but it can be more difficult in countries with cultures that condone (and even promote) conflict of interest. I seldom recommend going against management policy, but in that situation I do. If your management tries to place you in a conflict-of-interest situation, don't accept it. Trading your professional reputation for a job is not a good bargain. If you find yourself working for that sort of company, you are better off without it!

Personal drive

Construction management is a demanding profession that often generates long working hours, extensive travel, and frequent family relocation. Conflict resolution and problem solving add more stress to the job. All that is taking place within an environment of change, which further compounds the pressure. Many people won't consider those working conditions to be inducements for entering a field of work. Only people who look on that working environment as a challenge and have a strong personal drive to excel in it can be successful. On the plus side, a certain amount of authority, power, and respect comes with the job. The pay is usually attractive, and the personal satisfaction of a job well done is invigorating. Effective CMs usually are in short supply, so job security is good. The most recent exceptions to that statement occurred during the 1982 and 1991 recessions.

The mental side of the job demands a sound physical side to support it. CMs should keep in shape to handle their demanding schedules. This means having a regular exercise program and good dietary habits throughout your career. The fitness program and diet should be sensible and matched to your individual needs. In my opinion jogging, for example, is not an absolute necessity.

Relaxation is an equally important part of a CM's personal schedule. No one can live in a pressure cooker forever. Relaxing when time permits is necessary, but don't *play* as hard as you work. Some people have tried that unsuccessfully. Performing with a hangover the next morning does not improve your management abilities!

Multidiscipline capability

Construction people with experience across several disciplines are best suited for the job. Most of us who became CMs moved up from the technical ranks by working on capital projects. If you plan to get into construction management, seek work assignments that broaden your technical scope. That includes areas such as cost engineering, scheduling, procurement, and all areas of construction.

The most difficult cross-discipline transition is the one from nonprocess to process projects. The key here is the chemical engineering expertise required in the process environment. If you are making such a transition, a short course in chemical engineering for nonchemical engineers would be helpful. Obviously it's not necessary to become a chemical engineer, but the training will give you the necessary basics to better understand the inner workings of chemical process equipment and systems. That knowledge is especially useful for reading and understanding flow diagrams and for supervising the final plant checkout and acceptance by the owner. The same advice holds for training in any other discipline in which overcoming a knowledge gap will help you to do a better job and further your career.

The major weakness of most CMs is lack of skills on the business side of running projects. Earlier, I mentioned the missing humanity subjects in most technical college curricula. Also missing are business courses. The educational institutions leave it to us to get the business training necessary to supplement our technical management skills.

We can do that in several ways. The first is through on-the-job training, working as a field or area engineer under a construction manager/mentor for a few years. That approach can be effective, but time-consuming and costly because of the mistakes made during the learning process. The second is by taking an MBA degree. Preferably that is while working as an engineer rather than right after you earn a bachelor degree. That way you will get a better grasp of what the MBA work is about because you can base it on your work experience. The MBA is an added advantage in strengthening your resumé in addition to gaining knowledge. The MBA has only recently gained acceptance as a positive qualification for CMs on larger projects.

If you are past the age for standing up to the MBA course work, there is still hope. General management training is also available through courses such as the Alexander Hamilton Institute's Modern Business Program. This is a correspondence course requiring two years of study as part of either an individual or group training program. I used the course successfully as an individual and also later in some group training for my technical managers. In addition, I recommend a continuing program of training seminars and professional meetings to broaden and maintain your skills. Frequent self-analysis of your personal and business performance is valuable in evaluating your skills inventory . Be honest with yourself and cover both on- and off-the-job activities.

Human Relations

The area of human relations in construction management is broad because of the many interpersonal contacts involved in a largely conflict-oriented environment. In addition to the external human relations, there are the personal ones concerning the CM's career. Of course, the two are closely related. The rest of this chapter covers those crucial human factors and picks up the traits not covered in the previous lists. I have found the following areas to be the most critical to effective construction management:

- Client-contractor relations
- Contract administration
- Project relations
- Public relations
- Labor relations
- Leadership
- Common sense
- Keeping your cool
- Negotiating ability
- Patience

Client-contractor relations

When contractors' construction managers are polled about problem areas, client relations seem to come out on top every time. Ten years of leading project-management seminars has brought me in contact with about 500 client people who feel the same way about contractors. Obviously, there is something wrong when these key project groups have trouble working as a team.

In this discussion, I refer to the client-contractor relationship in its broadest sense. That includes construction managers who serve clients within their own companies as well as outside contractors and subcontractors. Some examples are central engineering departments and plant engineering groups that serve operating departments and divisions of their own companies. In each case it's difficult to get around the them-and-us syndrome.

Contractor-client relations involve two very important aspects in the business of meeting project goals. They are too valuable an asset to be sacrificed in a breakdown of human relations. Contractor-client relations can make or break

1. The success of the current project or, at the very least, meeting of the project goals
2. A contractor's opportunities for future work with the client

Those are two powerful incentives for making the client-contractor relationship work throughout the project. The project and construction managers can give up on meeting their personal goals, but they do not have the authority to give away their company's goals. They accepted those goals when they took on their respective project assignments.

It's impossible to write a scenario in which everyone has successful relations with a client or contractor. After all, a lot of personal chemistry is involved. A lot can happen over the complex course of a normal capital project. However, there are some things I can tell you that will improve the odds considerably. The primary necessity in any client-contractor scenario is having both parties realize that the relationship is essential to project success. Neither the project nor the construction manager can afford to have the project branded a failure, regardless of the personalities involved. Poor contractor-client relations can completely wipe out an otherwise outstanding performance. That makes an effective working relationship the number one priority throughout the project, both for your company and for your personal reputation in the business.

No two people will handle a given situation in the same way. You must handle human relations in a natural, unforced way to make any solution work. You will have to try various techniques with each contractor (or client) to see which one works for the given situation. Some key areas in the client-contractor relationship are discussed in the following sections.

Ethical conduct

Any failure in the key area of ethical conduct by either party will seriously undermine the relationship from the start. Neither party can condone unethical practices in the other. Fortunately, unethical conduct does not occur very often. When it does, squelch it immediately. PMs and CMs have a fiduciary responsibility to their respective managements to spend the project budget wisely. That doesn't leave any room for anyone on the project to show favoritism or accept kickbacks from vendors, suppliers, or subcontractors. Be sure that everyone on your project has a copy of the company's code of conduct and abides by it. If your company doesn't have a code of ethics, write one for the project to define your own ground rules.

Responsiveness

Contractors must be responsive to the client's needs first. This is a typical buyer/seller arrangement that places the customer in a preferred position. It does not, however, give the customer an automatic right to make unreasonable demands on the contractor or supplier.

Responsiveness is also a two-way street. Clients must be equally responsive to the contractor's needs for proper execution of the work. It's vital for clients to make their inputs and approvals in a professional and timely manner as defined in the contract. A strong effort by both client and contractor managers in this area will result in a highly successful working relationship.

Mental toughness

Weakness is seldom respected in any culture, not even by bullies. To keep control of their project, CMs should resist domination by their client counterparts. The buyer/seller relationship is not grounds for an uneven playing field. If there is any doubt in your mind, the contract should define the client-contractor relationship. I recommend that you gently, but firmly, cut off any one-upmanship activity whenever it occurs on your project. People who are prone to that type of activity will usually get back into line when pressed on the matter.

Contract administration

Contract administration is an area that sometimes produces friction in the client-contractor relationship. Even the best-crafted contract requires some interpretation from time to time. It's important that both parties take a proactive stance on meeting contractual requirements. That avoids friction over the need for one party to constantly remind the other of the contract requirements.

Also, there are areas of give-and-take in any contract that can make a contractual relationship run more smoothly. However, neither party can give away the store. Likewise, the other party should not ask for it. This area requires a lot of judgment for proper handling, so approach it with caution. Take a few smaller, calculated risks until you get the feel of it.

Relations of the project staffs

CMs also are responsible for the client-contractor relations throughout their field staffs. They must resolve any interstaff human relation problems not resolved at the individual level. The number of personality problems seems to be directly proportional to the size of the staff. The best way to reduce problems in this area is for the client and contractor project managers to set good examples as role models for the rest of the staff. Also, be sure that all members of your staff know the policies on how you expect them to handle their contractor (or client) relations.

Those are a few areas that I feel are critical to ensuring good contractor-client relations. If you learn how to handle them well, any minor areas should fall into place as well. Remember, try different approaches to suit different situations until you find the combination

that works for you. Also, nobody bats a thousand in the area of contractor-client relations. There will probably be a time where nothing
seems to work, so making a change is inevitable. If it happens more
than once or twice, there is cause for concern. You must review your
performance and correct the basic cause of the problem.

Internal project relations

The next most important human relations problem area is the field
team's working relationships with each other and with the rest of the
company. Eventually you and your key team leaders will have business contacts with everyone in the company, from general management down to the mail room. As the field team's leader, you are
responsible for seeing that these contacts are working smoothly.

The most frequent contacts are those with their peers in the firm's
home office, as well as subcontractors, vendors, government agencies,
and the like . The home office contacts can include department heads
and managers of construction, design, procurement, safety, and project-control groups. They may be the people who will be supplying the
members of your project team or having technical input to the work.
To a lesser degree you will have contact with business development,
corporate legal, accounting, and operating management. Certainly,
the first four items we covered under client-contractor relations also
apply to the relationships mentioned above. Those areas were ethical
conduct, responsiveness, mental toughness, and holding others to
their commitments. The one ingredient missing from the earlier discussion is the buyer-seller relationship since most of us are now working for the same organization.

That makes it more important to establish yourself as a knowledgeable manager with a mature outlook and respect for other people. It is
essential to command respect and cooperation from the people who
must perform well if your project is to succeed. That reputation is not
won by putting on your Superman suit every morning before leaving
for work. We build our respected reputations gradually, brick by brick
and stone by stone, much as we build our projects. Throwing the
stones and bricks at the other guys, even when they seem to be ask-
ing for it, will not build a solid reputation with our peers.

My philosophy is that it takes everybody pulling for you, in addition
to a good performance on your part, to have a successful construction
project. Having even a few people waiting around to pull the rug out
from under you increases the chances for failure.

You also should remember that project relations work in all directions—up, down, and sideways. Most managers are attentive to their
upward relations as a matter of personal survival. It is from there

that both the rewards and retribution flow. We certainly must maintain those relations if we are to meet our project and personal goals. Relations downward are sometimes another matter. Here we hold power and authority over the field staff. How we dispense that power and authority is a major factor in maintaining good project morale. Good project morale is the intangible factor that has a strong bearing on a successful performance. You are not likely to meet the project goals of quality, schedule, and budget without it.

Labor relations

Labor relations policy and practices are without doubt the most critical human relations area in the construction arena. Because labor relations impinge so heavily on day-to-day site operations and the ultimate success of the project, CMs have to understand their firm's basic labor relations policies and labor law. They need to have enough labor relations know-how to ensure effective choices when making field-level labor decisions.

Any contractor's overall labor relations policies must be based on their labor posture and the labor laws in effect at the construction site. The basic body of U.S. labor law is enacted by Congress, but there are many state labor laws that further affect the labor practices for construction activity in that state. The most notable of the latter type are the so-called "right to work laws," which permit open- or merit-shop construction operations in many states. The intent of most labor laws is to define and protect the rights of labor and management in an evenhanded manner. The laws have been around for a long time and are well supported by a broad base of case law. O'Brien and Zilly have an excellent review of labor laws and industrial relations in Chapter 4 of their *Contractor's Management Handbook.*[3]

In addition to the labor law directly affecting project working rules, there are the other laws covering affirmative action and equal opportunity for employment of minorities. Most contractors have the required plans in effect to meet those laws. The CM's role in this area is to see that the relevant field department heads are meeting the legal and corporate requirements of those laws in a positive manner. Just as with OSHA and safety, citations in this area can lead to suits or fines and bad publicity.

The main goal of any labor policy is to maintain high productivity and reduce work stoppages, which hamper job progress and increase labor costs. Local labor practice and usage play a large role in labor relations and make it difficult to develop specific guidelines covering all situations, particularly in a union-shop setting.

The basis for on-site labor relations is the union contract or the open-shop labor conditions that set the working rules for the field labor. Sometimes a supplemental agreement, called a *site agreement,* is used to define the special conditions in effect on a specific site for the duration of that project. The purpose of the contract or site agreement is to lay down a detailed set of rules governing the use of construction labor on the project. The agreement sets out working hours, pay scales, premium time, fringe benefits, grievance procedure, and so on, along with various management and labor policies. The CM uses the site agreement as the basis for administering the ongoing labor relations at that site. Those items that can't be resolved in the field are referred to the home office labor relations representatives for interpretation and resolution.

The relatively recent arrival of the open-shop construction concept has added several new dimensions to labor policy. The open-shop concept is most prevalent in the right-to-work states of the Sunbelt. The introduction of open-shop operations has created some competition for the construction labor once monopolized by the labor unions. The open-shop competition has forced the unions to relax some age-old working rules, fringes, and wage rates to become more cost-effective. Labor-management cooperation in the key cost and productivity areas was essential to keeping union-shop contractors competitive with their open-shop counterparts. Actually, the cost difference between open-shop and union-shop contractors has been closing in recent years, which has been a great benefit to owners and developers of new projects. Although there are no figures to prove the point, the advent of the open-shop construction contract has probably given the largest boost to construction cost-effectiveness in the last 30 years.

It is the nonroutine matters that cause the most problems on construction projects. Chief among these are jurisdictional disputes. A jurisdictional dispute is the result of two or more craft unions claiming the right to perform a certain type of work. This doesn't occur in open-shop work, where construction management decides which craft will do the work based on the skills required.

The debilitating effects of craft jurisdictional disputes are reduced by holding a prejob conference. The various craft business agents meet with construction management people to resolve their differences before the work starts. They discuss each major work activity in detail to agree on who will handle it. The agreements reached are in writing and are signed by all participants. You probably won't head off all jurisdictional disputes before construction starts, so you must eliminate the ill effects of those that do arise. Make an all-out effort to keep the rest of the work going while arguing any fine points of interpretation with the crafts in question. Take all possible legal steps within the site agreements to avoid setting up picket lines. Your negotiating skills will be severely tested in this area, so keep them sharp!

The contract usually includes time lost to labor stoppages in the force majeure clause, excusing any delays they may cause. As the owner's PM or CM, it's in your best interest to get the work restarted as quickly as possible. Therefore you should offer any support that the owner's organization can give to the contractor in settling the dispute. However, owners must never approach labor representatives directly. They should always go through the contractor to avoid muddying the waters of the ongoing labor negotiations.

If the project and home office labor relations personnel can't resolve a labor problem, it could be time to bring in the labor lawyers. That should be considered only as a last resort, however, because of the large amounts of time usually consumed in legal matters.

Public relations

The CM usually acts as the public relations (PR) representative for the construction project. CMs should mesh their project PR responsibilities into the overall corporate policies set by top management. It's natural to want your project to be presented in the most favorable light. A proactive approach to project PR is the best way to accomplish that. Bad PR can even get in the way of meeting the project goals. CMs ought to be alert to spot newsworthy events on their projects that could be of interest to the business or the general public. This is one of those subjects not covered in our technical educations, so developing a feel for newsworthy items is another self-taught skill.

It is even more important to take a proactive stance on items that are likely to be detrimental to your project or corporate image. Try to look ahead for any negative PR concerning the project that might be developing in the community. Make sure that any adverse publicity gets a fair counterpresentation the first time around. If an adverse or erroneous item gets into the media, it is virtually impossible to get a retraction later.

Establish the initial contacts with local government and community officials before opening the site. Tell them about your plans to handle such problem areas as traffic, dust, noise, and fumes, which could adversely impact the area. To keep the public on your side, stress the positive effects of the project on the community. The project or construction manager often acts as the project's technical representative in speaking to civic and government groups about environmental matters. The presentations require careful preparation, including clearance from top management. Be especially careful in handling the news media. Handing out a well-written and well-checked news release is safer than making an impromptu presentation.

Leadership

Leadership is an area that touches on most of the other human-factor areas in construction management. It is a crucial requirement for effective practice of *total construction management*. Building leadership and motivational skills is vital to becoming successful in construction management.

The following list of eight areas of importance in management illustrates my thoughts about the nature of leadership in project management. The list is a condensation of the material given by Henry Mintzberg in his book, *The Nature of Managerial Work*.[2]

1. Peer skills—the ability to set up and maintain a network of contacts with equals

2. Leadership skills—the ability to deal with subordinates and with the kinds of complications created by power, authority, and dependence

3. Conflict-resolution skills—the ability to mediate conflict and handle disturbances under psychological stress

4. Information-processing skills—the ability to build networks, extract and validate information, and transmit it effectively

5. Skills in unstructured decision making—the ability to find problems and solutions when alternatives, information, and objectives are ambiguous

6. Resource allocation skills—the ability to decide among alternative uses of time and other scarce resources

7. Entrepreneurial skills—the ability to take sensible risks and implement innovation

8. Skills of introspection—the ability to understand the position of a leader and the leader's impact on the organization

Each one of those areas has a direct bearing on the practice of construction management. Now let me tie them into our discussion.

Peer skills

We discussed this point earlier in connection with setting up effective working networks with key field staff, home office department heads, and other operating groups. The contacts with peer groups are vital for the necessary outside support to enable CMs to execute their projects. Those networks are built and maintained largely through a mature professional performance and a cooperative attitude with your peer groups.

Leadership skills

Management delegates to CMs a lot of the authority needed to fulfill the project goals. Knowing how to use that authority to build and motivate an effective project team is essential to successful leadership in the construction area. Wielding authority is the skill most involved with human relations!

Conflict-resolution skills

Construction management abounds with stressful conflict-resolution situations. Managers must learn how to cope with those emotionally charged situations quickly, calmly, and fairly, without damaging working relationships and project morale. Try to present decisions to the parties involved as win-win solutions. That will allow participants in the decision to keep their self-esteem and maintain enthusiasm.

Information-processing skills

Virtually all project communication passes over the CM's desk. In addition to assimilating that information, the CM must get out into the trenches to find out what is going on. That requires every communication skill: speaking, listening, reading, writing, and presenting information. The information gained is not worth anything until it's analyzed to determine its effect on project performance. This area is crucial to maintaining the good client-contractor and project relations mentioned earlier.

Skills in unstructured decision making

Positive leadership leaves no room for shilly-shallying. Decisiveness is one of the character traits listed earlier for successful CMs. "Shooting from the hip" on every decision is often hazardous, so I don't recommend that either. Make use of the time available for making the decision, but don't drag it out unnecessarily.

Resource allocation skills

Project and construction managers oversee the disposition of all project resources. That includes time, money, people, material, equipment, and systems. Each of them makes a contribution toward meeting the project goals, so allocate it wisely. That skill interacts closely with the decision making just discussed.

Entrepreneurial skills

This is my favorite! The project is your *business* (profit center) to run. Some projects require a Mom and Pop approach, while others are big business. In either case, you should take sound calculated risks when the payout looks good. By *internalizing* your project numbers, you should know what to expect when the project reports arrive. Most entrepreneurs are creative, so use some imagination in running your business.

Skills of introspection

In addition to understanding your position as a leader, skills of introspection mean periodic self-analysis of your total performance. Is everything possible being done to reach the project goals? Does your performance measure up to the personal standards you have set? Are your leadership skills getting effective results?

Applying introspection to your job environment is one of the best teaching tools available to you. Management schools and training courses can only point you in the right direction. Introspective practice of management theory is the quickest way to learn what really works for you!

Motivational skills

As good as these eight points of good management are, I was surprised to see that Mintzberg left out such a crucial area as motivational skills. We discussed those skills earlier under mobilizing the project and increasing productivity. Effective leadership in construction management is founded on having a fully motivated supervisory staff and labor force. Success in this key area will ensure a safer, more productive, and smoother-running job that will meet your project goals!

As you reviewed these nine areas of importance to management, I am sure you thought of several specific examples of recent project situations applicable to each one. Run through those examples in your own mind and rate yourself on how you actually handled them. If the answers are not good in some areas, try to mold your leadership skills to improve the outcome next time. Making a frank appraisal of your performance against the checklist a couple of times a year can improve your leadership skills immensely. Improvement in that area also will help you to mold the dominant-persuasive personality necessary for practicing *total construction project management.*

Common Sense

Common sense is one of those intangible attributes that some of us were lucky enough to be born with. Others, not so lucky, must acquire it. The dictionary defines common sense as "sound practical judgment that is independent of specialized knowledge, training or the like; normal native intelligence." To me, the operative words are "sound practical judgment." As a start, I recommend using sound practical judgment in areas where you *do* have specialized knowledge and training, such as construction management. Areas where you don't have specialized training and knowledge also call for common sense. You need common sense when people seek to promote the use of impractical ideas on your projects.

Suppose, for example, that a department head is trying to impose an unproven, overdetailed, costly scheduling technique on your small project. That is when common sense should tell you to ask some pertinent questions. Do we really need it on this type of job? Will it really work as well as you say? Can we afford to experiment with it on such a small job? Can we stand the extra cost?

The construction management system is really nothing but the application of common sense. The application of sound practical judgment in practicing total construction management is what I have been talking about throughout this book. One learns common sense by observation, practice, and experience.

Keeping Your Cool

Keeping your cool is another trait we can do something about. CMs can't survive by pushing the panic button. They have to learn to deal with panic situations calmly to avoid becoming nervous wrecks.

It's better to reserve your energies for clear thinking on how to solve the problem than to wallow in panic. If the project leader is running around in a state of panic, the panic will spread to the rest of the construction team. No serious problem ever was solved by creating a state of panic.

Panic is a symptom of extreme worry. If you have a tendency to worry, you also are likely to panic in difficult situations. One way I have used to overcome extreme worry is to analyze the problem to see what is the worst outcome that could happen. Many times it turns out to be less catastrophic than first imagined—you can live with it. Starting from that premise, coolly explore ways to improve on the worst-case scenario. Anything salvaged over and above the worst case is an improvement. In many cases you can turn the panic situation around, canceling out most of the adverse effects.

Furthermore, having successfully handled the panic coolly makes you look more professional in the eyes of your management and peers. That kind of performance builds your desired reputation as a mature, knowledgeable, and respected practitioner of *total construction project management.*

Negotiating Ability

CMs routinely find themselves in negotiating situations. Such situations arise with peers, clients, and subcontractors in almost every aspect of executing the project. They include areas such as staffing, estimating, and scheduling, in addition to such normal areas as contract negotiation, purchasing, and change orders.

Negotiation is one of the management arts that you can learn through training and practice. There are several good courses offered in the management training marketplace, and several good books on the subject are listed in the reference section at the end of the chapter. If you can't take a seminar in the subject, *The McGraw-Hill 36-Hour Negotiating Course*[5] will give you the equivalent material at a very nominal cost.

Remember, a negotiation is a special form of meeting, so a detailed meeting plan and agenda are critical to success. Before going into the negotiation, plan your basic strategy. That includes setting your short- and long-term goals, a profit strategy, and selection of the negotiating team. Next, set up your information-gathering and -processing systems and determine who makes the decisions. You should then be ready to study and implement your strategic and tactical approaches to finally arrive at a satisfactory close.

It also is important to control the negotiating meeting to your advantage even if you are the seller or plaintiff. It may surprise you to discover how often this can be done when your preparation is better than the opposition's. Review the discussion in Chapter 11 on planning, organizing, and controlling meetings.

Patience

Patience is a virtue—most of the time. However, don't confuse patience with allowing an existing problem area to continue, hoping that it will go away. Patience is something most of us develop naturally as we mature; we base it on common sense and practical judgment. That is why we often see effective project teams made up of mixtures of seasoned hands and young Turks, providing both patience and push.

Patience is akin to controlling one's temper in a difficult situation. When you lose your temper, you often lose the outcome of the situation. Patience is also a must in the client-contractor relationship discussed earlier. This is especially true in international projects where cultural differences are involved.

If you must have one admitted weakness in your construction management makeup, lack of patience is the most acceptable. I have found this admission to be a useful tool when undergoing annual appraisals. According to good management practice, the reviewer should ask what you think your weaknesses are. After a few moments of thought, admit to being a little impatient. The reviewer will immediately transpose your admitted weakness into a strength and go on to the next subject. Thus, you have neatly dodged the minefield of admitting your weaknesses. You are bound to increase your score by a couple of points!

Personal Human Factors

To wrap up human factors in construction management, there are a few items that affect the CM personally. The following tips can improve your performance and promote your career.

- Selling the organization
- Bending the rules
- Knowledge of the business
- Personal public relations

Selling the organization

Selling the organization is an important area that can relate to contractor-client or project relations. Never poor-mouth your company or any part of it in front of clients, contractors, vendors, or management. To do so is a common failing, especially after you have developed a close personal relationship with the other person. That doesn't mean you should try to convince anyone that a shoddy performance is great. When you run into a poor performance in the organization, get it corrected and make sure the right people hear about the corrections. But don't reinforce the ill effects by saying, "I always knew that so and so was a lousy department." If a member of your team makes an obvious blunder, get into the matter and straighten it out. Then go to the client or management and let them know how the matter was corrected and made right.

On the positive side, make sure people hear about a good performance in an unpretentious way. When that's done properly, one builds the image of the construction team and improves morale simultaneously. Proper handling in that area can make an average performance look good and a good one look excellent. It is more difficult to get a poor performance up to average, but it has been done.

Bending the rules

Bending the rules has to do with the art of construction management. I have spent most of the space in this volume laying down strict rules for you to follow in practicing total construction management. Now I want to talk about *bending* some rules. What's going on here!

Rules are necessary to keep organization in and chaos out. However, rules do offer the disadvantage of stifling creativity and *reasonable* risk taking. Creativity and risks can often pay large dividends if they succeed. We hate to miss out on any benefits of that type, so occasionally it pays to bend the rules. The only qualification I want to put on this point is that you should never consider bending the rules in the safety and risk-analysis areas. In other words, don't take chances on safety practices or insurance coverage.

The problem is knowing when and how much rule-bending is safe. When you spot an opportunity, analyze the situation carefully from all angles including a sound fallback position. Justify taking the risk by calculating the potential gains and losses to determine if the bending is really worthwhile. List the pros and cons and weigh each one to determine the chances of success. Make sure you have all the information necessary to make the decision and review the plan with your key peers. If the risk looks sound in relation to the reward, you may want to give it a try.

Learning when and where to bend the rules comes only with time and experimentation. I recommend you start in a small way and try to build on your experience. The cardinal rule is this: *Make sure the potential rewards warrant the risk.*

Knowledge of the business

You must have a thorough knowledge of the construction business, particularly the fields in which you are practicing. Keep current on general economic conditions and how they may affect your business. How are interest rates going? What is the outlook on inflation? What current events are likely to affect your project or the construction business as a whole?

You should take the time to read your industry trade journals so that you keep up with business and industry trends. Keep current on what the competition is doing, what big projects are breaking, and so on. All that background is necessary to participate intelligently in conversations with your management, clients, contractors, and peers. Being knowledgeable about your business builds your image as a competent and professional CM. Membership in professional societies and outside reading are key sources for that general background information. When you plan your time, leave some room for that sort of mind-expanding activity.

Personal public relations

Your personal PR program is something you have to do yourself— assuming you have not hired a press agent. One area of effort is getting letters of commendation at the end of the project. Naturally, you should expect them only after a good or outstanding performance. People often volunteer such letters, which is fine. However, if they don't volunteer, some timely hints may help.

Another personal PR activity is submitting newsworthy articles about unusual features of your project for house organs and trade journals. Such articles are image-builders. With some imagination, you should be able to come up with a few more ideas.

Summary

Good human relations skills in construction management are the key to personal and project success. The CM must develop a persona that will foster an image of an ethical, mature, competent, level-headed professional. CMs are responsible for the key areas of client, project, labor, and public relations. They must be effective, introspective leaders and role models for everyone on the project as well as their peers.

References

1. Fredrick Herzberg, *Work and the Nature of Man,* Thomas Y. Crowell, New York, 1966.
2. Henry Mintzberg, *The Nature of Managerial Work,* Harper and Row, New York, 1973.
3. James J. O'Brien and Robert G. Zilly, *Contractor's Management Handbook,* McGraw-Hill, New York, 1991.
4. Chester L. Karrass, *Give and Take,* Thomas Y. Crowell, New York, 1974.
5. Mark K. Schoenfeld and Rick M. Schoenfeld, *The McGraw-Hill 36-Hour Negotiating Course,* McGraw-Hill, New York, 1991.
6. Edward Levin, *Negotiating Tactics,* Fawcett Columbine, New York, 1980.
7. Robert J. Graham, *Project Management: Combining Technical and Behavioral Approaches for Effective Implementation,* Van Nostrand Reinhold, New York, 1985.

Case Study Instructions

1. Your corporate public relations department has asked for a list of project factors on your project that are likely to impact on the local area. It plans to use the information to indoctrinate local citizens groups, public officials, and labor leaders on the effect the new project will have on their community. Make a list of potential positive and negative effects your project will have on the community and the state.

2. After the 25 percent progress review, you are convinced that the contractor's CM is not performing to your satisfaction. As owner's CM/PM, how would you approach correcting the situation?

3. Perform a self-analysis of your personality as it relates to becoming a successful project manager. Include checking the inventory of the nine areas of importance in management.

4. As the contractor's CM, you are starting to have problems with your client relations. The owner is not living up to the contractual requirements in regard to information supply and drawing approvals. You already have requested improvement in the progress report, but it doesn't seem to be working. How would you approach that problem to ensure meeting your project goals?

13

Computers in Construction Management

As with virtually all other businesses, the computer age has penetrated the construction business. The construction business was somewhat slower to computerize than other businesses because of its diverse and fractured structure. Construction's innate reluctance to accept new ideas and to spend money on training were also delaying factors. Anyway, the 50 years necessary for the computer age to infiltrate business in general is only a blip on the time line in the history of construction, so the length of time doesn't really matter.

CMs must address the use of computers the same as any other technology that can benefit their project performance. Computers have arrived in the construction industry in both the construction technology and management segments. CMs that don't recognize that fact are not going to reach their full potentials. Modern CMs *must* become *computer-literate* to the degree that they can apply the computerized systems to gain the most benefit for their specific projects. Note the accent on *must* and *computer-literate*. By that I mean that CMs must at least know enough about computer systems and how they work to make the best use of the data the systems produce. They must know enough about computers just as they do about any other construction tool.

To gain a better understanding of my theme of using computers in construction management, let's look at what computers can do for us. The major operations that computers do fast and best is crunch numbers, operate databases, sort data, and print or plot information. Besides all that, computers can communicate data by modems, so they are already linked into our communications systems. Looking at it from that standpoint, all business, including construction, has been handed a management tool of epic proportions. Fortunately for America, the computer was developed here, or we would have lost our preeminent position in the business world by now. If the construction business doesn't embrace the computer, U.S. construction's preeminent position in the world is likely to be lost.

How do CMs coming into the business handle computer applications if they are not already computer-literate? First they should address the computer systems already in use. They have to ask themselves such key questions as, "How is this data produced?, How can I use it?, How accurate is it?, Do my people really know how to use it?", and the like. Talk to the people that use the systems to find out how it works. Don't be afraid to ask dumb questions just because you're the project leader. The best way to learn about computers is to buy your own PC for home use. If possible, join a local users group to learn as much as you can there. Also, take any available company training courses or outside seminars offered by vendors.

After you get up to speed and feel you have mastered the ongoing systems, start looking at other possible applications that could benefit your job performance. There will be areas that may require some seminar training. One example of that is learning to use a computerized CPM scheduling system so you don't have to depend completely on your field scheduler for information. The ability to sit down at the computer monitor yourself or with your scheduler and do "what if" exercises on the screen to solve knotty scheduling problems is a powerful tool. That way you can bring maximum scheduling expertise to the problem in the shortest time.

The first several chapters of Tidwell's *Microcomputer Applications for Field Construction Work*[2] give an excellent introduction to selecting available PC hardware for getting started. The book then goes into software applications that are useful in field construction operations. Although the clarity of software manuals has improved over recent years, you may still want to consider buying how-to books that give you shortcuts and a clearer picture of how to really run the program.

The Computer Revolution

I always like to take a look at the history of what happened to get us where we are today. That's based on the theory that "those who do not know about history are destined to relive it." Today's fast-paced world doesn't allow us to waste time on reliving history.

The electronic computer has been around only for about 50 years, having had its start with the first electronic-tube-based unit built for the U.S. Army during World War II. That's a very short time compared with its predecessor the abacus, which has been around for thousands of years. After the war, no one really thought that industry or commerce could afford one of those expensive monsters, so the computer industry evolved rather slowly. In the mid-1950s, improved and lower-cost mainframe computers were accepted into the commercial side of many *Fortune* 500 companies. Mostly those units were assigned to handle the more mundane general accounting and sales and marketing functions.

Meanwhile the electronic developments made the units smaller, more efficient, less costly, more user-friendly, and much easier to sell. In the early days, software to drive the hardware had developed hardly at all, so each computer operating center was supported by about an acre of programmers developing proprietary software for that company's applications. Actually, it was the cost of the support group that made the per-second cost to run the mainframes so high. A good economical alternative for smaller firms, who couldn't justify a computer, were the computer service agencies that offered computer services on an as-needed, time-share basis. Those firms are still active today and are still a viable option if they prove to be more efficient for your needs.

Eventually, a myriad of software development firms emerged to make and market generic programs for databases, spreadsheets, word processing, scheduling, and technical design and drawing programs. Today customers can select their software programs from a large slate of suppliers at a very reasonable cost. Many good programs are available for under $1000. That amount would run one of the old computer support groups for about 10 to 15 minutes.

All that happened before 1980, when the advent of the microchip and the personal computer (PC) caused another major revolution within the computer revolution. The PC revolution quickly resulted in distributing the computing function away from the mainframe to the point of use where the computer applications were really needed. Suddenly it was no longer necessary to run all computer applications through the knothole of the mainframe computer, to await your turn in the mainframe's interminable queue, or to argue about the priority of your work. The PC also sealed the fate of the acres of overhead people engaged in mainframe support. Mainframes became smaller and were generally relegated to the accounting functions and running occasional large programs that exceed the processing capability of a PC or minicomputer. The downsizing of mainframes and their market share has resulted in the present-day problems of computer giant IBM. It will be interesting to see how those problems play out.

The downside of the PC revolution was the time and cost of training the users necessary to reap the PC's benefits. To get full use from a PC, each user must gain some degree of computer literacy! There is no point in putting a PC on someone's desk if they aren't going to use it productively. There were plenty of examples of that form of inefficiency during the 1980s. Some of the construction industry's older hands have resisted the need for PC training, but they are now learning PCs or working their way out of the system through attrition. The modern-day construction manager in any sized firm has to become PC-literate to be competitive in the future.

In the past few years, major strides also have been taken in linking computers to communications systems by phone lines. Wireless communications for computers are already on our doorstep. Network applications vary from a simple local area network (LAN) inside an office to maximize use of computer peripherals, software, and data banks to wide-area networks (WAN) linking PCs around the world. In addition to transferring data, fax messages, and electronic mail, the new computer-communications systems, when combined with electronic pagers, instantaneously can put the home office in touch with key people on the job site, anywhere day or night. There is virtually nowhere CMs can go to hide for a few minutes to organize their thoughts. Many CMs and other business managers don't regard such high-tech developments as progress!

The PC has proven itself a very useful tool in all levels of the construction industry, from the smallest single-PC business to the largest top 100 contracting firms in the world. The flexibility of the PC permits a small firm to add hardware and software as its needs grow, and the largest firm can expand easily as new construction industry PC applications emerge. As of this writing, the cost of PC hardware has been dropping significantly even as the processing speed has been increasing by leaps and bounds. Microchips keep getting smaller and more powerful, so we now have palm-sized PCs that can easily be carried into the field in one's pocket. Data can be keyed in the field and transferred to the desktop PC in the office. As an author, I even hate to mention where computers stand today. With the computer revolution moving so fast, the author will surely look outdated by the time the book is published!

Aside from the PC as word processor replacing the typewriter, the single piece of software that cemented in the use of PCs in business was the electronic spreadsheet[2] as developed in Lotus *1-2-3*. Some prestigious business schools have claimed the electronic spreadsheet to be the most influential management tool discovered in the last 15 years. It actually made business democratic by bringing the spreadsheet's analytical powers to the lowest levels of the business organization. Its ability to organize input data and calculate various formulas makes it the workhorse of the control systems for business in general and construction operations in particular.

Before we get too far into the future, let's review the areas where computers have already become accepted in the construction industry. The following overview is designed to give an overall picture of the major developments in construction-related computer systems, so the reader should research the references listed at the end of the chapters and in other bibliographies for a more detailed treatment of the subject. Many of these applications already have been discussed in prior chapters and are mentioned again here simply to complete the computer applications picture.

Computer Applications in Design

When one considers the effect of computers on the construction industry, it's necessary to consider the design side of the business as well as the build side. After all, construction depends on the drawings and specifications to delineate what is to be built, and much of today's design documentation is produced on computers. Because design documentation is a major part of construction contracts, even CMs involved in the build-only side of the business need to know how construction documents are produced on the computer.

The largest application of computers in the design end of the business has been through the use of *computer-aided design (CAD)* for the production of construction drawings. Originally this process was dubbed *computer-aided design and drafting (CADD)* to cover both engineering design and drawing production. However, growth in both the design/drafting and engineering design areas was so explosive, the systems were divided into CAD for drawing production and CAE for engineering programs. That was only natural because each system used entirely different hardware and software systems. Most CAE software can be run on a standard PC at the engineer's desk, replacing the ubiquitous slide rule of former years. The use of CAE in the design office does not much affect the final design product seen by the construction people in the field. One fringe benefit for the field is that the design has a much better chance of being completed and issued for construction earlier than the former *manual* designs.

CAD systems

The more sophisticated CAD hardware and software systems are more complex and costly to install and maintain than desktop PCs. The drawing input terminals have been simplified in recent years, but the more complex computer-processing needs for handling the graphics require a larger computing capability than a PC can offer. Most larger CAD systems use a minicomputer such as a Digital Equipment Company VAX system to store and process the data. The VAX computer then transmits the data to a plotter to produce the drawing on reproducible media.

The advantages to the field of using CAD for drawing production are manifold, even though speed of production may not be one of them. The CAD system's greatest advantage is the availability of many electronic layers for the designer's use. The general layout can be drawn on one layer, with subsequent layers handling other trades like steel, electrical, plumbing, HVAC, and the like. The overlaying layers can then be computer-checked for interferences. That practice has gone a long way toward eliminating problems that occur in the field when two pieces of hardware wind up occupying the same space. Although the system is not yet 100 percent foolproof, CAD has come a long way in reducing costs by reducing the number of corrections of field interferences and other drawing errors.

Another CAD advantage is the possibility of taking off material through use of computer programs. Piping and insulation material takeoffs are examples of that capability. As the piping isometrics are produced on CAD, the program automatically produces a piping bill of material. The bills of materials are then processed to purchasing for procurement and on to the material-control function in the field warehouse. In addition to saving design hours, the system supplies more accurate data to procurement and the material-control functions.

The CAD system also offers the flexibility to complete some of the design work in the field, if that is more productive. A case in point are revamp jobs involving considerable amounts of process equipment and piping. It's possible to set up CAD units in the field office to make the new layouts, piping isometrics, and material takeoffs for rapid turnaround of the work.

The CAD system people have already developed some tools in the project design area that may represent even greater advantages for construction people in the future. A three-dimensional CAD has been around for almost 10 years now and has been developed into a practical working tool. It's still quite expensive, so its use is primarily within the larger firms where it can be economically justified. One area for the use of 3-D CAD is on complex, piping-intensive process projects such as petrochemical plants. Normally that type of project requires a hard-piped scale model to show the intricate piping and equipment layout. These models can be produced in 3-D CAD computer models, which can be viewed on CAD terminals at multiple locations including the field. The 3-D CAD models can be produced and updated much more cheaply than hard-piped models. Getting the complex layout information from the 3-D model out to the field crews doing the actual erection will require higher-tech supervision, foremen, and craftspeople. The recent development of digitized cameras to photograph the 3-D screens may offer a good avenue of approach in that area. The computer models also make it easier for owners to keep them up-to-date as plant changes are made over the years. Hard-piped scale models had a habit of disintegrating over time.

Three-D CAD can also benefit architectural-type projects by doing interior layouts that allow a pictorial walk-through of the project before the design has been completed. It is a powerful tool for working with owners and preconstruction analyses when they are used.

Since the development of my professional career and the computer revolution were roughly concurrent, I have had an excellent opportunity to observe and participate in it. I remember the brief life of the tape-driven electric typewriters that were the forerunners of today's word processors. We thought they were a great tool to use for building a reusable specification file. A little later, the computerized CPM engineering and construction schedules began to arrive on the scene.

CAD and CAE were close behind, in the mid-to-late 1970s. I well remember taking potential clients into the CAD room, where we had about five terminals set up alongside a computer and plotter. It was fun to see clients' eyes light up while watching the plotter methodically churn out perfect drawings before their eyes. They were rubbing their hands while imagining all the money they were going to save using the new technology. Unfortunately, it didn't work out that way right away, if ever. With CAD operating terminals costing about $45 per hour versus a draftsperson's then hourly rate of about $15 per hour, the CAD system had to produce an efficiency rating of three times better than manual. It was many years before that efficiency balance was reached, let alone surpassed. In any event, CAD offered so many other productivity advantages that contractors and clients went into CAD 100 percent without worrying about any direct economic payout.

With another world-class contractor, I had a first-hand opportunity to observe the folly of trying to develop a computer-aided design system in-house. In the mid-1970s the firm had spent $3 million in trying to develop a proprietary computerized piping and equipment layout drawing system for their own use. A team of piping designers and computer people had spent the money and had brought the system up to the point of trying it on a test project. While between project assignments, I had the job of bringing the system from the laboratory stage to the design room floor. It soon became obvious that the program was full of bugs and was not fully developed for use. My opinion was that the team members were not well enough qualified in software development to make it work. That, coupled with the natural resistance of the design office to accept the new technology, sealed its fate. Simultaneously with our in-house development, several software companies were working on similar systems and were successful in bringing them to market. Although I moved on to another firm, to my knowledge the in-house system was never used and the over $3 million was written off as a loss. I include this story to emphasize my principle that it's generally impractical and uneconomical to develop your own software in-house.

Computers in specification preparation

The advent of word processing and the Construction Specification Institute (CSI) standardization of construction specifications has greatly improved the production of specifications and special conditions for construction contracts. Pulling a print of the appropriate sections of the standards and marking them up to suit the new project makes easy work of the task. The modifications can be made by a word processing group from a marked-up set, or done quicker right on the computer if the specifications engineer is proficient with a word processor.

Computer Applications in the Home Office

The construction firm's main computer center is most commonly based in the home office. It usually serves as the base unit for any computer networks that tie together various operating units such as executive offices, heavy-equipment yards, warehouses, and construction sites. The home office computer's main function is to manage the firms's general accounting system.

The central office computer plays an important role in the preconstruction work of proposals, planning, budgeting, scheduling, staffing, safety programing, and the like, by acquiring and organizing the firm's available data bank. The preconstruction group develops their schedules and budget on PCs, which are then transferred to the field for execution. The final construction performance data completes the loop by being returned to the home office computer's database for storage or use on subsequent projects.

As the firm's central accounting office, the payrolls are usually processed in the home office based on the field timekeeping records.

An exception to that could be a mega-type project that could warrant a field payroll processing operation. Most of the job's materials are also paid for through the central office's computerized accounts payable program.

Computerized personnel records are a big plus when it comes to staffing field operations. Keeping an updated skills bank for supervisory and craft specialists is vitally important to staffing projects with well-qualified personnel. Open-shop personnel directors would be hard-pressed to function effectively without computerized personnel inventories. Keeping updated individual employee safety records is a valuable component of the computerized personnel records.

Computerized project-estimating procedures are a valuable adjunct to home office activities. As we said earlier, construction contractors live or die by the effectiveness of their estimating departments. Successfully won project estimates are than converted to computerized baseline project-control budgets that are used to control field costs. The improved efficiency of material takeoffs and organizing the estimate through computerization have significantly improved the quality and reduced the cost of preparing estimates.

The project procurement plan is formulated in the home office to delineate where the project procurement will be executed. Normally, the home office procurement portion starts before the field is opened. The computer programs to control the procurement documentation and material control are usually established in the home office.

Computer applications in procurement

Computer applications in procurement occur either in the home office or the field, depending on where the activity takes place. Computers in procurement often start with cutting the inquiries on the computer to set up a computerized project purchasing file. After the bids have been received and analyzed, the purchase order can be made on the computer to extend the order file to include price, delivery, change orders, and invoice payment. The figures from the procurement file can be automatically transferred to the budget to control the equipment and material costs charged against the project.

Other data can be transferred from the procurement file to the material-control report to simplify tracking the equipment and material from the vendor to the field. The field warehouse picks up the paper trail by creating the receiving and the over, short, and damage reports that go back to procurement for information and to accounting for use in paying the invoice. Once these computerized systems have been installed and are working, they can greatly facilitate the handling of a complex network of paperwork.

The computerized accounting system is set up in the home office to handle project invoicing and accounts payable. How the accounting program is tied into the field is important to ensure that the ongoing project financial commitments and payments are reflected in the project and field cost reports.

Computer Applications in the Field

The first use of computer systems in field operation was in the area of accounting and payrolls. In the old precomputer days, each week the field timekeepers collected the time cards to accumulate the hours worked per employee. That manual record was turned over to payroll clerks, who manually calculated the employees' wages, fringe benefits, and deductions to give the actual take-home pay. The money was taken out of the bank, counted out, and put into pay envelopes for each worker. On out-of-town jobs, that operation was all handled in the field with the help of the local bank. On larger projects the cash payroll was considerable, and sometimes was a handy target for an armed robbery.

With the advent of the computerized payroll practices in the home office, it was only a matter of time before the field payroll was relayed over phone lines to the home office for calculation and cutting the pay checks. In order to get faster turnaround on the payroll checks, the whole process was turned over to the field and run on a local computer. Fortunately, nowadays the CM and even the field office manager hardly know that the payroll operation is proceeding once it has been properly set up.

Computer use in field cost control

With the computerized payroll in the database, it is a simple matter to transfer labor costs into the field cost-control system. Automatically allotting the field labor costs to the proper cost accounts made evaluating field progress and craft productivity rates much easier and more accurate. The computer updated the labor cost and productivity spreadsheets, ending up with a variance column. By focusing on the off-target variances, the cost engineers and the CM can concentrate on correcting them and on continuing to monitor the trends. Doing that sort of activity by hand on large projects would require more cost-engineering staff and delay the production of cost reports.

Computerized CPM scheduling systems

In the scheduling area, we have already discussed the benefits of computers in generating a logic-based schedule like CPM. A small job using CPM could calculate the CPM network manually, but on a large project the task would be impossible. Again the computer permits a wide menu of data sorts to suit the varied needs of the construction team's operations. The main sort used by CMs is the one showing total float, which depicts the most critical activities heading the list.

The computer also allows the CM and field planners several options for evaluation of scheduling options during the course of the work as job conditions force a schedule change. Various "what-if" options can be tried on the field computer and the resulting logic diagram can be evaluated before making a final decision. Answering questions on the economics of "crashing" certain construction operations also can be studied on the logic diagram. Craft work-crew leveling is another computerized CPM scheduling resource available to the CM that we discussed in Chapter 4.

There is a plethora of computerized CPM programs available on the software market for both large and small projects.[2] CPM scheduling programs designed for smaller projects but handling up to several thousands work activities are available for as little as $500. That's more capacity than anyone would ever need for a smaller project. If you find yourself using over 500 activities on a small project, you are likely to suffer from computerized CPM overkill!

CPM programs for larger project systems can run from $10,000 to $30,000, just for software. Some of those also may require a minicomputer instead of a PC. James O'Brien has a chapter describing CPM computer programs and systems in the fourth edition of his *CPM in Construction Management.*[1] He covers the top 38 programs in common use in the construction industry today, including the hardware required for each.

Mr. O'Brien's chapter on CPM Costs starts with a paragraph that impressed me enough to reproduce it here. It sums up in three sentences what we need to know about selecting a CPM scheduling system:

> If you approach the application of CPM with a penny-wise and pound foolish attitude, you will doubtlessly get a poor bargain. If you hope to find something for nothing in CPM, you are well advised to forget it. It is, rather an investment which will return substantial and regular dividends.

Most CPM development firms have made progress in linking the treatment of CPM scheduling with cost. As I said earlier, that connection is difficult because the budget and schedule are broken down in separate systems and degrees of detail.[1] Some general relationships can be reached if the cost part of the equation is applied to the CPM schedule in a less-than-budget-quality breakdown. If a simple, inexpensive relationship of the cost of the CPM work activities can be made, it may be worthwhile to connect CPM schedule with cost. That arrangement gives both time and expenditure in a CPM printout. It must be understood, however, that the cost numbers thus generated are not accurate enough for the official project cost report. However, they can be used by management in evaluating the general cost trends on the project. CMs should evaluate the cost-benefit ratio of using such techniques on their specific projects.

Computerized material control

Computerization of modern field material-control systems pays dividends by ensuring that construction materials are available when needed. As we discussed earlier, field management's greatest contribution to improving field-labor productivity is having the materials on hand when the craft personnel need them. Closely tracking materials from procurement to installation with computerized methods is another proven business tool that has been adapted to construction work. Keeping the system as simple as possible to deliver the desired result for your particular type of construction work will maximize the cost-benefit ratio in your favor.

Computerized personnel records

On larger projects with a field personnel manager, the personnel files should be computerized for ready access to the firm's construction personnel resources. Although long-range planning is the best way to staff a project, there are many short-term staffing requirements that must be filled in a hurry. Waiting around for the home office to fill your needs may not be convenient or effective.

If the personnel people keep their records current, it is a good way to maintain the cumulative performance records of long-term employees. A solid data bank of proven construction performers can be a real asset to the project.

Computerized field engineering

Having the field engineering office computerize their files and records for technical data, inspection records, field surveys, change orders, correspondence, and the like can save time and money.[2] A CAD station tied into the home office system on design-build jobs also can be useful if the project is handling a lot of field changes. Having a few commonly used structural and civil design programs available can speed up the field engineer's response to requests for information in those areas.

Computerizing the construction field office

Having PCs in the CM's office and the administrative offices generally increases field office productivity and ensures professional-looking output. Handling the creation and flow of field-office paperwork by computers is virtually a must on today's information-driven projects. PCs equipped with modems have long since replaced the old teletype machines and are starting to cut into the need for separate facsimile machines. Continued combining of PCs with communications systems is expected to be the wave of the future for many computer applications.

Computerized Construction Management

It becomes readily apparent that there really isn't much left of field operations that hasn't been at least touched by computerization. It seems the next logical step is to bundle all the computer operations together into one self-contained, computerized construction management system. There is already one such system on the market, with others sure to follow.[2]

Some CPM developers regard the cost-schedule portion of computer applications as being the heart of project management, which it probably is. On that premise, those firms are developing computerized *total construction project management systems.* Design-build firms can carry that one step further to include the design portion and come up with a computerized *total project management system.* Such a system will offer computerization from project concept to project turnover, and beyond if required. Some may even refer to it as "cradle to the grave care," although *grave* may not be the right word to use in this context!

Among the construction contracting operations covered by one of the integrated construction management systems presently offered are the following major construction management activities:

- Tracking submittals
- Controlling project changes
- Cost and schedule control
- Contract administration
- Procurement and expediting
- Document control
- Project reporting
- Project control issues

This approach is a fairly quantum leap for computer applications in construction management. The program is based on all documentation being computer-generated and placed in a large relational data bank so that the various programs can tap into it for data and to update the database to keep it current. The complex nature of the system will require a thoroughly trained crew to operate it and a knowledgeable programmer to keep the software working. The systems will probably experience a rather painful growing experience through use on larger projects before they gain wider acceptance. There will be considerable training and development costs to get the new systems into effective operation and make them productive. In my opinion such comprehensive management programs represent the cutting edge of computerized construction management techniques at the moment, and eventually they will become common practice in the future.

Some Common-Sense Computer Rules

Having lived through the computer revolution, I have had plenty of opportunities to observe what does and doesn't work with computers. One must always remember that computer and software salespeoples' primary job is to sell the product. Sometimes they tend to get carried away with your application without knowing all the needs, so beware of their promises and claims. Be selective in your computer applications to ensure that they are cost-effective in your particular environment.

My first law of computers is to remember that they are only *a tool to help get the job done.* Until artificial intelligence (AI) is perfected, computers can only provide you with data, they cannot run the job. Development of practical AI systems, by the way, is still over the horizon according to many experts in the field. Using a computer today is roughly comparable to using a modern mobile crane in building the pyramids, versus the Egyptian method of inclined planes and sledges. Today you should be able to build pyramids more safely and efficiently, and with much less labor.

My second law of computers is that they can turn out more data than you can possibly consume and still do your job. Don't even try to absorb every scrap of data that the computer can generate. Be very selective and use only that part of the output that you need to make sound decisions and monitor your project's performance. One computer pundit recently likened PCs to washing machines: "Neither is really a labor-saving device. Instead, people just do more laundry." As we said in earlier chapters, select your computer output to maximize your management-by-exception techniques. Don't just do more laundry.

A corollary to the above laws is the rule that says, "Just because the output is computer-generated doesn't make it true." If the input or software is not accurate, the computer will not know the difference. Nothing says it better than the age-old computer adage, "Garbage in, garbage out!"

Another corollary to remember is that the input to the computer must be kept current. The project is a moving target for the computer, so the user of the output data must consider the cutoff point of the input data. Be sure that the system for updating the input is automated where possible and routinely posted where it is not. Usually clerical people are used for inputting the new data, so don't stint on having enough personnel dedicated to doing it.

A recurring theme in this book has been to ensure that people are properly trained in computer applications. That's also true for any other tools or equipment on the project. Actually, computer training is not as formidable as it appears. If the new hires have some basic training or experience, the cost of teaching them the applications for the work on your project should be manageable.

The nature of using computers in the field-construction environment is a little different from using them in normal continuous-business operations. The short-term nature of project work makes the data-processing effort relatively short-lived. The continuous, ongoing business in the construction industry takes place only in the home office. CMs must keep the short shelf life of the data in mind when applying computer systems to the project.

Some of the project data has to be saved to satisfy the contractual and legal requirements. Some other data like the estimating and final cost data will need to be transferred to update the historical files in the home office. Aside from that, the rest of the data may be of no use to the ongoing home office business activities.

On the other hand the advent of high-capacity electronic data storage equipment, such as tape and compact discs (CDs), makes storing all the electronic data more attractive. The whole project file probably could be stored on a single CD in read-only mode. The possibility of doing that rather than storing several cubic yards of paper file boxes in a moldy storage cave somewhere opens some attractive alternatives.

One also must consider the downside of computerization in the field. When one's operating system becomes computer-dependent, it becomes very critical to keep the system up. Given the temporary nature and interruptibility of some field electrical systems, it pays to ensure that the computer system stays up and that files are backed up regularly. A constant voltage power source, with standby power supply for the computer along with an automated file-backup system, should be considered.

The computer is really a multifaceted business and management tool, with hundreds of new possible applications being discovered daily. With the availability of PCs and a wide variety of software in most field-construction offices today, there will be many new uses developed on the job. It really comes down to the imagination of the field personnel that work with computers and the CM's affinity to computer applications as to how many new applications will be discovered on the job.

As to the downside of new uses, be sure that the new applications are truly cost-effective[2] or they will result in a loss of efficiency in managing the project. CMs also should guard against being tied to their computers to the detriment of being out in the trenches to follow the actual field work. The field work is done by humans, not computers. Common sense is the watchword to keep foremost in mind when melding computers into construction work.

Future Applications of Computers in the Construction Business

Not being a true computer guru places me on even shakier ground when it comes to predicting the future uses of a *high-tech* business like computers in a relatively *low-tech* industry like construction. When people are predicting the future, they often compare it to predicting the weather. The use of computers in weather forecasting has improved the forecasting accuracy to such a degree that the comparison is no longer valid.

Considering the computer predictions of the past reminds one that not one computer wizard predicted the PC revolution before its onset in 1980. The same applies to the astounding growth of software development. Obviously, if any of us had foreseen those two happenings, we would be a lot richer today! So on the basis of the above "weasel" clauses, I will make a few modest predictions which are likely to have a very insignificant effect on the construction industry.

I predict a moderate amount of success by the year 2000 for the self-contained, computerized, total-project-management software that I mentioned above. Again, it will be used primarily on larger-sized projects that can generate sufficient cost-benefit ratios. If that system eventually can be parlayed into artificial intelligence, that will be revolutionary.

The recent development of handheld, pen-based, and voice-actuated computers should find some new applications in the field in the next decade. The handheld portability of the unit, combined with the ease of pen or voice input, multiplies the opportunities for using that type of computer out in the field. Pen- and voice-based units require a minimum of computer skills and training to make data entries in a rigorous working environment. Similar handheld computers have already found a niche in many plant operations and maintenance applications. The data gathered in the plant is readily transferred to the office PC for further data logging or processing. Combining the portable units with bar code readers further extends their use to field warehousing and on-site fabricating shops.

Naturally there will be continuing evolutionary computer hardware improvements, to build faster processors with larger RAM and disk storage memories. I read recently that a microchip manufacturer has just succeeded in halving the distance between conducting strips on a microchip. That should lead to greater processing capability on yet smaller chips for even smaller computers. Obviously chip manufacturers eventually will arrive at the irreducible minimum blob, unless they come up with a revolutionary approach to laying out microchips. Corresponding continuation of software developments will lead to useful improvements in this area.

Future downside possibilities in computers

Despite all the euphoria as to future computer developments, there are a few clouds on the horizon that have to be considered. Most of these are not blockbusters, but they ought to be noted.

In analyzing the past success of the PC revolution, which led to the successful use of computers in construction, we have to look at the present direction of the industry. The original revolution was paced by many small start-up companies developing new hardware and software technologies at a rapid pace. Lately, the trend has been toward consolidation of the industry into monolithic giants that specialize in saturating markets with their products. That phenomenon tends to inhibit small companies from starting out with new ideas and making them grow. There has been some talk of the Federal Trade Commission (FTC) getting into the act to reverse that trend. Action paralleling the AT&T breakup could be in the wind. As of this writing (August 1993), the new administration in Washington has not been able to spend much time on the matter, but it may get to it later.

Another even more disturbing thought is that the trend toward computer networking may be forcing the PC revolution back to a *mainframe* mentality. That includes the resurgence of the old *man-*

agement information systems (MIS) department that for years successfully stifled the individual use of computing power. Here we are squarely facing the end of the PC revolution and the loss of all those hard-won gains that helped us to process what we wanted when we wanted it on our own machine.

The networking futurists are projecting the development of hyperconnected computing systems, similar to the telephone and electrical utility services we know today. One would plug a work station (a sophisticated telephone) into a wall socket and be immediately connected to any computer in the world. That opens the next stage of thought as to how to use all the unused memory available in the interconnected units. Some of the concepts are truly mind-boggling in both directions—some potentially beneficial and some not. I suppose the possibility of running an off-line PC as an independent operation would still be possible if the necessary hardware and software still were available.

What I have been addressing are primarily *evolutionary* improvements of already existing technology. I am sure there is at least one *revolutionary* event, equal to the computer itself, so far over the horizon that not even the high-tech pundits have had the vision to predict it. It will be interesting to see what develops in future construction-oriented generations!

This chapter has been done in a positive vein extolling the virtues of computer use in virtually all facets of construction operations. There are, I am sure, some construction people who don't entirely agree with that philosophy and can support their arguments. One of them is that the 1980s were the decade of the PC and its sweeping changes of the way business was done. By the same token the 1980s was also the decade of runaway federal budgets, the corporate buyout binge, junk bonds, insider trading, and two recessions. The construction business suffered worse from those two recessions than at any other time since the depression. In the same period, business spent $80 billion for information technology and had only a 1 percent increase in productivity; hardly an outstanding cost/benefit ratio.

I don't mean to imply that the economic problems of the 1980s were caused solely by computer use. What I do say is that all the labor-saving applications of the computer do not appear to have been used wisely. Some people gave too much credence to what the computer produced and got inaccurate information. Some applications just did more laundry instead of creating significant data. Some people bought computers (and then even used them) just to say they had them in the business. All of which can be summed up by repeating my First Law of Computers: The computer is only a tool, and it should be used accordingly to better run the business.

References

1. James J. O'Brien, *CPM in Construction Management,* 4th ed., McGraw-Hill, New York, 1993.
2. Mike C. Tidwell, *Microcomputer Applications for Field Construction Projects,* McGraw-Hill, New York, 1992.

Case Study Instructions

1. Make a list of the computer applications that you used on your project, from bidding through project closeout. Make a brief statement about what you hoped to accomplish with each application and give the estimated cost-benefit ratio.

2. Evaluate your computer literacy based on prior education, training, and experience. What is your program for expanding or maintaining your present computer qualifications?

3. What additional computer applications do you feel would benefit your performance in the field? Discuss your reasons why.

A

Business Roundtable Report List

The findings and recommendations of The Business Roundtable Construction Industry Cost Effectiveness Program are included in the reports listed below. Copies may be obtained at no cost by writing to The Business Roundtable, Attn: CICE, 200 Park Avenue, New York, NY 10166.

Project Management—Study Area A

A-1 Measuring Productivity in Construction

A-2 Construction Labor Motivation

A-3 Improving Construction Safety Performance

A-4 First and Second Level Supervisory Training

A-5 Management Education and Academic Relations

A-6 Modern Management Systems

A-7 Contractual Arrangements

Construction Technology—Study Area B

B-1 Integrating Construction Resource and Technology into Engineering

B-2 Technological Progress in the Construction Industry

B-3 Construction Technology Needs and Priorities

Labor Effectiveness—Study Area C

C-1 Exclusive Jurisdiction in Construction

C-2 Scheduled Overtime Effect on Construction Projects

C-3 Contractor Supervision in Unionized Construction

B

Job Description

Title: Construction Manager

Level: Various

1.0 SCOPE

Construction Managers (CMs) have full responsibility for the field operations associated with their assigned projects. They are responsible to the company management and/or the Project Director for effective and safe execution of the construction work to ensure completion of the project as specified, on time, and within budget.

The CM manages all aspects of construction execution in accordance with the contract. Key areas included are: field engineering. field procurement, construction activities, facility start-up assistance, and project closeout. Other important areas are contractual and financial considerations, personnel matters, safety, and client and public relations.

CMs are the focal point for all construction activities from project initiation to project closeout. They shall also be active in the precontract bidding and as the construction representative in precontract discussions with the client (or contractor).

Company management shall issue a construction management charter granting the CM sufficient authority to effectively execute the project within the parameters of this Job Description.

2.0 CONCEPT

Key CM functions are to effectively plan, organize, and control field operations. They must plan the work and organize the available people and methods, permitting normal field operations to proceed routinely. In addition they must be able to respond quickly and effectively to unusual or emergency situations as they arise. They shall also establish written goals and priorities for the field effort early in the project, and frequently review them with key field supervisors and home office management to ensure that the planned objectives are being met.

CMs must be true managers and organizers and not simply doers or supervisors. They must be effective delegators to project their expertise and management philosophy throughout the field organization. Training and developing potential new construction managers is an important consideration in that process.

The CM shall play a strong leadership role in fostering high morale, teamwork, and a motivated work force geared toward high productivity and meeting the project goals.

3.0   DUTIES AND RESPONSIBILITIES

3.1   Planning activities

a. Become completely familiar with all documents and special project and client (or contractor) requirements, and prepare and distribute pertinent construction information to those having a "need to know."

b. Develop the master plan for executing and controlling the construction activities in conjunction with company operating management and/or the project director.

c. Prepare and issue written field operating procedures governing all construction activities consistent with approved project plans, company policies and client's requirements.

d. Oversee preparation and approval of field budgets and establish procedures for controlling and reporting costs for all phases of the construction work.

e. Oversee preparation and approval of the construction schedule and establish procedures for monitoring, controlling, and reporting field progress through project completion.

f. Establish specific project goals and priorities for all facets of the construction work and issue them to the field staff. Incorporate these into an effective delegation and management-by-objectives system to ensure meeting the goals.

g. Participate with management in establishing the labor posture for field operations and preparing the site-labor agreements.

h. Establish a materials management plan for buying, receiving, storing, and issuing the project material resources. Include a field procurement section for field-procured goods and services if required.

i. Prepare a subcontracting plan for work to be subcontracted to others. The plan shall include taking bids, selecting, the contractors, and administering the subcontracts.

j. Review the project insurance requirements and risk-management plan for field operations. Ensure that the correct coverages are in effect throughout the project.

k. Review the heavy lifts and prepare a heavy-equipment schedule and a policy for small tools and light equipment.

l. Set project profitability goals, and plans for meeting them in conjunction with supervisory management.

3.2 Organizing

a. Develop a field organization chart showing the lines of authority and interrelationships of key personnel and their activities.

b. Prepare job descriptions detailing the duties, responsibilities, and objectives for key field personnel. Incorporate the job descriptions into a management-by-objectives system, setting goals with key personnel and a regular performance-monitoring schedule.

c. Initiate and participate in the selection of key field supervisors, and establish the requirements for hiring temporary local personnel.

d. Implement a total field personnel plan in conjunction with key project supervisors and the personnel department in accordance with the project schedule. Review the plan and eliminate any sharp peaks in personnel loading.

e. Finalize policies and procedures for assignment, transfer, and expense reimbursement of field personnel to the job site.

f. Continually review the field organization and adjust to suit actual project needs, especially near the end of the project.

g. Arrange internal and client (or contractor) construction kickoff meetings to initiate the execution of the construction work in an orderly manner.

h. Issue the Field Procedure Manual (FPM) within 30 days of notice to proceed (using "holds" where necessary). Use the FPM for field staff orientation sessions.

i. Keep the client field representative informed of changes in the project organizational structure and manning, and secure contractual approvals as required.

j. Set up a job-site public relations program to promote the project and company image in the local area.

k. Organize and promote the site-safety program and monitor it closely to ensure a safe site in compliance with company, client, and government requirements and regulations.

3.3 Controlling

a. Closely monitor field activities for conformance to contract scope requirements and establish a change-order procedure for scope and field revisions. Monitor contractual requirements and recommend adjustments when required.

b. Administer and enforce compliance with the terms of the contract, the construction master plan, the field procedures, and management directives, paying particular attention to quality, guarantee, and warranty requirements.

c. Regularly monitor the systems to control field costs, schedule, and quality and to ensure that they are effectively meeting project objectives. All control systems should forecast field activities to project completion.

d. Maintain effective communications with client, subcontractors, and key project participants by:

- Conducting construction progress review meetings on a regular basis

- Issuing quality field reports covering the status of physical progress versus schedule, cost versus budget, material status, etc., and explaining off-target conditions

- Discussing any problem areas along with your recommendations for solving them

- Documenting the field work with minutes of meetings, correspondence, telephone call confirmations, daily diaries, etc. to build working files and a project history.

e. Review field personnel requirements regularly, ensuring that the human resources are properly matched to the work load and schedule.

f. Establish construction quality-control procedures, ensuring quality construction in accordance with the plans and specifications.

g. Monitor flow of design documentation, vendor data, and project information, ensuring that all parts of the field work can progress smoothly.

h. Monitor all field invoices and payments to ensure adherence to the cashflow plan. Periodically review the escalation and contingency accounts to see that they are being properly allocated.

i. Promptly inform company management and the project director of any unusual construction events or problems to keep them current on unforeseen field events.

j. Review and approve all outside communications, ensuring that the field and company images are being presented fairly.

k. Practice control by exception and give immediate attention and corrective action to off-target items.

l. Review with company management on a monthly basis the actual project profitability results versus plan.

4.0 AUTHORITY

To have strong and effective control of field operations, the CM's authority must be established in writing and supported by top management policy and actions. The CM shall have authority to

4.1 Participate in the selection of personnel who will be assigned to the field and shall be consulted prior to any proposed changes in assignment of field personnel.

4.2 Act as company's representative on all matters relating to labor relations in the field. The CM handles all jurisdictional disputes and grievances in accordance with overall company and site-agreement policies.

4.3 Request the presence of any departmental personnel whose services are required to serve the field operations.

4.4 Arbitrate interdepartmental and interdiscipline differences on matters pertaining to field operations. Upper management may be required to approve the decision.

4.5 Approve all field expenditures and commitments for the project within any limits that may be set by upper management.

5.0 WORKING RELATIONSHIPS

To successfully complete a construction project, the CM must have the full cooperation of all departments of the company in addition to the authority granted by top management. To gain this cooperation, the CM must maintain good working relationships within and across all organizational lines of the project and the company. At a minimum, the Construction Manager must

5.1 Cooperate with the project director and design office staff to meet the overall project goals.

5.2 Cooperate with corporate staff members, department heads, and other management personnel in matters relating to their assigned areas of responsibility.

5.3 Cooperate with other operating units, management centers, and/or affiliates so that the best interests of the company and the project are served at all times.

5.4 Keep company operating management and department heads current on all project matters that could affect their operations.

5.5 Be responsive to requests for information and services from client, company, and project operating groups.

5.6 Provide routine and/or special reports required by the company procedures, operating management, or the owner.

6.0 LEADERSHIP QUALITIES

A successful CM must have strong leadership qualities to ensure that the field organization is performing at top efficiency at all times. To develop into a true leader, the CM must

6.1 Direct all field operations to meet project and company contractual obligations at all times while maintaining high project team morale.

6.2 Develop and maintain a system of decision making within the field organization whereby decisions are made at the lowest possible level.

6.3 Promote the development and career growth of key field supervisors and encourage them to do likewise with their people.

6.4 Establish written project objectives and performance goals for all key members of the construction team and review them periodically via an MBO system.

6.5 Promote an atmosphere of team spirit with the field and client (or contractor) staffs.

6.6 Conduct himself or herself at all times in an exemplary manner, setting a good example for all team members to follow.

6.7 Be a good listener and fairly resolve any problems or differences between project personnel, owners, department heads, subcontractors, etc. that may arise.

6.8 Anticipate and minimize potential problems before they arise by maintaining frequent contact with and current knowledge of all field activities, project status, owner and contractor attitudes, and outside factors that might affect the project.

6.9 Maintain a positive attitude toward the field and project staffs, clients, subcontractors, management, and peers at all times.

6.10 Attack problem areas quickly no matter how distasteful they may be. All problems must be brought into the open and resolved as soon as possible.

C

Case Study Selection

These case studies are hypothetical projects that have been created for use with the Case Study Instructions given at the end of each chapter. A broad spectrum of project types is given to cover the various types of working environments as shown in Table 1.1. In the event that you do not find a project on this list, you can create a do-it-yourself selection by filling in the generic format at the end of the list.

Since it is not possible to give you every detail of a project in this limited space, you will have to make reasonable assumptions for any of the missing details you will need to create a viable solution to the case study.

You may choose to assume the position of construction or project manager for either the owner or contractor as best suits your working environment or career goals.

PROJECT 1

Your firm has just been awarded the contract to construct a 55-story office building in a busy metropolitan downtown area, and you have been appointed as site construction manager. The building will occupy the complete site, with a six-level underground parking garage. The lower five above-ground levels will be hotel and commercial rental space. The remaining 50 stories will be office space, consisting of 20 floors of preleased space including interior partitioning and finishes and the remaining floors without partitioning and floor finishes. As more floors are leased, the owner reserves the right to request change-order pricing on the interior partitioning and finishes of the leased spaces. Before the contract has been completed, the owner leases 10 more floors and wants the work covered by change order to the original contract.

The contractor has a prime contract and is responsible for all procurement, site safety (including public street access), compliance with local codes, receiving materials, negotiating and supervising all subcontractors, and supervision of all construction activities. The design has been completed, including the interior design. The A&E firm and the interior designer will act as owner's inspectors and approve the final construction. The owner will have an on-site resident representative and the A&E firm will have on-site inspectors.

The contract will be based on a current issue of an AIA standard construction agreement format. The contract basis is lump-sum price, with progress payments and 10 percent retention on billings. The schedule is considered to be tight but feasible. The contract calls for liquidated damages of $5000 per day for the first two weeks and $10,000 per day thereafter until substantial completion of the lower 25 floors and parking garage.

PROJECT 2

Your firm is going to build a grassroots paper mill in northern British Columbia, Canada. The preliminary cost estimate is US$600 million for the initial plant, with one paper machine being installed now and two machines added later as sales warrant. The site is remote, but is served by a deep-water port with limited facilities. The area is populated by an indigenous native population and has little or no infrastructure. The provincial government has agreed to furnish some of the infrastructure to encourage development in the area.

The schedule for the project is 48 months from conceptual design approval to plant start-up. The owner plans that the work will be performed on a turnkey basis by a single major engineering-construction contractor selected from a worldwide slate of contracting firms. The owner has prepurchased the paper machine and continuous digester system because of the long delivery, and will turn the orders over to the successful contractor for completion of the delivery. The continuous digester has been purchased as an installed package from a Swedish firm. A basic-design package for a paper-coating process has been licensed by the owner from a firm in Finland.

Financing will be from internal funding and government loans, so cashflow and budgets will be very tight. Cost and schedule overruns will be catastrophic for the financial plan for the project; therefore, good project control is vital to the success of the project. You are to start the project at the proposal stage and run through project start-up and operator training.

You may select the role of owner's construction manager, contractor's project director, or field construction manager. Don't forget to include your participation in taking and/or preparing the bids and/or selecting the contractor.

PROJECT 3

The Department of Energy (DOE) has been chosen as the agency to construct a new supercollider for the National Science Foundation on a site south of Dallas, Texas. The project has a preliminary price tag of $8.25 billion and is urgently required to reestablish the preeminence of the United States in superconductivity research.

The supercollider will take seven years to build and will contain the following major parts: an oval, 53-mile particle accelerator tunnel, a ground-level 350 acre campus of 15 buildings, 4 lab buildings inside the oval, and smaller support buildings every 5 miles on the tunnel. The remaining area inside the oval not occupied by buildings will be left in its native state. Access roads and a security system will be required to service the total facility.

Financing is going to be furnished by the federal and state governments as well as foreign governments interested in participating in the venture. It is estimated that about 50,000 contractors and suppliers will be participating in the construction of the facility and its equipment. Your group will be responsible only for the construction of the capital facilities portion of the project.

Three years into the project, DOE calls for the work on the project to be suspended, because Congress would not approve more funds. After a delay of 12 months, work is resumed and the project is completed.

You may select the position of project construction manager for the DOE, contractor's construction manager, or a major contractor on any portion of the project in your area of expertise.

PROJECT 4

Your firm is bidding on the construction of a single-story sheltered workshop to carry out light assembly work for a county board of mental retardation. Separate prime contracts are to be let for the mechanical and electrical portions of the work. As general prime contractor, your firm is responsible for overall project coordination and scheduling. The contract will be with the county on a lump-sum basis, using an AIA construction contract format. Each prime will have liquidated damages for late completion—yours will be $500 per day you are late on substantial completion. The schedule is 12 months from notice to proceed. The A&E design firm will coordinate and inspect the construction work for the owner.

The area of the building is 36,000 sq ft on a three-acre site, with the improvements estimated at $3.5 million for the complete job. The facility includes a school bus parking lot with security fencing and a limited bus-maintenance shop. The space includes a lunch facility and several classrooms. The furniture and tooling will be by owner.

Assume the role of contractor's construction manager responsible for bidding and executing the project as the low bidder. The contract stipulates that there be at least 15 percent of the contract value in set-asides for minority subcontractors, and they are to be named in the bid.

PROJECT 5

Your firm is bidding on 35 miles of new interstate highway through a section of the Rocky Mountains. The project includes exits and a half-mile-long four-lane tunnel through a mountain. The work will take place in a virgin site and include all clearing, grubbing, grading, sub-base, finished paving, markings, reflectors, and signage. The schedule including the tunnel is 24 months, with liquidated damages of $1000 per day the road is not available for traffic.

The work will interface on each end with other contractors finishing the whole highway. The complete design has been furnished by the Department of Transportation. The field work will be subject to inspection by state and federal DOTs with on-site inspectors. There is a 30% minority-subcontractor set-aside with named contractors in the bid. You may select the position of owner or contractor construction manager.

PROJECT 6

This is a *generic project* format that you may use to construct a do-it-yourself sample to suit your particular area of expertise in the capital project field. Assume that you start the project at the proposal stage and carry it through start-up and turnover.

Owner:	_____ government, _____ industry, _____ _____ developer, _____ other
Facility:	___ institution, ___ plant, ___ building, ___ laboratory, ___ commercial _____ your choice
Complexity:	___ addition, ___ green field site, ___ ___ revamp, ___ high-tech, _____ other
Size:	_____ sq ft, _____ capacity, ____ stories, $_____ estimated cost, _____ labor hrs, _____ other
Location:	_____country, ____ urban, ___ remote, ___ developed country, ___ third world, _____ other
Infrastructure:	___ existing, ___ nonexisting, ___ partial, _____ transportation, ___ utilities, _____ labor supply, _____ other
Services:	___ project development ___ design, ___ procurement, ___ construction, ___ inspection, ___ start-up, ___ extras, _____ other
Financing:	___ internal, ___ construction loan, ___ participative, ___ government loan, _____other
Schedule:	___ loose, ___ tight, ___ medium, ___ fast track, ___ impossible, _____other
Budget:	___ liberal, ___ tight, ___ average, ___ improbable, _____ other
Design Basis:	___ well-defined, ___ preliminary, ___ loose, ___ by owner, ___ by contractor, ____ other environmental permits _____
Execution plan:	___ in-house, ___ prime contract, ___ separate subcontracts, ___ mixed-basis, _____other
Long-delivery items:	equipment _____ materials _____ other factors _____
Other factors:	_____ _____

Index

ABOUT THE AUTHOR

George J. Ritz is a leading expert in construction and project management, with 40 years' experience in executing projects in the U.S. and abroad. He has worked on a broad spectrum of construction projects, ranging from schools for the physically challenged to world-class petrochemical plants. Mr. Ritz is a registered Professional Engineer and a frequent lecturer on project management and related topics. He is the author of *Total Engineering Project Management*, also published by McGraw-Hill.